# Foundations of Free Noncommutative Function Theory

**Mathematical Surveys and Monographs**

**Volume 199**

# Foundations of Free Noncommutative Function Theory

Dmitry S. Kaliuzhnyi-Verbovetskyi
Victor Vinnikov

**American Mathematical Society**
Providence, Rhode Island

2010 *Mathematics Subject Classification.* Primary 17A50, 40A05, 46-02, 46L52, 46L54, 47A60, 47L25.

This work was partially supported by the US–Israel Binational Science Foundation (BSF) grant 2010432. The first author was partially supported by the US National Science Foundation (NSF) grant DMS 0901628 and by the Center of Advanced Studies in Mathematics of Ben-Gurion University of the Negev. The second author was partially supported by the Israel Science Foundation (ISF). Part of this work was carried out during the authors' visit to the Banff International Research Station (BIRS) under the program Research in Teams and at Mathematisches Forschungsinstitut Oberwolfach (MFO) under the program Research in Pairs.

---

For additional information and updates on this book, visit
**www.ams.org/bookpages/surv-199**

---

**Library of Congress Cataloging-in-Publication Data**

Kaliuzhnyi-Verbovetskyi, Dmitry S., 1965–
Foundations of free noncommutative function theory / Dmitry S. Kaliuzhnyi-Verbovetskyi, Victor Vinnikov.
pages cm. – (Mathematical surveys and monographs ; volume 199)
Includes bibliographical references and index.
ISBN 978-1-4704-1967-3 (alk. paper)
1. Functional analysis. 2. Noncommutative algebras. 3. Noncommutative function spaces. I. Vinnikov, V. (Victor). II. Title.

QA320.K25 2014
515′.7–dc23 2014025799

---



♾ The paper used in this book is acid-free and falls within the guidelines
established to ensure permanence and durability.
Visit the AMS home page at `http://www.ams.org/`

10 9 8 7 6 5 4 3 2 1 19 18 17 16 15 14

# Contents

CHAPTER 1

# Introduction

The goal of this work is to develop, in a systematic way and in a full natural generality, the foundations of a theory of functions of free[1] noncommuting variables. This theory offers a unified treatment for many free noncommutative objects appearing in various branches of mathematics.

Analytic functions of $d$ noncommuting variables originate in the pioneering work of J. L. Taylor on noncommutative spectral theory [**99, 100**]. The underlying idea is that a function of $d$ noncommuting variables is a function on $d$-tuples of square matrices of all sizes that respects simultaneous intertwinings (or equivalently — as we will show — direct sums and simultaneous similarities). Taylor showed that such functions admit a good differential (more precisely, difference-differential) calculus, all the way to the noncommutative counterpart of the classical (Brook) Taylor formula. Of course a $d$-tuple of matrices (say over $\mathbb{C}$) is the same thing as a matrix over $\mathbb{C}^d$, so we can view a noncommutative function as defined on square matrices of all sizes over a given vector space. This puts noncommutative function theory in the framework of operator spaces [**37, 78, 79**]. Also, noncommutative functions equipped with the difference-differential operator form an infinitesimal bialgebra [**90, 3**].[2] The theory has been pushed forward by Voiculescu [**105, 106, 107**], with an eye towards applications in free probability [**102, 103, 104, 108**]. We mention also the work of Hadwin [**46**] and Hadwin–Kaonga–Mathes [**47**], of Popescu [**85, 86, 88, 89**], of Helton–Klep–McCullough [**55, 51, 52, 53**], and of Muhly–Solel [**69, 71**]. The (already nontrivial) case of functions of a single noncommutative variable[3] was considered by Schanuel [**93**] (see also Schanuel–Zame [**94**]) and by Niemiec [**76**].

In a purely algebraic setting, polynomials and rational functions in $d$ noncommuting indeterminates and their evaluations on $d$-tuples of matrices of an arbitrary fixed size (over a commutative ring $\mathcal{R}$) are central objects in the theory of polynomial and rational identities; see, e.g., [**91, 41**]. A deep and detailed study of the ring of noncommutative polynomials and the skew field of noncommutative rational

[1]We consider only the case of free noncommuting variables, namely a free algebra or more generally the tensor algebra of a module; we will therefore say simply "noncommutative" instead of "free noncommutative".

[2]More precisely, to use our terminology, we have to consider noncommutative functions with values in the noncommutative space over an algebra with a directional noncommutative difference-differential operator as a comultiplication, and it is a topological version of the bialgebra concept where the range of the comultiplication is a completed tensor product. See Section 2.3.4 for the Leibnitz rule, and Section 3.4 for the coassociativity of the comultiplication. We will not pursue the infinitesimal bialgrebra viewpoint explicitly.

[3]See Remark 5.11.

functions has been pursued in the work of P. M. Cohn [**29, 31**]. The noncommutative difference-differential operator in the setting of noncommutative polynomials is well known as the universal derivation on the free algebra, see [**67, 25, 33**].

In systems and control, noncommutative rational functions and formal power series appear naturally as recognizable formal power series of the theory of automata and formal languages [**66, 95, 96, 38, 39, 40, 26**] and as transfer functions of multidimensional systems with evolution along the free monoid [**20, 15, 17, 16, 8, 18**]. In particular, transfer functions of conservative noncommutative multidimensional systems are characterized as formal power series whose values on a certain class of $d$-tuples of operators are contractive (matrix evaluations actually suffice — see [**9**]). Such classes of formal power series appear as the noncommutative generalization of the classical Schur class of contractive analytic functions on the unit disc [**2, 48, 14**] in the operator model theory for row contractions and more general noncommuting operator tuples [**81, 82, 83, 84, 87**] and in the representation theory of the Cuntz algebra [**27, 32**]; see also the generalized Hardy algebras associated to a $W^*$-correspondence [**68, 70, 12**].

Coming from a different direction, it turns out that most optimization problems appearing in systems and control are dimension-independent, i.e., the natural variables are matrices, and the problem involves rational expressions in these matrix variables which have therefore the same form independent of matrix sizes; see [**50**]. This leads to exploring such techniques as Linear Matrix Inequalities (LMIs) — see, e.g., [**74, 73, 98**] — in the context of noncommutative convexity and noncommutative real semialgebraic geometry, where one considers polynomials and rational functions in $d$ noncommuting indeterminates evaluated on $d$-tuples of matrices over $\mathbb{R}$ [**49, 56, 59, 58, 54, 57**].

A key feature of noncommutative functions that we establish in this work is very strong analyticity under very mild assumptions. In an algebraic setting, this means that a noncommutative function which is polynomial in matrix entries when it is evaluated on $n \times n$ matrices, $n = 1, 2, \ldots$, of bounded degree, is a noncommutative polynomial. In an analytic setting, local boundedness implies the existence of a convergent noncommutative power series expansion.

Difference-differential calculus for noncommutative rational functions, its relation to matrix evaluations, and applications were considered in [**64, 65**].

Recent papers [**80, 21, 22**] used the results of the present work on noncommutative function theory to study noncommutative infinite divisibility and limit theorems in operator-valued free probability. For instance, in the scalar-valued case a measure is free infinitely divisible if and only if its so-called $R$-transform has positive imaginary part on the complex upper half-plane ([**23**], see also [**75**] — this is the free analogue of the classical Levi–Hinçin Theorem). One of the main results of [**80**] is a similar statement in the operator-valued case except that the $R$-transform of an operator-valued distribution is a noncommutative function.

In a recent paper [**1**], a general fixed point theorem for noncommutative functions has been proved, and, in particular, the corresponding variation of the Banach contraction mapping theorem has been obtained. This result was applied then to prove the existence and uniqueness theorem for ODEs in noncommutative spaces. In addition, a noncommutative version of the principle of nested closed sets has been established.

The recent papers [**5, 6, 7**] made further progress in noncommutative function theory. Specifically, [**5**] established an analogue of the realization theorem of [**10**] and [**13**] for noncommutative functions on a domain defined by a matrix noncommutative polynomial (the result was originally established for special cases in [**17**] in the framework of noncommutative power series, see Section 1.3.5 below for a further discussion). [**5**] used this realization result to establish noncommutative analogues of the Oka–Weil approximation theorem and of the Carleson corona theorem. [**6**] applied these ideas to a noncommutative version of the Nevanlinna–Pick interpolation problem, and [**7**] gave an application to symmetric functions of two noncommuting variables.

We proceed now to give some motivating examples of noncommutative functions followed by their definition; we then discuss the difference-differential calculus and present some of the main results of the theory, and we finish the introduction with a detailed overview. We postpone a review of and a comparison to some of the earlier work on the subject to the short chapter at the end of the book.

## 1.1. Noncommutative (nc) functions: examples and genesis

Let $\mathcal{R}$ be a unital commutative ring, and let $\mathcal{R}\langle x_1,\ldots,x_d\rangle$ be the ring of nc polynomials (the free associative algebra) over $\mathcal{R}$. Here $x_1,\ldots,x_d$ are nc indeterminates, and $f\in\mathcal{R}\langle x_1,\ldots,x_d\rangle$ is of the form

$$f=\sum_{w\in\mathcal{G}_d} f_w x^w, \tag{1.1}$$

where $\mathcal{G}_d$ denotes the free monoid on $d$ generators (letters) $g_1,\ldots,g_d$ with identity $\emptyset$ (the empty word), $f_w\in\mathcal{R}$, $x^w$ are nc monomials in $x_1,\ldots,x_d$ ($x^w=x_{j_1}\cdots x_{j_m}$ for $w=g_{j_1}\cdots g_{j_m}\in\mathcal{G}_d$ and $x^\emptyset=1$), and the sum is finite. $f$ can be evaluated in an obvious way on $d$-tuples of square matrices of all sizes over $\mathcal{R}$: for $X=(X_1,\ldots,X_d)\in\left(\mathcal{R}^{n\times n}\right)^d$,

$$f(X)=\sum_{w\in\mathcal{G}_d} f_w X^w=\sum_{w\in\mathcal{G}_d} X^w f_w\in\mathcal{R}^{n\times n}. \tag{1.2}$$

We can also consider nc formal power series or nc rational functions. The ring $\mathcal{R}\langle\langle x_1,\ldots,x_d\rangle\rangle$ of nc formal power series over $\mathcal{R}$ is the (formal) completion of the ring of nc polynomials; $f\in\mathcal{R}\langle\langle x_1,\ldots,x_d\rangle\rangle$ is of the same form as in (1.1), except that the sum is in general infinite. There are two settings in which we can define the evaluation of $f$ on $d$-tuples of square matrices:

- Assume that $X=(X_1,\ldots,X_d)\in\left(\mathcal{R}^{n\times n}\right)^d$ is a jointly nilpotent $d$-tuple, i.e., $X^w=0$ for all $w\in\mathcal{G}_d$ with $|w|\ge k$ for some $k$, where $|w|$ denotes the length of the word $w$; when $\mathcal{R}=\mathbb{K}$ is a field, this simply means that $X$ is jointly similar to a $d$-tuple of strictly upper triangular matrices. Then we can define $f(X)$ as in (1.2), since the sum is actually finite.
- Assume that $\mathcal{R}=\mathbb{K}$ is the field of real or complex numbers and that $f$ has a positive nc multiradius of convergence, i.e., there exists a $d$-tuple $\rho=(\rho_1,\ldots,\rho_d)$ of strictly positive numbers such that

$$\limsup_{k\to\infty}\sqrt[k]{\sum_{|w|=k}|f_w|\rho^w}\le 1.$$

Then we can define $f(X)$ as in (1.2), where the infinite series converges absolutely and uniformly on any *nc polydisc*

$$\coprod_{n=1}^{\infty} \left\{ X \in \left(\mathbb{K}^{n\times n}\right)^d : \|X_j\| < r_j,\ j = 1, \ldots, d \right\}$$

of multiradius $r = (r_1, \ldots, r_d)$ with $r_j < \rho_j$, $j = 1, \ldots, d$.

The skew field of nc rational functions over a field $\mathbb{K}$ is the universal skew field of fractions of the ring of nc polynomials over $\mathbb{K}$. This involves some nontrivial details since unlike the commutative case, a nc rational function does not admit a canonical coprime fraction representation; see [**11, 24, 28, 30**] for some of the original constructions, and [**91**, Chapter 8] and [**29, 31**] for good expositions and background. The following quick description follows [**64, 65**], to which we refer for both details and further references. We first define (scalar) nc rational expressions by starting with nc polynomials and then applying successive arithmetic operations — addition, multiplication, and inversion. A nc rational expression $r$ can be evaluated on a $d$-tuple $X$ of $n \times n$ matrices in its *domain of regularity*, $\operatorname{dom} r$, which is defined as the set of all $d$-tuples of square matrices of all sizes such that all the inverses involved in the calculation of $r(X)$ exist. (We assume that $\operatorname{dom} r \neq \emptyset$, in other words, when forming nc rational expressions we never invert an expression that is nowhere invertible.) Two nc rational expressions $r_1$ and $r_2$ are called *equivalent* if $\operatorname{dom} r_1 \cap \operatorname{dom} r_2 \neq \emptyset$ and $r_1(Z) = r_2(Z)$ for all $d$-tuples $Z \in \operatorname{dom} r_1 \cap \operatorname{dom} r_2$. We define a *nc rational function* $\mathfrak{r}$ to be an equivalence class of nc rational expressions; notice that it has a well-defined evaluation on $\bigcup_{r\in\mathfrak{r}} \operatorname{dom} r$ (in fact, on a somewhat larger set called the extended domain of regularity of $\mathfrak{r}$ ).

We notice that in all these cases the evaluation of a formal algebraic object $f$ (a nc polynomial, formal power series, or rational function) on $d$-tuples of matrices possesses two key properties.

- $f$ *respects direct sums*: $f(X \oplus Y) = f(X) \oplus f(Y)$, where
$$X \oplus Y = (X_1 \oplus Y_1, \ldots, X_d \oplus Y_d) = \left( \begin{bmatrix} X_1 & 0 \\ 0 & Y_1 \end{bmatrix}, \ldots, \begin{bmatrix} X_d & 0 \\ 0 & Y_d \end{bmatrix} \right)$$
(we assume here that $X = (X_1, \ldots, X_d)$, $Y = (Y_1, \ldots, Y_d)$ are such that $f(X)$, $f(Y)$ are both defined).
- $f$ *respects simultaneous similarities*: $f(TXT^{-1}) = Tf(X)T^{-1}$, where
$$TXT^{-1} = (TX_1T^{-1}, \ldots, TX_dT^{-1})$$
(we assume here that $X = (X_1, \ldots, X_d)$ and $T$ are such that $f(X)$ and $f(TXT^{-1})$ are both defined).

More generally, one can consider a $p\times q$ matrix nc polynomial $f$ with coefficients $f_w \in \mathbb{K}^{p\times q}$ and with evaluation

$$f(X) = \sum_{w\in\mathcal{G}_d} X^w \otimes f_w \in \mathbb{K}^{n\times n} \otimes \mathbb{K}^{p\times q} \cong \mathbb{K}^{np\times nq} \tag{1.3}$$

for $X \in \left(\mathbb{K}^{n\times n}\right)^d$, and similarly for nc formal power series with matrix coefficients and matrix nc rational functions.[4] It is still true that the evaluation of $f$ on matrices respects direct sums and simultaneous similarities (where for similarities we replace

[4] Notice that, unlike in [**64**] and [**65**], we write the coefficients $f_w$ on the right.

$Tf(X)T^{-1}$ by $(T \otimes I_p)f(X)(T \otimes I_q)^{-1})$. In the case $\mathbb{K} = \mathbb{C}$ or $\mathbb{K} = \mathbb{R}$, one can also consider operator nc polynomials and formal power series.

*Quasideterminants* [**44, 45, 42**] and *nc symmetric functions* [**43**] are important examples of nc rational functions, whereas *formal Baker–Campbell–Hausdorff series* [**36**] are an important example of nc formal power series. Let us also mention here *nc continued fractions* [**109**].

## 1.2. NC sets, nc functions, and nc difference-differential calculus

Both for the sake of potential applications and for the sake of developing the theory in its natural generality, it turns out that the proper setting for the theory of nc functions is that of matrices of all sizes over a given vector space or a given module. In the special case when the module is $\mathcal{R}^d$, $n \times n$ matrices over $\mathcal{R}^d$ can be identified with $d$-tuples of $n \times n$ matrices over $\mathcal{R}$, and we recover nc functions of $d$ variables, key examples of which appeared in Section 1.1.

Let $\mathcal{M}$ be a module over a unital commutative ring $\mathcal{R}$; we call

$$\mathcal{M}_{\rm nc} = \coprod_{n=1}^{\infty} \mathcal{M}^{n \times n}$$

the *nc space over* $\mathcal{M}$. A subset $\Omega \subseteq \mathcal{M}_{\rm nc}$ is called a *nc set* if it is closed under direct sums, i.e., we have

$$X \oplus Y = \begin{bmatrix} X & 0 \\ 0 & Y \end{bmatrix} \in \Omega_{n+m}$$

for all $n, m \in \mathbb{N}$ and all $X \in \Omega_n$, $Y \in \Omega_m$, where we denote $\Omega_n = \Omega \cap \mathcal{M}^{n \times n}$. NC sets are the only reasonable domains for nc functions, but additional conditions on the domain are needed for the development of the nc difference-differential calculus. Essentially, we need the domain to be closed under formation of upper-triangular block matrices with an arbitrary upper corner block, but this is too stringent a requirement (e.g., this is false for nc polydiscs or nc balls — see Section 1.2.3 below). The proper notion turns out to be as follows: a nc set $\Omega \subseteq \mathcal{M}_{\rm nc}$ is called *right admissible*[5] if for all $X \in \Omega_n$, $Y \in \Omega_m$ and all $Z \in \mathcal{M}^{n \times m}$ there exists an invertible $r \in \mathcal{R}$ such that

$$\begin{bmatrix} X & rZ \\ 0 & Y \end{bmatrix} \in \Omega_{n+m}.$$

Our primary examples of right admissible nc sets are as follows:

**1.2.1.** The set $\Omega = \operatorname{Nilp}(\mathcal{M})$ of nilpotent matrices over a module $\mathcal{M}$. Here $X \in \mathcal{M}^{n \times n}$ is called nilpotent if $X^{\odot k} = 0$ for some $k$, where $X^{\odot k}$ denotes the power of $X$ as a matrix over the tensor algebra

$$\mathbf{T}(\mathcal{M}) = \bigoplus_{j=0}^{\infty} \mathcal{M}^{\otimes j}$$

of $\mathcal{M}$ (this is often called the "faux" product in operator space theory, when $\mathcal{M} = \mathcal{V}$ is an operator space); in the case where $\mathcal{R} = \mathbb{K}$ is a field, this means that there exists an invertible $T \in \mathbb{K}^{n \times n}$ such that $TXT^{-1}$ is strictly upper triangular.

[5] "Upper admissible" could have been a more appropriate terminology, however we stick with "right admissible" since it is related to the right difference-differential operator, see below. A similar comment applies to the analogous notion of "left admissible" which could have been called "lower admissible".

**1.2.2.** Assume that $\mathcal{V}$ is a Banach space (so $\mathbb{K} = \mathbb{C}$ or $\mathbb{K} = \mathbb{R}$) and that so are the spaces $\mathcal{V}^{n\times n}$, $n = 2, 3, \dots$. We usually need the topologies on $\mathcal{V}^{n\times n}$ to be compatible in the following sense: we require that the corresponding system of matrix norms $\|\cdot\|_n$ is *admissible*, i.e., for every $n, m \in \mathbb{N}$ there exist $C_1(n,m)$, $C_1'(n,m) > 0$ such that for all $X \in \mathcal{V}^{n\times n}$ and $Y \in \mathcal{V}^{m\times m}$,

$$\begin{aligned}(1.4)\quad C_1(n,m)^{-1}\max\{\|X\|_n, \|Y\|_m\} &\le \|X \oplus Y\|_{n+m} \\ &\le C_1'(n,m)\max\{\|X\|_n, \|Y\|_m\},\end{aligned}$$

and for every $n \in \mathbb{N}$ there exists $C_2(n) > 0$ such that for all $X \in \mathcal{V}^{n\times n}$ and $S, T \in \mathbb{K}^{n\times n}$,

$$\|SXT\|_n \le C_2(n)\|S\|\,\|X\|_n\|T\|, \tag{1.5}$$

where $\|\cdot\|$ denotes the operator norm of $\mathbb{K}^{n\times n}$ with respect to the standard Euclidean norm of $\mathbb{K}^n$. If $\Omega \subseteq \mathcal{V}_{\rm nc}$ is open in the sense that $\Omega_n \subseteq \mathcal{V}^{n\times n}$ is open for all $n$, then $\Omega$ is right admissible.

**1.2.3.** Assume that $\mathcal{V}$ is an *operator space*, see, e.g., [**37, 78, 79, 92**]. Recall that this means (by Ruan's Theorem) that there exists a system of norms $\|\cdot\|_n$ on $\mathcal{V}^{n\times n}$, $n = 1, 2, \dots$, satisfying

$$\|X \oplus Y\|_{n+m} = \max\{\|X\|_n, \|Y\|_m\} \quad \text{for all } X \in \mathcal{V}^{n\times n},\ Y \in \mathcal{V}^{m\times m}, \tag{1.6}$$

and

$$\|TXS\|_n \le \|T\|\|X\|_n\|S\| \quad \text{for all } X \in \mathcal{V}^{n\times n},\ T, S \in \mathbb{C}^{n\times n}. \tag{1.7}$$

Clearly, this system of norms is admissible, with the constants in (1.4) and (1.5) satisfying

$$C_1(n,m) = C_1'(n,m) = C_2(n) = 1, \quad n, m \in \mathbb{N}.$$

For $Y \in \mathcal{V}^{s\times s}$ and $r > 0$, define a *nc ball centered at $Y$ of radius $r$* as

$$B_{\rm nc}(Y,r) = \coprod_{m=1}^{\infty} B\Big(\bigoplus_{\alpha=1}^{m} Y, r\Big) = \coprod_{m=1}^{\infty}\Big\{X \in \mathcal{V}^{sm\times sm} \colon \Big\|X - \bigoplus_{\alpha=1}^{m} Y\Big\|_{sm} < r\Big\}.$$

NC balls form a basis for a topology on $\mathcal{V}_{\rm nc}$, that we call the *uniformly-open topology*, which is weaker than the disjoint union topology of Section 1.2.2. In particular, every uniformly-open nc set is right admissible.

Let $\mathcal{M}$ and $\mathcal{N}$ be modules over a unital commutative ring $\mathcal{R}$, and let $\Omega \subseteq \mathcal{M}_{\rm nc}$ be a nc set. A function $f\colon \Omega \to \mathcal{N}_{\rm nc}$ with $f(\Omega_n) \subseteq \mathcal{N}^{n\times n}$ is called a *nc function* if:

- *$f$ respects direct sums*: $f(X \oplus Y) = f(X) \oplus f(Y)$ for all $X \in \Omega_n$, $Y \in \Omega_m$.
- *$f$ respects similarities*: $f(TXT^{-1}) = Tf(X)T^{-1}$ for all $X \in \Omega_n$ and invertible $T \in \mathcal{R}^{n\times n}$ such that $TXT^{-1} \in \Omega_n$.

It turns out that these two conditions are equivalent to a single one: *$f$ respects intertwinings*, namely if $XS = SY$ then $f(X)S = Sf(Y)$, where $X \in \Omega_n$, $Y \in \Omega_m$, and $S \in \mathcal{R}^{n\times m}$. This condition originates in the pioneering work of Taylor [**100**]. We denote the module of nc functions on $\Omega$ with values in $\mathcal{N}_{\rm nc}$ by $\mathcal{T}(\Omega; \mathcal{N}_{\rm nc})$.

The main idea behind the nc difference-differential calculus is to evaluate a nc function on block upper triangular matrices. Let $f$ be a nc function on a right admissible nc set $\Omega \subseteq \mathcal{M}_{\rm nc}$ with values in $\mathcal{N}_{\rm nc}$. Let $X \in \Omega_n$, $Y \in \Omega_m$, and

$Z \in \mathcal{M}^{n\times m}$, and let $r \in \mathcal{R}$ be invertible and such that $\begin{bmatrix} X & rZ \\ 0 & Y \end{bmatrix} \in \Omega_{n+m}$. Then it turns out that

$$f\left(\begin{bmatrix} X & rZ \\ 0 & Y \end{bmatrix}\right) = \begin{bmatrix} f(X) & \Delta_R f(X,Y)(rZ) \\ 0 & f(Y) \end{bmatrix},$$

where the mapping $Z \mapsto \Delta_R f(X,Y)(Z)$ from $\mathcal{M}^{n\times m}$ to $\mathcal{N}^{n\times m}$ is $\mathcal{R}$-linear.

We will call $\Delta = \Delta_R$ the *right nc difference-differential operator.* (The left nc difference-differential operator $\Delta_L$ can be defined analogously via evaluations on block lower triangular matrices.) Its main property is that for all $n, m \in \mathbb{N}$, $X \in \Omega_n$, $Y \in \Omega_m$, and $S \in \mathcal{R}^{m\times n}$ one has

$$Sf(X) - f(Y)S = \Delta f(Y,X)(SX - YS). \tag{1.8}$$

In particular, when $n = m$ and $S = I_n$, we have the following formula of finite differences:

$$f(X) - f(Y) = \Delta f(Y,X)(X - Y) \quad (= \Delta f(X,Y)(X - Y)). \tag{1.9}$$

Thus, the linear mapping $\Delta f(Y,Y)(\cdot)$ plays the role of a nc differential.

Let $\mathcal{R} = \mathbb{K}$ be a field of real or complex numbers. Setting $X = Y + tZ$ (with $t \in \mathbb{K}$), we obtain from (1.9) that

$$f(Y + tZ) - f(Y) = t\Delta f(Y, Y + tZ)(Z).$$

Under appropriate continuity conditions, it follows that $\Delta f(Y,Y)(Z)$ is the directional derivative of $f$ at $Y$ in the direction $Z$.

In the case of $\mathcal{M} = \mathcal{R}^d$, (1.9) turns into

$$f(X) - f(Y) = \sum_{i=1}^{d} \Delta_i f(Y,X)(X_i - Y_i), \quad X, Y \in \Omega_n, \tag{1.10}$$

where $\Delta_i f(Y,X)(C) = \Delta f(Y,X)(0,\dots,0,C,0,\dots,0)$, and $C \in \mathcal{R}^{n\times n}$ is at the $i$-th position. The linear mapping $\Delta_i f(Y,Y)(\cdot)$ plays the role of an $i$-th partial nc differential at the point $Y$.

For $\mathcal{M}_0$, $\mathcal{M}_1$, $\mathcal{N}_0$, $\mathcal{N}_1$ modules over a unital commutative ring $\mathcal{R}$, and $\Omega^{(0)} \subseteq \mathcal{M}_{0,\mathrm{nc}}$, $\Omega^{(1)} \subseteq \mathcal{M}_{1,\mathrm{nc}}$ nc sets, we define a *nc function of order 1* to be a function $f$ on $\Omega^{(0)} \times \Omega^{(1)}$ so that for $X^0 \in \Omega^{(0)}_{n_0}$ and $X^1 \in \Omega^{(1)}_{n_1}$, $f(X^0, X^1)\colon {\mathcal{N}_1}^{n_0\times n_1} \to {\mathcal{N}_0}^{n_0\times n_1}$ is a linear mapping, and so that $f$ respects, in a natural way, direct sums and similarities in each argument. A typical nc function of order 1 is $f(X,Y)(Z) = f_0(X)(Zf_1(Y))$, where $f_0 \in \mathcal{T}(\Omega^{(0)}; \mathcal{N}_{0,\mathrm{nc}})$ and $f_1 \in \mathcal{T}(\Omega^{(1)}; \mathcal{N}^*_{1,\mathrm{nc}})$. We denote the class of nc functions of order 1 by $\mathcal{T}^1(\Omega^{(0)}, \Omega^{(1)}; \mathcal{N}_{0,\mathrm{nc}}, \mathcal{N}_{1,\mathrm{nc}})$.

It turns out that for $f \in \mathcal{T}(\Omega; \mathcal{N}_{\mathrm{nc}})$ one has $\Delta f \in \mathcal{T}^1(\Omega, \Omega; \mathcal{N}_{\mathrm{nc}}, \mathcal{M}_{\mathrm{nc}})$. More generally, one can define *nc functions of order k*,

$$\mathcal{T}^k(\Omega^{(0)}, \dots, \Omega^{(k)}; \mathcal{N}_{0,\mathrm{nc}}, \dots, \mathcal{N}_{k,\mathrm{nc}}),$$

where $\Omega^{(0)} \subseteq \mathcal{M}_{0,\mathrm{nc}}$, $\dots$, $\Omega^{(k)} \subseteq \mathcal{M}_{k,\mathrm{nc}}$, to be functions of $k+1$ arguments in $\Omega^{(0)}_{n_0}, \dots, \Omega^{(k)}_{n_k}$ whose values are $k$-linear mappings

$${\mathcal{N}_1}^{n_0\times n_1} \times \dots \times {\mathcal{N}_k}^{n_{k-1}\times n_k} \longrightarrow {\mathcal{N}_0}^{n_0\times n_k},$$

and that respect, in a natural way, direct sums and similarities in each argument. There is a difference-differential operator $\Delta\colon \mathcal{T}^k \to \mathcal{T}^{k+1}$, so that by iteration we obtain for $f \in \mathcal{T}(\Omega; \mathcal{N}_{\mathrm{nc}})$ that $\Delta^\ell f \in \mathcal{T}^\ell(\Omega, \dots, \Omega; \mathcal{M}_{\mathrm{nc}}, \mathcal{N}_{\mathrm{nc}}, \dots, \mathcal{N}_{\mathrm{nc}})$. (We view usual nc functions as nc functions of order 0: $\mathcal{T}(\Omega; \mathcal{N}_{\mathrm{nc}}) = \mathcal{T}^0(\Omega; \mathcal{N}_{\mathrm{nc}})$.)

Rather than using iterations, we can also calculate $\Delta^\ell f(X^0,\dots,X^\ell)(Z^1,\dots,Z^\ell)$ directly, by evaluating $f$ on a block matrix having $X^0,\dots,X^\ell$ on the main diagonal, $Z^1,\dots,Z^\ell$ just above the main diagonal, and all the other block entries zero, and taking the $(1,\ell+1)$-th block entry.

The difference-differential operators $\Delta$ (and $\Delta_i$) are obviously linear. They also satisfy a version of the chain rule (for a composition of nc functions), and — in the case when a product operation is defined — a version of the Leibnitz rule.

Using these higher order difference-differential operators, iterating the first order finite difference formula (1.9), and using the fact that direct sums are respected, leads to the following nc counterpart of the classical Taylor formula (see, e.g., [**97**] for the commutative multivariable version), that we call the *Taylor–Taylor formula* (or simply the *TT formula*), in honor of Brook Taylor and of Joseph L. Taylor. Let $f\colon\Omega\to\mathcal{N}_{\rm nc}$ be a nc function on a right admissible nc set $\Omega\subseteq\mathcal{M}_{\rm nc}$, let $s\in\mathbb{N}$, and let $Y\in\Omega_s$; then for all $m\in\mathbb{N}$, $X\in\Omega_{ms}$, and $N=0,1,\dots$ one has

$$(1.11)\quad f(X)=\sum_{\ell=0}^{N}\Big(X-\bigoplus_{\alpha=1}^{m}Y\Big)^{\odot_s\ell}\Delta^\ell f(\underbrace{Y,\dots,Y}_{\ell+1\text{ times}}) + \Big(X-\bigoplus_{\alpha=1}^{m}Y\Big)^{\odot_s N+1}\Delta^{N+1}f(\underbrace{Y,\dots,Y}_{N+1\text{ times}},X).$$

Here $(X-\bigoplus_{\alpha=1}^{m}Y)^{\odot_s\ell}$ denotes the $\ell$-th power of $X-\bigoplus_{\alpha=1}^{m}Y$ viewed as a $m\times m$ matrix over the tensor algebra $\mathbf{T}(\mathcal{M}^{s\times s})$; this power is a $m\times m$ matrix over $(\mathcal{M}^{s\times s})^{\otimes\ell}$, to which the mapping $\Delta^\ell f(Y,\dots,Y)$ viewed as a linear mapping on $(\mathcal{M}^{s\times s})^{\otimes\ell}$ is applied entrywise, yielding a $m\times m$ matrix over $\mathcal{N}^{s\times s}$, i.e., an element of $\mathcal{N}^{ms\times ms}$. The remainder term has a similar meaning.

The TT formula (1.11) makes it natural to consider the infinite TT series of $f$ around $Y\in\Omega_s$,

$$(1.12)\qquad \sum_{\ell=0}^{\infty}\Big(X-\bigoplus_{\alpha=1}^{m}Y\Big)^{\odot_s\ell}\Delta^\ell f(Y,\dots,Y).$$

In the case $\mathcal{M}=\mathcal{R}^d$ and $Y=\mu\in\mathbb{K}^d$ is a scalar point (so $s=1$), (1.12) can be rewritten as an ordinary nc power series using the $d$-tuple of partial nc difference-differential operators $\Delta=(\Delta_1,\dots,\Delta_d)$ (with an obvious abuse of notation),

$$(1.13)\qquad \sum_{w\in\mathcal{G}_d}(X-\mu I_m)^w\Delta^{w^\top}f(\mu,\dots,\mu),$$

where $\Delta^{w^\top}=\Delta_{i_\ell}\cdots\Delta_{i_1}$ for a word $w=g_{i_1}\cdots g_{i_\ell}\in\mathcal{G}_d$. There is a version of (1.13) for a general matrix center $Y$:

$$(1.14)\qquad \sum_{w\in\mathcal{G}_d}\Big(X-\bigoplus_{\alpha=1}^{m}Y\Big)^{\odot_s w}\Delta^{w^\top}f(Y,\dots,Y).$$

For $f\in\mathcal{T}(\Omega;\mathcal{N}_{\rm nc})$ and $Y\in\Omega_s$, the sequence of $\ell$-linear mappings $f_\ell:=\Delta^\ell f(Y,\dots,Y)\colon(\mathcal{M}^{s\times s})^\ell\to\mathcal{N}^{s\times s}$, $\ell=0,1,\dots$, satisfies the following conditions:

$$(1.15)\qquad Sf_0-f_0S=f_1(SY-YS),$$

and for $\ell = 1, \dots$,

$$(1.16)\qquad Sf_\ell(Z^1, \dots, Z^\ell) - f_\ell(SZ^1, Z^2, \dots, Z^\ell) = f_{\ell+1}(SY - YS, Z^2, \dots, Z^\ell),$$

$$(1.17)\qquad f_\ell(Z^1, \dots, Z^{j-1}, Z^j S, Z^{j+1}, \dots, Z^\ell) - f_\ell(Z^1, \dots, Z^j, SZ^{j+1}, Z^{j+2}, \dots, Z^\ell) \\ = f_{\ell+1}(Z^1, \dots, Z^j, SY - YS, Z^{j+1}, \dots, Z^\ell),$$

$$(1.18)\qquad f_\ell(Z^1, \dots, Z^{\ell-1}, Z^\ell S) - f_\ell(Z^1, \dots, Z^\ell)S = f_{\ell+1}(Z^1, \dots, Z^\ell, SY - YS),$$

for every $S \in \mathcal{R}^{s \times s}$. (Notice, that in the case $s = 1$, conditions (1.15)–(1.18) are trivial.) Conversely, given a sequence of $\ell$-linear mappings $f_\ell\colon (\mathcal{M}^{s\times s})^\ell \to \mathcal{N}^{s\times s}$, $\ell = 0, 1, \dots$, satisfying conditions (1.15)–(1.18), the sum of the series

$$(1.19)\qquad f(X) = \sum_{\ell=0}^{\infty} \Big(X - \bigoplus_{\alpha=1}^{m} Y\Big)^{\odot_s \ell} f_\ell,$$

whenever it makes sense, defines a nc function. Similar direct and inverse statement can also be formulated in the case where $\mathcal{M} = \mathcal{R}^d$ for the series along $\boldsymbol{\mathcal{G}}_d$.

## 1.3. Applications of the Taylor–Taylor formula

The Taylor–Taylor formula is the main tool for the study of local behavior of nc functions. Here are some sample results.

**1.3.1.** Any nc function on $\mathrm{Nilp}_d(\mathcal{R}) := \mathrm{Nilp}(\mathcal{R}^d)$ (the set of jointly nilpotent $d$-tuples of matrices over a unital commutative ring $\mathcal{R}$) is given by a nc power series

$$f(X) = \sum_{w \in \boldsymbol{\mathcal{G}}_d} X^w \Delta^{w^\top} f(0, \dots, 0),$$

where the sum is finite. More generally, let $Y \in (\mathcal{R}^{s\times s})^d$. Denote by $\mathrm{Nilp}_d(\mathcal{R}, Y)$ the set of $d$-tuples of $sm \times sm$ matrices over $\mathcal{R}$, $m = 1, 2, \dots$, such that $X - \bigoplus_{\alpha=1}^m Y \in \mathrm{Nilp}_d(\mathcal{R})$. Then any nc function on $\mathrm{Nilp}_d(\mathcal{R}, Y)$ can be written as in (1.14) where, again, the sum is finite.

**1.3.2.** Let $\mathbb{K}$ be an infinite field, and let $f$ be a nc function on $(\mathbb{K}^d)_{\rm nc}$ such that for every $n$ each matrix entry of $f(X_1, \dots, X_d)$ is a polynomial in matrix entries of $X_1, \dots, X_d$ of a uniformly (in $n$) bounded degree. Then $f$ is a nc polynomial.

The condition of uniform boundedness of the degrees is necessary. However, we can get a sharp result without this condition as well: namely, $f$ belongs to the completion of the ring of nc polynomials with respect to the decreasing sequence of ideals $\{\boldsymbol{\mathcal{I}}_n\}_{n=1}^\infty$, where $\boldsymbol{\mathcal{I}}_n$ denotes the ideal of identities [**91**] for $n \times n$ matrices (over $\mathbb{K}$ and in $d$ variables).

**1.3.3.** Let $\mathbb{K} = \mathbb{C}$, let $\mathcal{V}$ and $\mathcal{W}$ be Banach spaces equipped with admissible systems of matrix norms over $\mathcal{V}$ and over $\mathcal{W}$ — see Section 1.2.2, and let $\Omega \subseteq \mathcal{V}_{\rm nc}$ be an open nc set. If a nc function $f$ on $\Omega$ is locally bounded, then it is analytic, i.e., $f|_{\Omega_n}$ is an analytic function on $\Omega_n \subseteq \mathcal{V}^{n\times n}$ with values in $\mathcal{W}^{n\times n}$ (see, e.g., [**61, 72**] for the general theory of analytic functions between Banach spaces), and the TT series (1.12) converges to $f$ locally uniformly. More precisely, if $f$ is bounded on $B(\bigoplus_{\alpha=1}^m Y, \delta)$, then the series (1.12) converges absolutely and uniformly on $B(\bigoplus_{\alpha=1}^m Y, r)$ for all $r < \delta$. In the case $\mathcal{V} = \mathbb{C}^d$ and $\mathcal{W} = \mathbb{C}$, analyticity simply means that for every $n$ each matrix entry of $f(X_1, \dots, X_d)$ is an analytic function of the matrix entries of $X_1, \dots, X_d$; in this case it is enough to require that for every $n$, $f|_{\Omega_n}$ is locally bounded on slices, that is, for any fixed $d$-tuples of matrices $X = (X_1, \dots, X_d) \in \Omega_s$ and $Z = (Z_1, \dots, Z_d) \in (\mathbb{C}^{s\times s})^d$, $f(X+tZ)$ is bounded for all $t$: $|t| < \epsilon$, with some $\epsilon > 0$. In fact, if a nc function $f\colon \Omega \to \mathcal{W}_{\mathrm{nc}}$ (for a general vector space $\mathcal{V}$ and a Banach space $\mathcal{W}$ as before) is locally bounded on slices, then $f$ is Gâteaux differentiable and its TT series converges.

**1.3.4.** Let $\mathcal{V}$ and $\mathcal{W}$ be operator spaces, and let $\Omega \subseteq \mathcal{V}_{\mathrm{nc}}$ be a uniformly-open nc set (see Section 1.2.3). If a nc function $f$ on $\Omega$ is locally bounded in the uniformly-open topology, then the TT series (1.12) converges to $f$ locally uniformly in that topology. More precisely, if $f$ is bounded on $B_{\mathrm{nc}}(Y, \delta)$, then the series converges absolutely and uniformly on $B_{\mathrm{nc}}(Y, r)$ for all $r < \delta$. $f$ is called a uniformly analytic nc function.

In the case where $\mathcal{V} = \mathbb{C}^d$ with some operator space structure given by a sequence of norms $\|\cdot\|_n$, it is further true that the series (1.13) — without grouping together terms of the same degree into homogeneous nc polynomials — converges absolutely and uniformly on somewhat smaller sets, namely on every *open nc diamond about* $Y$,

$$\diamondsuit_{\mathrm{nc}}(Y, r) := \coprod_{m=1}^{\infty} \left\{ X \in \Omega_{sm} \colon \sum_{j=1}^{d} \|e_j\|_1 \left\| X_j - \bigoplus_{\alpha=1}^{m} Y_j \right\| < r \right\}$$

with $r < \delta$; here $e_1, \dots, e_d$ denote the standard basis for $\mathbb{C}^d$.

**1.3.5.** TT series expansions allow us to reformulate some results that were originally established for nc power series as results about nc functions. This includes, in particular, realizations of nc power series as transfer functions of noncommutative multidimensional systems (systems with evolution along the free monoid $\mathcal{G}_d$[6]). These systems were first studied in [**20**] in the conservative infinite-dimensional setting, in the context of operator model theory for row contractions due mostly to Popescu [**81, 82, 83, 84**] and of representation theory of the Cuntz algebra [**27, 32**]. A comprehensive study of nc realization theory, in both finite-dimensional and infinite-dimensional setting, appears in [**15, 17, 16**]; these papers give a unified framework of structured nc multidimensional linear systems for different kinds of realization formulae. We also mention the paper [**18**] where an even more general class of nc systems (given though in a frequency domain) was described and the corresponding dilation theory was developed.

We give (a somewhat imprecise version of) the main conservative realization theorem from [**17**] restated as a theorem about nc functions. Let $Q_1, \dots, Q_d$ be $p\times q$ complex matrices with a certain additional structure that comes from a bipartite graph. Let $\mathcal{U}$ and $\mathcal{Y}$ be Hilbert spaces, and let $\mathcal{L}(\mathcal{U}, \mathcal{Y})$ be the space of bounded linear operators from $\mathcal{U}$ to $\mathcal{Y}$. The corresponding *nc Schur–Agler class* consists of contraction-valued $f \in \mathcal{T}(B_{\mathrm{nc}}(0,1); \mathcal{L}(\mathcal{U}, \mathcal{Y})_{\mathrm{nc}})$, where $B_{\mathrm{nc}}(0,1) \subseteq (\mathbb{C}^d)_{\mathrm{nc}}$, and $\mathbb{C}^d$ is equipped with the operator space structure defined by the system of matrix norms

$$\|X\|_n = \|Q(X)\|_{\mathcal{L}(\mathbb{C}^n \otimes \mathbb{C}^q, \mathbb{C}^n \otimes \mathbb{C}^p)},$$

where $Q(X) := X_1 \otimes Q_1 + \cdots + X_d \otimes Q_d$. Then the following are equivalent:

[6] Referred to often as systems with evolution along the free semigroup.

(1) $f$ is in the nc Schur–Agler class.
(2) There exists an *Agler decomposition*

$$I_{\mathbb{C}^n \otimes \mathcal{Y}} - f(X)f(Y)^* = H(X)(I_{\mathbb{C}^n \otimes \mathbb{C}^p \otimes \mathcal{G}} - (Q(X) \otimes I_{\mathcal{G}})(Q(Y)^* \otimes I_{\mathcal{G}}))H(Y)^*$$

for an auxiliary Hilbert space $\mathcal{G}$ and some $H \in \mathcal{T}(B_{\rm nc}(0,1); \mathcal{L}(\mathbb{C}^p \otimes \mathcal{G}, \mathcal{Y})_{\rm nc})$ which is bounded on every nc ball $B_{\rm nc}(0,r)$ of radius $r < 1$.
(3) There exists a conservative realization:

$$f(X) = I_n \otimes D + (I_n \otimes C)(I_{\mathbb{C}^n \otimes \mathbb{C}^p \otimes \mathcal{G}} - Q(X)(I_n \otimes A))^{-1} Q(X)(I_n \otimes B).$$

Here the colligation matrix

$$\begin{bmatrix} A & B \\ C & D \end{bmatrix} : \begin{bmatrix} \mathbb{C}^p \otimes \mathcal{G} \\ \mathcal{U} \end{bmatrix} \to \begin{bmatrix} \mathbb{C}^q \otimes \mathcal{G} \\ \mathcal{Y} \end{bmatrix}$$

is unitary.

(The statement in [**17**] is considerably more precise in identifying the state spaces $\mathbb{C}^p \otimes \mathcal{G}$ and $\mathbb{C}^q \otimes \mathcal{G}$ of the unitary colligation in terms of the bipartite graph generating the matrices $Q_1, \ldots, Q_d$.) This is the nc version of the commutative multidimensional realization theorem of [**10**] and [**13**] for the linear nc function $Q$ defining the domain of $f$, which in turn generalizes Agler's seminal result on the polydisc [**4**], see also [**19**]. Such generalized transfer-function realizations occur also in the context of the generalized Hardy algebras of Muhly–Solel (elements of which are viewed as functions on the unit ball of representations of the algebra—see [**68, 70, 12**]) and of the generalized Schur–Agler class associated with an admissible class of test functions (see [**34, 35**]).

We note that the formulation of the realization theorem for the nc Schur–Agler class given above suggests generalizations to an arbitrary finite- or infinite-dimensional operator space (instead of $\mathbb{C}^d$ equipped with the system of norms associated with the linear nc function $Q$) and to more general nc domains (a nc counterpart of the commutative domains of [**10**] and [**13**], where the defining nc function $Q$ is a nc polynomial). In the finite-dimensional case such a generalization has been carried out recently in [**5**]; we refer to both this paper and [**6, 7**] for applications.

## 1.4. An overview

In Chapter 2, we introduce nc functions and their difference-differential calculus. In Section 2.1, we define nc spaces, nc sets, and nc functions, and prove that respecting direct sums and similarities is equivalent to respecting intertwinings (Proposition 2.1). In Section 2.2, we introduce the right and left difference-differential operators $\Delta_R$ and $\Delta_L$ via evaluation of nc functions on block upper triangular matrices and then taking the upper right (left) corner block of the value, and show that $\Delta_R f(X,Y)(Z)$ and $\Delta_L f(X,Y)(Z)$ are linear in $Z$ (Propositions 2.4 and 2.6). We proceed to formulate basic nc calculus rules in Section 2.3 and to establish the first-order finite difference formulae in Section 2.4. Then we prove in Section 2.5 that $\Delta_R f(X,Y)(Z)$ and $\Delta_L f(X,Y)(Z)$ respect direct sums and similarities in $X$ and in $Y$ (in the terminology of Chapter 3, $\Delta_R f$ and $\Delta_L f$ are nc functions of order 1). We conclude Chapter 2 by discussing in Section 2.6 directional and (in the special case $\mathcal{M} = \mathcal{R}^d$) partial nc difference-differential operators.

In Chapter 3, we first introduce higher order nc functions, and then extend the difference-differential operators to these functions, which leads to a higher order calculus for ordinary and higher order nc functions. In Section 3.1, we define nc functions of order $k$ and introduce the classes $\mathcal{T}^k$ of those functions (so that the original nc functions belong to the class $\mathcal{T}^0$). The values of nc functions of order $k$ are $k$-linear mappings of matrices over modules, which can also be interpreted as elements of tensor products of $k+1$ matrix modules (Remarks 3.4 and 3.5). This allows us to define a natural mapping $\mathcal{T}^k \otimes \mathcal{T}^\ell \to \mathcal{T}^{k+\ell+1}$, $k, \ell = 0, 1, \ldots$, and consequently — a mapping from the tensor product of $k+1$ classes $\mathcal{T}^0$ (perhaps over different operator spaces) to $\mathcal{T}^k$ (Remark 3.6) and interpret a higher order nc function in terms of nc functions of order 0. We then extend in Section 3.2 the right difference-differential operator $\Delta_R$ to a mapping of classes $\mathcal{T}^k \to \mathcal{T}^{k+1}$, $k = 0, 1, \ldots$. In this definition, the last argument of $f(X^0, \ldots, X^k)$, where $f$ is a nc function of order $k$, is taken in the form of a block matrix $X^k = \begin{bmatrix} X^{k\prime} & Z \\ 0 & X^{k\prime\prime} \end{bmatrix}$, and then we take the upper right corner block of the matrix value. The extended operator $\Delta_R$ is also linear as a function of $Z$ (Propositions 3.8 and 3.9). Starting from this point, we concentrate on the right version of calculus, since the left version can be developed analogously or simply obtained from the right one by symmetry.

The iterated operator $\Delta_R^\ell \colon \mathcal{T}^k \to \mathcal{T}^{k+\ell}$ can also be computed directly, by evaluating a nc function on block upper bidiagonal matrices and then taking the upper right corner block of its matrix value (Therems 3.11 and 3.12). The definition of the extended operator $\Delta_R$ becomes more transparent when we interpret higher order nc functions using tensor products of nc functions of order 0 (Remark 3.17).

We can also take $X^j$ in $f(X^0, \ldots, X^k)$ in the block upper triangular form, which leads to operators ${}_j\Delta_R \colon \mathcal{T}^k \to \mathcal{T}^{k+1}$, $j = 0, 1, \ldots$ (so that $\Delta_R = {}_k\Delta_R$), that also extend the original operator $\Delta_R \colon \mathcal{T}^0 \to \mathcal{T}^1$ (Remark 3.18). In Section 3.3, we establish the first order difference formulae for higher order nc functions (Theorem 3.19) and its generalized version for the case of matrices $X$ and $Y$ where we evaluate the higher order nc function in these formulae, of possibly different size (Theorem 3.20). Similar formulae are valid for operators ${}_j\Delta_R$ (Remark 3.22).

In Section 3.4, we show that for a nc function $f \in \mathcal{T}^k$ which is $k$ times integrable, i.e., can be represented as $\Delta_R^k g$ with $g \in \mathcal{T}^0$, one has ${}_i\Delta_R\, {}_j\Delta_R f = {}_j\Delta_R\, {}_i\Delta_R f$, for all $i, j = 0, \ldots, k$. Finally, we discuss in Section 3.5 higher order directional nc difference-differential operators; in particular, when the underlying module is $\mathcal{R}^d$, we discuss higher order partial difference-differential operators $\Delta_R^w$, $w \in \mathcal{G}_d$.

In Chapter 4, we establish the Taylor–Taylor formula (Theorem 4.1 or more general Theorem 4.2; in the case $\mathcal{M} = \mathcal{R}^d$ — Corollary 4.4 and, in tensor product interpretation of $\Delta_R^{w^\top} f$, Theorem 4.6). We show in Remark 4.3 that the coefficients in the TT formula, the multilinear mappings $f_\ell = \Delta_R f(Y, \ldots, Y)$, $\ell = 0, 1, \ldots$, satisfy conditions (1.15)–(1.18) that we mentioned in Section 1.2. The counterpart of these condition for the sequence $f_w = \Delta_R^{w^\top} f(Y, \ldots, Y)$, $w \in \mathcal{G}_d$, in the case $\mathcal{M} = \mathcal{R}^d$ is established in Remark 4.5.

As a first application of the TT formula, we describe in Chapter 5 nc functions on the set $\mathrm{Nilp}(\mathcal{M})$ (or, more generally, on $\mathrm{Nilp}(\mathcal{M}; Y)$) of nilpotent matrices (or nilpotent matrices about $Y$) over a module $\mathcal{M}$ — see Section 1.2.1 for the definition. In Section 5.1 we show that every such a nc function is a sum of its TT series (Theorems 5.2, 5.4, 5.6, and 5.8). Then we show that the coefficients of a nc power series representing the nc function are uniquely determined. In fact, this series is unique and is equal to the corresponding TT series (Theorems 5.9 and 5.10).

In Section 5.2 we show that, conversely, given a sequence of multilinear mappings $f_\ell\colon (\mathcal{M}^{s\times s})^\ell \to \mathcal{N}^{s\times s}$, $\ell = 0,1,\ldots$, satisfying conditions (1.15)–(1.18), the sum of the nc power series $\sum_{\ell=0}^\infty (X - \bigoplus_{\alpha=1}^m Y)^{\odot_s \ell} f_\ell$ is a nc function on $\mathrm{Nilp}(\mathcal{M};Y)$ with values in $\mathcal{N}_{\mathrm{nc}}$ (Theorem 5.13), and a version of this result, Theorem 5.15, for the case $\mathcal{M} = \mathcal{R}^d$ and a sequence of multilinear mappings $f_w\colon (\mathcal{M}^{s\times s})^\ell \to \mathcal{N}^{s\times s}$, $w \in \mathcal{G}_d$, satisfying the conditions analogous to (1.15)–(1.18). These results are algebraic in their nature, since the sum of the corresponding series is finite at every point of $\mathrm{Nilp}(\mathcal{M};Y)$.

In Chapter 6, we give some other algebraic applications of the TT formula. We prove that a nc function on $(\mathbb{K}^d)_{\mathrm{nc}}$, with $\mathbb{K}$ an infinite field, which is polynomial in matrix entries when evaluated on $n \times n$ matrices, $n = 1,2,\ldots$, of bounded degree, is necessarily a nc polynomial (Theorem 6.1). The degree boundedness condition is essential and cannot be omitted (Example 6.3). However, if we do not require this condition, then the following is true: an arbitrary nc function on $(\mathbb{K}^d)_{\mathrm{nc}}$, which is polynomial of degree $M_n$ in matrix entries when evaluated on $n \times n$ matrices, is a sum of a nc polynomial and an infinite series of homogeneous polynomials $f_j$ vanishing on all $d$-tuples of $n \times n$ matrices for $j > M_n$ (Theorem 6.4). We also obtain a generalization of Theorem 6.1 to nc functions on (much more general) nc spaces, which are polynomial on slices (Theorem 6.8).

In Chapter 7 we study analytic nc functions and the convergence of their TT series. We consider three different topologies on a nc space $\mathcal{V}_{\mathrm{nc}}$ (over a complex vector space $\mathcal{V}$), or three types of convergence of nc power series, and, correspondingly, three types of analyticity of nc functions: finitely open topology and analyticity on slices, norm topology and analyticity on $n\times n$ matrices over $\mathcal{V}$ for every $n = 1,2,\ldots$, and uniformly-open topology and uniform (in $n$) analyticity. The main feature of analytic nc functions established in Chapter 7 is that local boundedness implies analyticity, in each of the three settings. In the first part of Section 7.1, we introduce the finitely open topology on $\mathcal{V}_{\mathrm{nc}}$ and prove that a nc function which is locally bounded on slices is analytic on slices and discuss the convergence of the corresponding TT series (Theorem 7.2). In the second part of Section 7.1, we show that a nc function which is locally bounded (with respect to the norm topology on $\mathcal{V}^{n\times n}$, for every $n$) is analytic, with the TT series convergent uniformly and absolutely on certain open complete circular nc sets about the center $Y \in \Omega_s$ (Theorem 7.4) or on open balls centered at $Y$ (Corollary 7.5). We also establish the uniqueness of the convergent nc power series expansions, in both of the above topologies (Theorem 7.9). In Section 7.2, we introduce and study the uniformly-open topology on a nc space $\mathcal{V}_{\mathrm{nc}}$ over an operator space $\mathcal{V}$. In Section 7.3, we show that a uniformly locally bounded nc function is uniformly analytic, and its TT series converges uniformly and absolutely on certain uniformly-open complete circular nc sets about $Y$ (Theorem 7.21) or on certain uniformly-open matrix circular nc sets (Theorem 7.23) or on uniformly-open nc balls (Corollary 7.26). We also prove a version of this result for the case $\mathcal{V} = \mathbb{C}^d$ and the TT series convergent along $\mathcal{G}_d$ (Theorem 7.29 and Corollary 7.31). In Section 7.4 we introduce and study analytic higher order nc functions, also in the three different settings. The main results are analogous to those in Sections 7.1 and 7.3.

In Chapter 8, we study the convergence of nc power series in the three different topologies, as in Chapter 7, to an (analytic in the corresponding sense) nc function. The results are the converse of the results of Sections 7.1 and 7.3. We discuss the convergence of nc power series in the finitely open topology (resp., in the norm topology and in the uniformly-open topology) in Section 8.1 (resp., 8.2 and 8.3). In each topology, we have Cauchy–Hadamard type estimates for radii of convergence and sharp estimates for the size of convergence domains of various shapes. We also characterize the maximal nc sets where the sum of the series is an analytic on slices (resp., analytic and uniformly analytic) nc function and give many examples illustrating the differences between various types of convergence domains.

In Chapter 9, we define and study so-called direct summands extensions of nc sets and nc functions. If a nc set $\Omega$ is invariant under similarities, then one can extend $\Omega$ to a larger nc set, $\Omega_{\text{d.s.e.}}$, which contains, together with a matrix which is decomposable into a direct sum of matrices, every direct summand of the decomposition. This extension preserves many properties of $\Omega$ (Proposition 9.1). We then can extend a nc function $f$ on $\Omega$ to a nc function $f_{\text{d.s.e.}}$ on $\Omega_{\text{d.s.e.}}$, which inherits most important properties of $f$, in particular, analyticity (Proposition 9.2). Similarly, we define and establish the properties of the direct summands extensions of higher order nc functions (Proposition 9.3). We also define and study direct summands extensions of sequences $f_\ell$, $\ell = 0, 1, \ldots$ (resp., $f_w$, $w \in \mathcal{G}_d$) of multilinear mappings satisfying conditions (1.15)–(1.18) (resp., their counterparts in the case $\mathcal{M} = \mathcal{R}^d$) in Proposition 9.4 (resp., in Proposition 9.6).

As we mentioned in Section 1.2, we need a nc set $\Omega$ to be right (left) admissible in order to define right (left) difference-differential operators via evaluations of nc functions $f$ on $\Omega$ at block upper (lower) triangular matrices. However, for a (say, upper) block triangular matrix $\begin{bmatrix} X & Z \\ 0 & Y \end{bmatrix}$ we need to scale $Z$, so that $\begin{bmatrix} X & rZ \\ 0 & Y \end{bmatrix} \in \Omega$. Then we can define $\Delta_R f(X,Y)(rZ)$ and extend this definition to arbitrary $Z$ by linearity of $\Delta_R f(X,Y)(\cdot)$. Although the properties of $\Delta_R$ ($\Delta_L$) can be established using these scalings, this creates cumbersome technicalities in the proofs. To bypass those technicalities, we define the so-called similarity invariant envelope $\widetilde{\Omega}$ of $\Omega$, which is the smallest nc set containing $\Omega$ and invariant under similarities. Then it turns out that $\widetilde{\Omega}$ is also invariant under formation of block triangular matrices (Proposition A.2). Our next step is to extend a nc function $f$ on $\Omega$ to a nc function $\widetilde{f}$ on $\widetilde{\Omega}$. Such an extension is unique (Proposition A.3). A similar extension is constructed for higher order nc functions (Proposition A.5). It turns out that $(\Delta_R \widetilde{f})|_{\Omega\times\Omega} = \Delta_R f$ (Remark 2.5), which allows us to prove various statements about the difference-differential operators throughout the monograph assuming, without loss of generality, that the underlying nc set $\Omega$ is similarity invariant. The results on similarity invariant envelopes and the corresponding extensions of nc functions are presented in Appendix A, which is written in collaboration with Shibananda Biswas.

## Acknowledgements

We wish to thank Mihai Putinar for directing our attention to the work of J. L. Taylor, and Shibananda Biswas for a careful reading of the manuscript and valuable suggestions. It is also a pleasure to thank J. A. Ball, S. Belinschi, M. Dritschel, J. W. Helton, I. Klep, S. McCullough, P. Muhly, M. Popa, and B. Solel for fruitful discussions. Last but not least, we would like to thank the three anonymous referees for their many helpful remarks.

CHAPTER 2

# NC functions and their difference-differential calculus

## 2.1. Definition of nc functions

Let $\mathcal{R}$ be a commutative ring with identity. For $\mathcal{M}$ a module over $\mathcal{R}$, we define the *nc space over* $\mathcal{M}$

$$\mathcal{M}_{\rm nc} = \coprod_{n=1}^{\infty} \mathcal{M}^{n\times n}.$$

For $X \in \mathcal{M}^{n\times n}$ and $Y \in \mathcal{M}^{m\times m}$ we define their *direct sum*

$$X \oplus Y = \begin{bmatrix} X & 0 \\ 0 & Y \end{bmatrix} \in \mathcal{M}^{(n+m)\times(n+m)}.$$

Notice that matrices over $\mathcal{R}$ act from the right and from the left on matrices over $\mathcal{M}$ by the standard rules of matrix multiplication and by the action of $\mathcal{R}$ on $\mathcal{M}$: if $X \in \mathcal{M}^{p\times q}$ and $T \in \mathcal{R}^{r\times p}$, $S \in \mathcal{R}^{q\times s}$, then

$$TX \in \mathcal{M}^{r\times q}, \quad XS \in \mathcal{M}^{p\times s}.$$

A subset $\Omega \subseteq \mathcal{M}_{\rm nc}$ is called a *nc set* if it is closed under direct sums; explicitly, denoting $\Omega_n = \Omega \cap \mathcal{M}^{n\times n}$, we have $X \oplus Y \in \Omega_{n+m}$ for all $X \in \Omega_n$, $Y \in \Omega_m$. We will sometimes use the convention that $\mathcal{M}^{n\times 0}$, $\mathcal{M}^{0\times n}$, and $\Omega_0 = \mathcal{M}^{0\times 0}$ consist each of a single (zero) element, the empty matrix "of appropriate size", and that $X \oplus Y = Y \oplus X = X \in \Omega_n$ for $n \in \mathbb{N}$, $X \in \Omega_n$, and $Y \in \Omega_0$.

In the case of $\mathcal{M} = \mathcal{R}^d$, we identify matrices over $\mathcal{M}$ with $d$-tuples of matrices over $\mathcal{R}$:

$$\left(\mathcal{R}^d\right)^{p\times q} \cong \left(\mathcal{R}^{p\times q}\right)^d.$$

Under this identification, for $d$-tuples $X = (X_1, \ldots, X_d) \in \left(\mathcal{R}^{n\times n}\right)^d$ and $Y = (Y_1, \ldots, Y_d) \in \left(\mathcal{R}^{m\times m}\right)^d$,

$$X \oplus Y = \left( \begin{bmatrix} X_1 & 0 \\ 0 & Y_1 \end{bmatrix}, \ldots, \begin{bmatrix} X_d & 0 \\ 0 & Y_d \end{bmatrix} \right) \in \left(\mathcal{R}^{(n+m)\times(n+m)}\right)^d;$$

and for a $d$-tuple $X = (X_1, \ldots, X_d) \in \left(\mathcal{R}^{p\times q}\right)^d$ and matrices $T \in \mathcal{R}^{r\times p}$, $S \in \mathcal{R}^{q\times s}$,

$$TX = (TX_1, \ldots, TX_d) \in \left(\mathcal{R}^{r\times q}\right)^d, \quad XS = (X_1S, \ldots, X_dS) \in \left(\mathcal{R}^{p\times s}\right)^d.$$

Let $\mathcal{M}$ and $\mathcal{N}$ be modules over $\mathcal{R}$, and let $\Omega \subseteq \mathcal{M}_{\rm nc}$ be a nc set. A mapping $f \colon \Omega \to \mathcal{N}_{\rm nc}$, with $f(\Omega_n) \subseteq \mathcal{N}^{n\times n}$, $n = 0, 1, \ldots$, is called a *nc function* if $f$ satisfies the following two conditions:

- $f$ *respects direct sums*:

$$f(X \oplus Y) = f(X) \oplus f(Y) \tag{2.1}$$

  for all $X, Y \in \Omega$.

- *$f$ respects similarities*: if $X \in \Omega_n$ and $S \in \mathcal{R}^{n\times n}$ is invertible with $SXS^{-1} \in \Omega_n$, then

$$f(SXS^{-1}) = Sf(X)S^{-1}. \tag{2.2}$$

The two conditions in the definition of a nc function can be actually replaced by a single one.

PROPOSITION 2.1. *A mapping $f\colon \Omega \to \mathcal{N}_{\rm nc}$, with $f(\Omega_n) \subseteq \mathcal{N}^{n\times n}$, $n = 0, 1, \ldots$, respects direct sums and similarities if and only if it* respects intertwinings*: for any $X \in \Omega_n$, $Y \in \Omega_m$, and $T \in \mathcal{R}^{n\times m}$, one has*

$$XT = TY \Longrightarrow f(X)T = Tf(Y). \tag{2.3}$$

Condition (2.3) was used by J. L. Taylor in [**100**].

PROOF OF PROPOSITION 2.1. Assume that $f$ respects direct sums and similarities. If $X \in \Omega_n$, $Y \in \Omega_m$, and $T \in \mathcal{R}^{n\times m}$ are such that $XT = TY$, then

$$\begin{bmatrix} X & 0 \\ 0 & Y \end{bmatrix} = \begin{bmatrix} I_n & -T \\ 0 & I_m \end{bmatrix} \begin{bmatrix} X & 0 \\ 0 & Y \end{bmatrix} \begin{bmatrix} I_n & T \\ 0 & I_m \end{bmatrix}.$$

Since

$$\begin{bmatrix} I_n & -T \\ 0 & I_m \end{bmatrix} = \begin{bmatrix} I_n & T \\ 0 & I_m \end{bmatrix}^{-1},$$

we can apply first (2.2) and then (2.1) to get

$$\begin{bmatrix} f(X) & 0 \\ 0 & f(Y) \end{bmatrix} = \begin{bmatrix} I_n & -T \\ 0 & I_m \end{bmatrix} \begin{bmatrix} f(X) & 0 \\ 0 & f(Y) \end{bmatrix} \begin{bmatrix} I_n & T \\ 0 & I_m \end{bmatrix}.$$

Equating the $(1,2)$ block entries on the two sides of this equality, we get $0 = f(X)T - Tf(Y)$, i.e., (2.3) holds.

Conversely, assume that $f$ respects intertwinings. It is then clear that $f$ respects similarities. To prove that $f$ respects direct sums, let $X \in \Omega_n$, $Y \in \Omega_m$. Obviously,

$$\begin{bmatrix} X & 0 \\ 0 & Y \end{bmatrix} \begin{bmatrix} I_n \\ 0 \end{bmatrix} = \begin{bmatrix} I_n \\ 0 \end{bmatrix} X.$$

Therefore

$$f\left(\begin{bmatrix} X & 0 \\ 0 & Y \end{bmatrix}\right) \begin{bmatrix} I_n \\ 0 \end{bmatrix} = \begin{bmatrix} I_n \\ 0 \end{bmatrix} f(X). \tag{2.4}$$

Denote

$$f\left(\begin{bmatrix} X & 0 \\ 0 & Y \end{bmatrix}\right) = \begin{bmatrix} A & B \\ C & D \end{bmatrix}.$$

Then (2.4) becomes

$$\begin{bmatrix} A \\ C \end{bmatrix} = \begin{bmatrix} f(X) \\ 0 \end{bmatrix}.$$

In the same way the identity

$$\begin{bmatrix} X & 0 \\ 0 & Y \end{bmatrix} \begin{bmatrix} 0 \\ I_m \end{bmatrix} = \begin{bmatrix} 0 \\ I_m \end{bmatrix} Y$$

implies that

$$\begin{bmatrix} B \\ D \end{bmatrix} = \begin{bmatrix} 0 \\ f(Y) \end{bmatrix}.$$

Thus $A = f(X)$, $B = 0$, $C = 0$, $D = f(Y)$, i.e., (2.1) holds. □

## 2.2. NC difference-differential operators

Our next goal is to introduce right and left difference-differential operators, $\Delta_R$ and $\Delta_L$, on nc functions. This will be done by evaluating a nc function $f$ on block upper (respectively, lower) triangular matrices. The idea is to show that, for $X \in \Omega_n$ and $Y \in \Omega_m$,

$$f\left(\begin{bmatrix} X & Z \\ 0 & Y \end{bmatrix}\right) = \begin{bmatrix} f(X) & \Delta_R f(X,Y)(Z) \\ 0 & f(Y) \end{bmatrix}. \tag{2.5}$$

Here $Z \mapsto \Delta_R f(X,Y)(Z)$ is a $\mathcal{R}$-linear operator from $\mathcal{M}^{n\times m}$ to $\mathcal{N}^{n\times m}$ which plays the role of a right nc difference operator when $Y \neq X$, and of a right nc differential operator when $Y = X$. Analogously,

$$f\left(\begin{bmatrix} X & 0 \\ Z & Y \end{bmatrix}\right) = \begin{bmatrix} f(X) & 0 \\ \Delta_L f(X,Y)(Z) & f(Y) \end{bmatrix}, \tag{2.6}$$

where $Z \mapsto \Delta_L f(X,Y)(Z)$ is a $\mathcal{R}$-linear operator from $\mathcal{M}^{m\times n}$ to $\mathcal{N}^{m\times n}$ which plays the role of a left nc difference operator when $Y \neq X$, and of a left nc differential operator when $Y = X$. The assumption that $\begin{bmatrix} X & Z \\ 0 & Y \end{bmatrix} \in \Omega_{n+m}$ for all $X \in \Omega_n$, $Y \in \Omega_m$, and $Z \in \mathcal{M}^{n\times m}$ (respectively, that $\begin{bmatrix} X & 0 \\ Z & Y \end{bmatrix} \in \Omega_{n+m}$ for all $X \in \Omega_n$, $Y \in \Omega_m$, and $Z \in \mathcal{M}^{m\times n}$) would be however too strong, and we will proceed with a weaker assumption on a nc set $\Omega$, which is naturally suggested by the following proposition.

PROPOSITION 2.2. *Let $f$ be a nc function on a nc set $\Omega$. Let $X \in \Omega_n$, $Y \in \Omega_m$, and $Z \in \mathcal{M}^{n\times m}$ be such that $\begin{bmatrix} X & Z \\ 0 & Y \end{bmatrix} \in \Omega_{n+m}$. Then (2.5) holds, where the off-diagonal block entry $\Delta_R f(X,Y)(Z)$ is determined uniquely by (2.5) and has the following property. If $r \in \mathcal{R}$ is such that $\begin{bmatrix} X & rZ \\ 0 & Y \end{bmatrix} \in \Omega_{n+m}$, then*

$$\Delta_R f(X,Y)(rZ) = r\Delta_R f(X,Y)(Z). \tag{2.7}$$

PROOF. Denote

$$f\left(\begin{bmatrix} X & Z \\ 0 & Y \end{bmatrix}\right) = \begin{bmatrix} A & \Delta_R f(X,Y)(Z) \\ C & D \end{bmatrix}.$$

We have

$$\begin{bmatrix} X & Z \\ 0 & Y \end{bmatrix}\begin{bmatrix} I_n \\ 0 \end{bmatrix} = \begin{bmatrix} I_n \\ 0 \end{bmatrix} X.$$

By Proposition 2.1,

$$f\left(\begin{bmatrix} X & Z \\ 0 & Y \end{bmatrix}\right)\begin{bmatrix} I_n \\ 0 \end{bmatrix} = \begin{bmatrix} I_n \\ 0 \end{bmatrix} f(X).$$

This implies that $A = f(X)$ and $C = 0$. We have also

$$Y\begin{bmatrix} 0 & I_m \end{bmatrix} = \begin{bmatrix} 0 & I_m \end{bmatrix}\begin{bmatrix} X & Z \\ 0 & Y \end{bmatrix}.$$

By Proposition 2.1,

$$f(Y)\begin{bmatrix} 0 & I_m \end{bmatrix} = \begin{bmatrix} 0 & I_m \end{bmatrix} f\left(\begin{bmatrix} X & Z \\ 0 & Y \end{bmatrix}\right),$$

which implies that $D = f(Y)$ and that again $C = 0$. Thus, (2.5) holds.

Next, we have

$$\begin{bmatrix} rI_n & 0 \\ 0 & I_m \end{bmatrix} \begin{bmatrix} X & Z \\ 0 & Y \end{bmatrix} = \begin{bmatrix} X & rZ \\ 0 & Y \end{bmatrix} \begin{bmatrix} rI_n & 0 \\ 0 & I_m \end{bmatrix}.$$

By Proposition 2.1,

$$\begin{bmatrix} rI_n & 0 \\ 0 & I_m \end{bmatrix} f\left(\begin{bmatrix} X & Z \\ 0 & Y \end{bmatrix}\right) = f\left(\begin{bmatrix} X & rZ \\ 0 & Y \end{bmatrix}\right) \begin{bmatrix} rI_n & 0 \\ 0 & I_m \end{bmatrix},$$

which is by (2.5) equivalent to

$$\begin{bmatrix} rI_n & 0 \\ 0 & I_m \end{bmatrix} \begin{bmatrix} f(X) & \Delta f(X,Y)(Z) \\ 0 & f(Y) \end{bmatrix} = \begin{bmatrix} f(X) & \Delta f(X,Y)(rZ) \\ 0 & f(Y) \end{bmatrix} \begin{bmatrix} rI_n & 0 \\ 0 & I_m \end{bmatrix}.$$

Writing out the $(1,2)$ block in the product matrix on both sides of this equality, we obtain (2.7). □

The "left" counterpart of Proposition 2.2 (which has a similar proof) is the following.

PROPOSITION 2.3. *Let $f$ be a nc function on a nc set $\Omega$. Let $X \in \Omega_n$, $Y \in \Omega_m$, and $Z \in \mathcal{M}^{m \times n}$ be such that $\begin{bmatrix} X & 0 \\ Z & Y \end{bmatrix} \in \Omega_{n+m}$. Then (2.6) holds, where the off-diagonal block entry $\Delta_L f(X,Y)(Z)$ is determined uniquely by (2.6) and has the following property. If $r \in \mathcal{R}$ is such that $\begin{bmatrix} X & 0 \\ rZ & Y \end{bmatrix} \in \Omega_{n+m}$, then*

$$\Delta_L f(X,Y)(rZ) = r\Delta_L f(X,Y)(Z). \tag{2.8}$$

We will say that the nc set $\Omega \subseteq \mathcal{M}_{\rm nc}$ is *right admissible* if for every $X \in \Omega_n$, $Y \in \Omega_m$, and $Z \in \mathcal{M}^{n \times m}$ there exists an invertible $r \in \mathcal{R}$ such that $\begin{bmatrix} X & rZ \\ 0 & Y \end{bmatrix} \in \Omega_{n+m}$. We will say that the nc set $\Omega \subseteq \mathcal{M}_{\rm nc}$ is *left admissible* if for every $X \in \Omega_n$, $Y \in \Omega_m$, and $Z \in \mathcal{M}^{m \times n}$ there exists an invertible $r \in \mathcal{R}$ such that $\begin{bmatrix} X & 0 \\ rZ & Y \end{bmatrix} \in \Omega_{n+m}$.

Let the nc set $\Omega$ be right admissible, and let $f$ be a nc function on $\Omega$. Then for every $X \in \Omega_n$, $Y \in \Omega_m$, and $Z \in \mathcal{M}^{n \times m}$, and for an invertible $r \in \mathcal{R}$ such that $\begin{bmatrix} X & rZ \\ 0 & Y \end{bmatrix} \in \Omega_{n+m}$ we define first $\Delta_R f(X,Y)(rZ)$ as the $(1,2)$ block of the matrix $f\left(\begin{bmatrix} X & rZ \\ 0 & Y \end{bmatrix}\right)$ (see (2.5)), and then define

$$\Delta_R f(X,Y)(Z) = r^{-1}\Delta_R f(X,Y)(rZ). \tag{2.9}$$

Similarly, let the nc set $\Omega$ be left admissible, and let $f$ be a nc function on $\Omega$. Then for every $X \in \Omega_n$, $Y \in \Omega_m$, and $Z \in \mathcal{M}^{m \times n}$, and for an invertible $r \in \mathcal{R}$ such that $\begin{bmatrix} X & 0 \\ rZ & Y \end{bmatrix} \in \Omega_{n+m}$ we define first $\Delta_L f(X,Y)(rZ)$ as the $(2,1)$ block of the matrix $f\left(\begin{bmatrix} X & 0 \\ rZ & Y \end{bmatrix}\right)$ (see (2.6)), and then define

$$\Delta_L f(X,Y)(Z) = r^{-1}\Delta_L f(X,Y)(rZ). \tag{2.10}$$

By Proposition 2.2 (respectively, by Proposition 2.3), the right-hand side of (2.9) (respectively, of (2.10)) is independent of $r$, hence the left-hand side is defined unambiguously. Moreover, we obtain the following.

PROPOSITION 2.4. *Let $f$ be a nc function on a nc set $\Omega$. The mappings $Z \mapsto \Delta_R f(X,Y)(Z)$ and $Z \mapsto \Delta_L f(X,Y)(Z)$ are* homogeneous *as operators acting from $\mathcal{M}^{n\times m}$ to $\mathcal{N}^{n\times m}$ and from $\mathcal{M}^{m\times n}$ to $\mathcal{N}^{m\times n}$, respectively. In other words, if $\Omega$ is right admissible then for every $X \in \Omega_n$, $Y \in \Omega_m$, and $Z \in \mathcal{M}^{n\times m}$, and every $r \in \mathcal{R}$, (2.7) holds; respectively, if $\Omega$ is left admissible then for every $X \in \Omega_n$, $Y \in \Omega_m$, and $Z \in \mathcal{M}^{m\times n}$, and every $r \in \mathcal{R}$, (2.8) holds.*

PROOF. Let us prove one of these homogeneity properties — the proof of the other one is analogous. Let $\Omega$ be right admissible, $X \in \Omega_n$, $Y \in \Omega_m$, $Z \in \mathcal{M}^{n\times m}$, and $r \in \mathcal{R}$. Then there exist invertible $r_1, r_2 \in \mathcal{R}$ such that $\begin{bmatrix} X & r_1 Z \\ 0 & Y \end{bmatrix} \in \Omega_{n+m}$ and $\begin{bmatrix} X & r_2 r Z \\ 0 & Y \end{bmatrix} \in \Omega_{n+m}$. Then

$$\Delta_R f(X,Y)(Z) = r_1^{-1} \Delta_R f(X,Y)(r_1 Z)$$

and

$$\Delta_R f(X,Y)(rZ) = r_2^{-1} \Delta_R f(X,Y)(r_2 r Z).$$

Moreover, by Proposition 2.2,

$$\Delta_R f(X,Y)(r_2 r Z) = r_2 r r_1^{-1} \Delta_R f(X,Y)(r_1 Z).$$

These three equalities imply (2.7). □

REMARK 2.5. The proof of Proposition 2.4 can be simplified with a use of the similarity invariant envelope $\widetilde{\Omega}$ of a right (resp., left) admissible nc set $\Omega$ and the canonical extension $\widetilde{f} \colon \widetilde{\Omega} \to \mathcal{N}_{\rm nc}$ of a nc function $f \colon \Omega \to \mathcal{N}_{\rm nc}$; see Appendix A for the definitions. We will show this for a right admissible nc set $\Omega$. By Proposition A.2, $\begin{bmatrix} \widetilde{X} & Z \\ 0 & \widetilde{Y} \end{bmatrix} \in \widetilde{\Omega}_{n+m}$ for all $\widetilde{X} \in \widetilde{\Omega}_n$, $\widetilde{Y} \in \widetilde{\Omega}_m$, and $Z \in \mathcal{M}^{n\times m}$, and by Proposition A.3, $\widetilde{f}$ is a nc function. Therefore, a formula analogous to (2.5) defines $\Delta_R \widetilde{f}(\widetilde{X},\widetilde{Y})(Z)$ for all $n, m \in \mathbb{N}$, $\widetilde{X} \in \widetilde{\Omega}_n$, $\widetilde{Y} \in \widetilde{\Omega}_m$, and $Z \in \mathcal{M}^{n\times m}$, and by Proposition 2.2 applied to $\widetilde{f}$, one has

$$\Delta_R \widetilde{f}(\widetilde{X},\widetilde{Y})(rZ) = r \Delta_R \widetilde{f}(\widetilde{X},\widetilde{Y})(Z)$$

for every $r \in \mathcal{R}$. In particular, if $X \in \Omega_n$, $Y \in \Omega_m$, $Z \in \mathcal{M}^{n\times m}$, and $r \in \mathcal{R}$ is such that $\begin{bmatrix} X & rZ \\ 0 & Y \end{bmatrix} \in \Omega_{n+m}$, then

$$\Delta_R f(X,Y)(rZ) = \Delta_R \widetilde{f}(X,Y)(rZ) = r \Delta_R \widetilde{f}(X,Y)(Z).$$

If, in addition, $r$ is invertible, then from (2.9) we obtain

$$\Delta_R f(X,Y)(Z) = \Delta_R \widetilde{f}(X,Y)(Z). \tag{2.11}$$

Consequently, the mapping $Z \mapsto \Delta_R f(X,Y)(Z)$ is homogeneous, and the statement of Proposition 2.4 for a right admissible nc set $\Omega$ follows. An analogous argument works for a left admissible nc set $\Omega$.

We see that the homogeneity of the mapping $Z \mapsto \Delta_R f(X,Y)(Z)$ follows directly from the homogeneity of the mapping $Z \mapsto \Delta_R \widetilde{f}(\widetilde{X},\widetilde{Y})(Z)$, and the proof for the latter is less involved, since it does not require manipulating with appropriate invertible elements $r \in \mathcal{R}$ so that the relevant block triangular matrices get into the nc set $\Omega$. In the sequel, we will often use similarity envelopes of nc sets and canonical extensions of nc functions to simplify the proofs.

PROPOSITION 2.6. *Let $f$ be a nc function on a nc set $\Omega$. The mappings $Z \mapsto \Delta_R f(X,Y)(Z)$ and $Z \mapsto \Delta_L f(X,Y)(Z)$ are* additive *as operators acting from $\mathcal{M}^{n\times m}$ to $\mathcal{N}^{n\times m}$ and from $\mathcal{M}^{m\times n}$ to $\mathcal{N}^{m\times n}$, respectively. In other words, if $\Omega$ is right admissible, then for every $X \in \Omega_n$, $Y \in \Omega_m$, and $Z_I, Z_{II} \in \mathcal{M}^{n\times m}$,*

$$\Delta_R f(X,Y)(Z_I + Z_{II}) = \Delta_R f(X,Y)(Z_I) + \Delta_R f(X,Y)(Z_{II}); \tag{2.12}$$

*respectively, if $\Omega$ is left admissible then for every $X \in \Omega_n$, $Y \in \Omega_m$, and $Z_I, Z_{II} \in \mathcal{M}^{m\times n}$,*

$$\Delta_L f(X,Y)(Z_I + Z_{II}) = \Delta_L f(X,Y)(Z_I) + \Delta_L f(X,Y)(Z_{II}). \tag{2.13}$$

PROOF. Let us prove one of these additivity properties — the proof of the other one is analogous. Let $\Omega$ be right admissible, $X \in \Omega_n$, $Y \in \Omega_m$, $Z_I, Z_{II} \in \mathcal{M}^{n\times m}$. By (2.11), we may assume, without loss of generality, that $\Omega$ is similarity invariant, i.e., $\widetilde{\Omega} = \Omega$; see Remark 2.5 and Appendix A. Then, by Proposition A.2,

$$\begin{bmatrix} X & Z_I \\ 0 & Y \end{bmatrix} \in \Omega_{n+m}, \quad \begin{bmatrix} X & Z_{II} \\ 0 & Y \end{bmatrix} \in \Omega_{n+m}, \quad \begin{bmatrix} X & Z_I + Z_{II} \\ 0 & Y \end{bmatrix} \in \Omega_{n+m},$$

and

$$\begin{bmatrix} X & 0 & Z_I \\ 0 & X & Z_{II} \\ 0 & 0 & Y \end{bmatrix} \in \Omega_{2n+m}$$

(for the latter, we use the fact that if $X \in \Omega_n$ then $\begin{bmatrix} X & 0 \\ 0 & X \end{bmatrix} \in \Omega_{2n}$). We have

$$\begin{bmatrix} I_n & 0 & 0 \\ 0 & 0 & I_m \end{bmatrix} \begin{bmatrix} X & 0 & Z_I \\ 0 & X & Z_{II} \\ 0 & 0 & Y \end{bmatrix} = \begin{bmatrix} X & Z_I \\ 0 & Y \end{bmatrix} \begin{bmatrix} I_n & 0 & 0 \\ 0 & 0 & I_m \end{bmatrix}.$$

By Proposition 2.1,

$$\begin{bmatrix} I_n & 0 & 0 \\ 0 & 0 & I_m \end{bmatrix} f\left(\begin{bmatrix} X & 0 & Z_I \\ 0 & X & Z_{II} \\ 0 & 0 & Y \end{bmatrix}\right) = f\left(\begin{bmatrix} X & Z_I \\ 0 & Y \end{bmatrix}\right) \begin{bmatrix} I_n & 0 & 0 \\ 0 & 0 & I_m \end{bmatrix}.$$

Clearly, we have

$$f\left(\begin{bmatrix} X & 0 & Z_I \\ 0 & X & Z_{II} \\ 0 & 0 & Y \end{bmatrix}\right) = \begin{bmatrix} f(X) & 0 & A \\ 0 & f(X) & B \\ 0 & 0 & f(Y) \end{bmatrix},$$

with some matrices $A$ and $B$, and also we have

$$f\left(\begin{bmatrix} X & Z_I \\ 0 & Y \end{bmatrix}\right) = \begin{bmatrix} f(X) & \Delta_R f(X,Y)(Z_I) \\ 0 & f(Y) \end{bmatrix}.$$

Therefore,

$$\begin{bmatrix} I_n & 0 & 0 \\ 0 & 0 & I_m \end{bmatrix} \begin{bmatrix} f(X) & 0 & A \\ 0 & f(X) & B \\ 0 & 0 & f(Y) \end{bmatrix} = \begin{bmatrix} f(X) & \Delta_R f(X,Y)(Z_I) \\ 0 & f(Y) \end{bmatrix} \begin{bmatrix} I_n & 0 & 0 \\ 0 & 0 & I_m \end{bmatrix}.$$

Comparing the $(1,3)$ block entries in a product matrix on both sides, we get

$$A = \Delta_R f(X,Y)(Z_I).$$

Similarly, we proceed with the intertwining

$$\begin{bmatrix} 0 & I_n & 0 \\ 0 & 0 & I_m \end{bmatrix} \begin{bmatrix} X & 0 & Z_I \\ 0 & X & Z_{II} \\ 0 & 0 & Y \end{bmatrix} = \begin{bmatrix} X & Z_{II} \\ 0 & Y \end{bmatrix} \begin{bmatrix} 0 & I_n & 0 \\ 0 & 0 & I_m \end{bmatrix}.$$

By Proposition 2.1,

$$\begin{bmatrix} 0 & I_n & 0 \\ 0 & 0 & I_m \end{bmatrix} f\left(\begin{bmatrix} X & 0 & Z_I \\ 0 & X & Z_{II} \\ 0 & 0 & Y \end{bmatrix}\right) = f\left(\begin{bmatrix} X & Z_{II} \\ 0 & Y \end{bmatrix}\right) \begin{bmatrix} 0 & I_n & 0 \\ 0 & 0 & I_m \end{bmatrix},$$

or equivalently,

$$\begin{bmatrix} 0 & I_n & 0 \\ 0 & 0 & I_m \end{bmatrix} \begin{bmatrix} f(X) & 0 & A \\ 0 & f(X) & B \\ 0 & 0 & f(Y) \end{bmatrix} = \begin{bmatrix} f(X) & \Delta_R f(X,Y)(Z_{II}) \\ 0 & f(Y) \end{bmatrix} \begin{bmatrix} 0 & I_n & 0 \\ 0 & 0 & I_m \end{bmatrix}.$$

Comparing the $(1,3)$ block entries in a product matrix on both sides, we get

$$B = \Delta_R f(X,Y)(Z_{II}).$$

Next, we use the intertwining relation

$$\begin{bmatrix} I_n & I_n & 0 \\ 0 & 0 & I_m \end{bmatrix} \begin{bmatrix} X & 0 & Z_I \\ 0 & X & Z_{II} \\ 0 & 0 & Y \end{bmatrix} = \begin{bmatrix} X & Z_I + Z_{II} \\ 0 & Y \end{bmatrix} \begin{bmatrix} I_n & I_n & 0 \\ 0 & 0 & I_m \end{bmatrix}.$$

By Proposition 2.1,

$$\begin{bmatrix} I_n & I_n & 0 \\ 0 & 0 & I_m \end{bmatrix} f\left(\begin{bmatrix} X & 0 & Z_I \\ 0 & X & Z_{II} \\ 0 & 0 & Y \end{bmatrix}\right) = f\left(\begin{bmatrix} X & Z_I + Z_{II} \\ 0 & Y \end{bmatrix}\right) \begin{bmatrix} I_n & I_n & 0 \\ 0 & 0 & I_m \end{bmatrix},$$

or equivalently,

$$\begin{aligned} &\begin{bmatrix} I_n & I_n & 0 \\ 0 & 0 & I_m \end{bmatrix} \begin{bmatrix} f(X) & 0 & \Delta_R f(X,Y)(Z_I) \\ 0 & f(X) & \Delta_R f(X,Y)(Z_{II}) \\ 0 & 0 & f(Y) \end{bmatrix} \\ &\qquad = \begin{bmatrix} f(X) & \Delta_R f(X,Y)(Z_I + Z_{II}) \\ 0 & f(Y) \end{bmatrix} \begin{bmatrix} I_n & I_n & 0 \\ 0 & 0 & I_m \end{bmatrix}. \end{aligned}$$

Comparing the $(1,3)$ block entries in a product matrix on both sides, we get

$$\Delta_R f(X,Y)(Z_I) + \Delta_R f(X,Y)(Z_{II}) = \Delta_R f(X,Y)(Z_I + Z_{II}),$$

i.e., (2.12) is true. □

The transposition of matrices induces naturally a transposition operation on nc functions and enables us to relate right and left nc difference-differential operators.

PROPOSITION 2.7. (1) *If $\Omega \subseteq \mathcal{M}_{\rm nc}$ is a nc set, then so is*

$$\Omega^\top := \{X^\top : X \in \Omega\} \subseteq \mathcal{M}_{\rm nc}.$$

(2) *If $f\colon \Omega \to \mathcal{N}_{\rm nc}$ is a nc function, then so is $f^\top\colon \Omega^\top \to \mathcal{N}_{\rm nc}$, where $f^\top(X) := f\left(X^\top\right)^\top$.*

(3) *If $\Omega$ is right (respectively, left) admissible, then $\Omega^\top$ is left (respectively, right) admissible. In this case,*

$$\Delta_L f^\top(X,Y)(Z) = \left(\Delta_R f\left(X^\top, Y^\top\right)\left(Z^\top\right)\right)^\top$$

*(respectively, $\Delta_R f^\top(X,Y)(Z) = \left(\Delta_L f\left(X^\top, Y^\top\right)\left(Z^\top\right)\right)^\top$).*

The proof is straightforward. Proposition 2.7 allows us to deduce results for $\Delta_L$ from the corresponding results for $\Delta_R$ and vice versa, using a transposition argument.

In the case where a nc set $\Omega$ is both right and left admissible, there is also a simple symmetry relation between right and left nc difference-differential operators.

PROPOSITION 2.8. *Let $f\colon \Omega \to \mathcal{N}_{\rm nc}$ be a nc function on a nc set $\Omega \subseteq \mathcal{M}_{\rm nc}$ which is both right and left admissible. Then*

$$\Delta_R f(X,Y)(Z) = \Delta_L f(Y,X)(Z)$$

*for all $n,m \in \mathbb{N}$, $X \in \Omega_n$, $Y \in \Omega_m$, and $Z \in \mathcal{M}^{n\times m}$.*

PROOF. By (2.11) and its left analogue, we may assume, without loss of generality, that $\Omega$ is similarity invariant, i.e., $\widetilde{\Omega} = \Omega$; see Remark 2.5 and Appendix A. Then by Proposition A.2, for arbitrary $n,m \in \mathbb{N}$, $X \in \Omega_n$, $Y \in \Omega_m$, and $Z \in \mathcal{M}^{n\times m}$ one has

$$\begin{bmatrix} X & Z \\ 0 & Y \end{bmatrix}, \begin{bmatrix} Y & 0 \\ Z & X \end{bmatrix} \in \Omega_{n+m}.$$

Clearly,

$$\begin{bmatrix} X & Z \\ 0 & Y \end{bmatrix}\begin{bmatrix} 0 & I_n \\ I_m & 0 \end{bmatrix} = \begin{bmatrix} 0 & I_n \\ I_m & 0 \end{bmatrix}\begin{bmatrix} Y & 0 \\ Z & X \end{bmatrix}.$$

By Proposition 2.1,

$$f\left(\begin{bmatrix} X & Z \\ 0 & Y \end{bmatrix}\right)\begin{bmatrix} 0 & I_n \\ I_m & 0 \end{bmatrix} = \begin{bmatrix} 0 & I_n \\ I_m & 0 \end{bmatrix} f\left(\begin{bmatrix} Y & 0 \\ Z & X \end{bmatrix}\right),$$

i.e.,

$$\begin{bmatrix} f(X) & \Delta_R f(X,Y)(Z) \\ 0 & f(Y) \end{bmatrix}\begin{bmatrix} 0 & I_n \\ I_m & 0 \end{bmatrix} = \begin{bmatrix} 0 & I_n \\ I_m & 0 \end{bmatrix}\begin{bmatrix} f(Y) & 0 \\ \Delta_L f(Y,X)(Z) & f(X) \end{bmatrix}.$$

Comparing the (1,1) block entries in the matrix products on the left-hand side and on the right-hand side, we obtain

$$\Delta_R f(X,Y)(Z) = \Delta_L f(Y,X)(Z).$$

□

## 2.3. Basic rules of nc difference-differential calculus

We proceed to identify basic building blocks which allow one to compute $\Delta_R f$ and $\Delta_L f$ much as in the classical commutative calculus; here and below, whenever it does not cause confusion, we often write $\Delta_R f$ and $\Delta_L f$ without arguments.

**2.3.1.** If $f\colon \mathcal{M}_{\rm nc} \to \mathcal{N}_{\rm nc}$ is a *constant nc function*, i.e., there is a $c \in \mathcal{N}$ such that $f(X) = cI_n$ $(= \operatorname{diag}[c,\ldots,c])$ for all $n \in \mathbb{N}$ and all $X \in \mathcal{M}^{n\times n}$, then $\Delta_R f(X,Y)(Z) = 0$ and $\Delta_L f(X,Y)(Z) = 0$ for all $X,Y,Z$ of appropriate sizes.

PROOF. Trivial. □

**2.3.2.** If $f, g\colon \Omega \to \mathcal{N}_{\rm nc}$ are nc functions on a right (respectively, left) admissible nc set $\Omega \subseteq \mathcal{M}_{\rm nc}$ and $a, b \in \mathcal{R}$, then $\Delta_R(af+bg) = a\Delta_R f + b\Delta_R g$ (respectively, $\Delta_L(af+bg) = a\Delta_L f + b\Delta_L g$).

PROOF. Trivial. □

**2.3.3.** Let $l\colon \mathcal{M} \to \mathcal{N}$ be a $\mathcal{R}$-linear mapping. We extend $l$ to a linear mapping from $\mathcal{M}^{n\times m}$ to $\mathcal{N}^{n\times m}$, $n, m \in \mathbb{N}$ by $l([\mu_{ij}]) = [l(\mu_{ij})]$. Then $l$ is a nc function from $\mathcal{M}_{\rm nc}$ to $\mathcal{N}_{\rm nc}$, and $\Delta_R l(X,Y)(Z) = l(Z)$ and $\Delta_L l(X,Y)(Z) = l(Z)$ for all $X, Y, Z$ of appropriate sizes. In particular, if $l_j\colon (\mathcal{R}^d)_{\rm nc} \to \mathcal{R}_{\rm nc}$ is the *j-th coordinate nc function*, i.e.,

$$l_j(X) = l_j(X_1, \ldots, X_d) = X_j,$$

then $\Delta_R l_j(X,Y)(Z) = Z_j$ and $\Delta_L l_j(X,Y)(Z) = Z_j$ for all $d$-tuples $X, Y, Z$ of matrices over $\mathcal{R}$ of appropriate sizes.

PROOF. Trivial. □

**2.3.4.** Let $f\colon \Omega \to \mathcal{N}_{\rm nc}$, $g\colon \Omega \to \mathcal{O}_{\rm nc}$ be nc functions on a right (respectively, left) admissible nc set $\Omega \subseteq \mathcal{M}_{\rm nc}$. Assume that we are given a product operation $(x, y) \mapsto x \cdot y$ on $\mathcal{N} \times \mathcal{O}$ with values in a module $\mathcal{P}$ over $\mathcal{R}$ which is left and right distributive and commutes with the action of $\mathcal{R}$ (i.e., we are given a $\mathcal{R}$-linear map from $\mathcal{N} \otimes_{\mathcal{R}} \mathcal{O}$ to $\mathcal{P}$). We extend the product operation to matrices over $\mathcal{N}$ and over $\mathcal{O}$ of appropriate sizes. It is easy to check that $f \cdot g\colon \Omega \to \mathcal{P}_{\rm nc}$ is a nc function.
Then

$$\Delta_R(f\cdot g)(X,Y)(Z) = f(X)\cdot \Delta_R g(X,Y)(Z) + \Delta_R f(X,Y)(Z)\cdot g(Y)$$

for all $n, m \in \mathbb{N}$, $X \in \Omega_n$, $Y \in \Omega_m$, and $Z \in \mathcal{M}^{n\times m}$ (respectively,

$$\Delta_L(f\cdot g)(X,Y)(Z) = \Delta_L f(X,Y)(Z)\cdot g(X) + f(Y)\cdot \Delta_L g(X,Y)(Z)$$

for all $n, m \in \mathbb{N}$, $X \in \Omega_n$, $Y \in \Omega_m$, and $Z \in \mathcal{M}^{m\times n}$).

PROOF. We shall now give a proof of the first equality. By (2.11), we may assume, without loss of generality, that $\Omega$ is similarity invariant, i.e., $\widetilde{\Omega} = \Omega$; see Remark 2.5 and Appendix A. Then by Proposition A.2, for arbitrary $X \in \Omega_n$, $Y \in \Omega_m$, and $Z \in \mathcal{M}^{n\times m}$ one has $\begin{bmatrix} X & Z \\ 0 & Y \end{bmatrix} \in \Omega_{n+m}$. On one hand,

$$(f\cdot g)\left(\begin{bmatrix} X & Z \\ 0 & Y \end{bmatrix}\right) = \begin{bmatrix} (f\cdot g)(X) & \Delta_R(f\cdot g)(X,Y)(Z) \\ 0 & (f\cdot g)(Y) \end{bmatrix}. \tag{2.14}$$

On the other hand,

$$\begin{aligned} f\left(\begin{bmatrix} X & Z \\ 0 & Y \end{bmatrix}\right)\cdot g\left(\begin{bmatrix} X & Z \\ 0 & Y \end{bmatrix}\right) &= \begin{bmatrix} f(X) & \Delta_R f(X,Y)(Z) \\ 0 & f(Y) \end{bmatrix}\cdot \begin{bmatrix} g(X) & \Delta_R g(X,Y)(Z) \\ 0 & g(Y) \end{bmatrix} \\ &= \begin{bmatrix} f(X)\cdot g(X) & f(X)\cdot\Delta_R g(X,Y)(Z) + \Delta_R f(X,Y)(Z)\cdot g(Y) \\ 0 & f(Y)\cdot g(Y) \end{bmatrix}. \end{aligned} \tag{2.15}$$

Comparing the $(1,2)$ blocks on the right hand sides of (2.14) and (2.15), we obtain the required formula for $\Delta_R(f\cdot g)(X,Y)(Z)$.

The formula for $\Delta_L(f\cdot g)(X,Y)(Z)$ is proved analogously. In the case where $\Omega$ is both right and left admissible, we can also use a symmetry argument (Proposition 2.8). $\square$

REMARK 2.9. We can define also an opposite product operation $(y,x)\mapsto y\cdot_{\rm op} x$ on $\mathcal{O}\times\mathcal{N}$ with values in $\mathcal{P}$ by $y\cdot_{\rm op} x := x\cdot y$ (i.e., we are using the canonical isomorphism $\mathcal{N}\otimes_{\mathcal{R}}\mathcal{O}\cong\mathcal{O}\otimes_{\mathcal{R}}\mathcal{N}$). Extending the product operation "$\cdot_{\rm op}$" to matrices over $\mathcal{O}$ and over $\mathcal{N}$ of appropriate sizes, we clearly have $(X\cdot Y)^\top = X^\top\cdot_{\rm op} Y^\top$, hence also $(f\cdot g)^\top = f^\top\cdot_{\rm op} g^\top$. In particular, we can use Proposition 2.7 to derive the formula for $\Delta_L(f\cdot g)(X,Y)(Z)$ from the formula for $\Delta_R\left(f^\top\cdot_{\rm op} g^\top\right)\left(X^\top,Y^\top\right)\left(Z^\top\right)$.

**2.3.5.** Let $f\colon\Omega\to\mathcal{A}_{\rm nc}$ be a nc function on a right (respectively, left) admissible nc set $\Omega\subseteq\mathcal{M}_{\rm nc}$, where $\mathcal{A}$ is a unital algebra over $\mathcal{R}$. Let

$$\Omega^{\rm inv} := \coprod_{n=1}^{\infty}\{X\in\Omega_n\colon f(X) \text{ is invertible in } \mathcal{A}^{n\times n}\}.$$

Then $\Omega^{\rm inv}$ is a right (respectively, left) admissible nc set, $f^{-1}\colon\Omega^{\rm inv}\to\mathcal{A}_{\rm nc}$ defined by $f^{-1}(X) := f(X)^{-1}$ is a nc function, and

$$\Delta_R f^{-1}(X,Y)(Z) = -f(X)^{-1}\Delta_R f(X,Y)(Z) f(Y)^{-1}$$

for all $n,m\in\mathbb{N}$, $X\in\Omega_n^{\rm inv}$, $Y\in\Omega_m^{\rm inv}$, and $Z\in\mathcal{A}^{n\times m}$ (respectively,

$$\Delta_L f^{-1}(X,Y)(Z) = -f(Y)^{-1}\Delta_L f(X,Y)(Z) f(X)^{-1}$$

for all $n,m\in\mathbb{N}$, $X\in\Omega_n^{\rm inv}$, $Y\in\Omega_m^{\rm inv}$, and $Z\in\mathcal{A}^{m\times n}$).

PROOF. Observe that $\Omega^{\rm inv}$ is a nc set since $f$ respects direct sums.

We show next that if $\Omega$ is right admissible, then so is $\Omega^{\rm inv}$. Recall first that a block upper triangular matrix $\begin{bmatrix} A & C\\ 0 & B\end{bmatrix}$ is invertible if and only if both $A$ and $B$ are invertible, and in this case

$$\begin{bmatrix} A & C\\ 0 & B\end{bmatrix}^{-1} = \begin{bmatrix} A^{-1} & -A^{-1}CB^{-1}\\ 0 & B^{-1}\end{bmatrix}. \tag{2.16}$$

Now, let $X\in\Omega_n^{\rm inv}\subseteq\Omega_n$, $Y\in\Omega_m^{\rm inv}\subseteq\Omega_m$, and $Z\in\mathcal{M}^{n\times m}$. Then there exists an invertible $r\in\mathcal{R}$ such that $\begin{bmatrix} X & rZ\\ 0 & Y\end{bmatrix}\in\Omega_{n+m}$. We have

$$f\left(\begin{bmatrix} X & rZ\\ 0 & Y\end{bmatrix}\right) = \begin{bmatrix} f(X) & \Delta_R f(X,Y)(rZ)\\ 0 & f(Y)\end{bmatrix}.$$

Since $f(X)$ and $f(Y)$ are invertible, so is the right-hand side. Therefore, $\begin{bmatrix} X & rZ\\ 0 & Y\end{bmatrix}\in\Omega_{n+m}^{\rm inv}$, and $\Omega^{\rm inv}$ is right admissible.

By (2.16),

$$\Delta_R f^{-1}(X,Y)(rZ) = -f(X)^{-1}\Delta_R f(X,Y)(rZ) f(Y)^{-1},$$

which implies the required formula for $\Delta_R f^{-1}(X,Y)(Z)$ by linearity.

The "left" statements can be proved analogously, or can be deduced from the "right" statements by a transposition argument (see Proposition 2.7) using the fact that $\left(f^{-1}\right)^\top = \left(f^\top\right)^{-1}$. In the case where $\Omega$ is both right and left admissible, we can also use a symmetry argument (Proposition 2.8).

Alternatively, we could also use equality (2.11) and its left analogue, and Proposition A.2 which guarantees that the similarity envelope $\widetilde{\Omega}$ of the right (resp., left) admissible nc set $\Omega$ is right (resp., left) admissible; see Remark 2.5 and Appendix A. Then, similar to the argument in the proof above, so is $\widetilde{\Omega}^{\rm inv} = \widetilde{\Omega^{\rm inv}}$, and then the rest of the proof above works with $r = 1$. □

**2.3.6.** Let $f\colon \Omega \to \mathcal{N}_{\rm nc}$ and $g\colon \Lambda \to \mathcal{O}_{\rm nc}$ be nc functions on right (respectively, left) admissible nc sets $\Omega \subseteq \mathcal{M}_{\rm nc}$ and $\Lambda \subseteq \mathcal{N}_{\rm nc}$, such that $f(\Omega) \subseteq \Lambda$. Then the composition $g \circ f\colon \Omega \to \mathcal{O}_{\rm nc}$ is a nc function, and

$$\Delta_R(g \circ f)(X,Y)(Z) = \Delta_R g(f(X), f(Y))(\Delta_R f(X,Y)(Z))$$

for all $n, m \in \mathbb{N}$, $X \in \Omega_n$, $Y \in \Omega_m$, and $Z \in \mathcal{M}^{n \times m}$ (respectively,

$$\Delta_L(g \circ f)(X,Y)(Z) = \Delta_L g(f(X), f(Y))(\Delta_L f(X,Y)(Z))$$

for all $n, m \in \mathbb{N}$, $X \in \Omega_n$, $Y \in \Omega_m$, and $Z \in \mathcal{M}^{m \times n}$).

PROOF. It is easy to see that $g \circ f$ is a nc function.

By (2.11) applied to $f$, $g$, and $g \circ f$, we may assume, without loss of generality, that $\Omega$ is similarity invariant, i.e., $\widetilde{\Omega} = \Omega$; see Remark 2.5 and Appendix A. Then, by Proposition A.2, for arbitrary $n, m \in \mathbb{N}$, $X \in \Omega_n$, $Y \in \Omega_m$, and $Z \in \mathcal{M}^{n \times m}$ one has $\begin{bmatrix} X & Z \\ 0 & Y \end{bmatrix} \in \Omega_{n+m}$. On one hand,

$$(g \circ f)\left(\begin{bmatrix} X & Z \\ 0 & Y \end{bmatrix}\right) = \begin{bmatrix} (g \circ f)(X) & \Delta_R(g \circ f)(X,Y)(Z) \\ 0 & (g \circ f)(Y) \end{bmatrix}. \tag{2.17}$$

On the other hand,

$$\begin{aligned} g\left(f\left(\begin{bmatrix} X & Z \\ 0 & Y \end{bmatrix}\right)\right) &= g\left(\begin{bmatrix} f(X) & \Delta_R f(X,Y)(Z) \\ 0 & f(Y) \end{bmatrix}\right) \\ &= \begin{bmatrix} g(f(X)) & \Delta_R g(f(X), f(Y))(\Delta_R f(X,Y)(Z)) \\ 0 & g(f(Y)) \end{bmatrix}. \end{aligned} \tag{2.18}$$

Comparing the $(1,2)$ blocks on the right hand sides of (2.17) and (2.18), we obtain the required formula for $\Delta_R(g \circ f)(X,Y)(Z)$.

The formula for $\Delta_L(g \circ f)(X,Y)(Z)$ can be proved analogously, or by a transposition argument, or (when $\Omega$ is both right and left admissible) by a symmetry argument. □

## 2.4. First order difference formulae

We establish now formulae which justify the claim made in Section 2.2 that $\Delta_R$ and $\Delta_L$ play the role of nc finite difference operators of the first order. We will see in the sequel that these formulae are nc analogues of zeroth order Brook Taylor expansions.

THEOREM 2.10. *Let $f\colon \Omega \to \mathcal{N}_{\rm nc}$ be a nc function on a right admissible nc set $\Omega$. Then for all $n \in \mathbb{N}$, and arbitrary $X, Y \in \Omega_n$,*

$$f(X) - f(Y) = \Delta_R f(Y, X)(X - Y) \tag{2.19}$$

*and*

$$f(X) - f(Y) = \Delta_R f(X, Y)(X - Y). \tag{2.20}$$

*Analogously, let* $f\colon \Omega \to \mathcal{N}_{\rm nc}$ *be a nc function on a left admissible nc set* $\Omega$. *Then for all* $n \in \mathbb{N}$, *and arbitrary* $X, Y \in \Omega_n$,

$$f(X) - f(Y) = \Delta_L f(Y, X)(X - Y) \tag{2.21}$$

*and*

$$f(X) - f(Y) = \Delta_L f(X, Y)(X - Y). \tag{2.22}$$

In fact, Theorem 2.10 is a special case of the following more general finite difference formulae.

THEOREM 2.11. *Let* $f\colon \Omega \to \mathcal{N}_{\rm nc}$ *be a nc function on a right admissible nc set* $\Omega$. *Then for all* $n, m \in \mathbb{N}$, *and arbitrary* $X \in \Omega_n$, $Y \in \Omega_m$ *and* $S \in \mathcal{R}^{m\times n}$,

$$Sf(X) - f(Y)S = \Delta_R f(Y, X)(SX - YS). \tag{2.23}$$

*Analogously, let* $f\colon \Omega \to \mathcal{N}_{\rm nc}$ *be a nc function on a left admissible nc set* $\Omega$. *Then for all* $n, m \in \mathbb{N}$, *and arbitrary* $X \in \Omega_n$, $Y \in \Omega_m$ *and* $S \in \mathcal{R}^{m\times n}$,

$$Sf(X) - f(Y)S = \Delta_L f(X, Y)(SX - YS). \tag{2.24}$$

PROOF. We prove the "right" statement first. By (2.11) and Proposition A.3, we may assume, without loss of generality, that $\Omega$ is similarity invariant, i.e., $\widetilde{\Omega} = \Omega$; see Remark 2.5 and Appendix A. Then, by Proposition A.2, for arbitrary $n, m \in \mathbb{N}$, $X \in \Omega_n$, $Y \in \Omega_m$, and $S \in \mathcal{R}^{m\times n}$ one has $\begin{bmatrix} Y & SX - YS \\ 0 & X \end{bmatrix} \in \Omega_{m+n}$. Clearly,

$$\begin{bmatrix} Y & SX - YS \\ 0 & X \end{bmatrix} \begin{bmatrix} S \\ I_n \end{bmatrix} = \begin{bmatrix} S \\ I_n \end{bmatrix} X.$$

Therefore,

$$f\left(\begin{bmatrix} Y & SX - YS \\ 0 & X \end{bmatrix}\right) \begin{bmatrix} S \\ I_n \end{bmatrix} = \begin{bmatrix} S \\ I_n \end{bmatrix} f(X),$$

i.e.,

$$\begin{bmatrix} f(Y) & \Delta_R f(Y, X)(SX - YS) \\ 0 & f(X) \end{bmatrix} \begin{bmatrix} S \\ I_n \end{bmatrix} = \begin{bmatrix} S \\ I_n \end{bmatrix} f(X).$$

Comparing the (1,1) block entries in the matrix products in the right-hand side and in the left-hand side, we obtain

$$f(Y)S + \Delta_R f(Y, X)(SX - YS)) = Sf(X),$$

which yields (2.23).

The "left" statements can be proved analogously using the intertwining

$$\begin{bmatrix} S & -I_m \end{bmatrix} \begin{bmatrix} X & 0 \\ SX - YS & Y \end{bmatrix} = Y \begin{bmatrix} S & -I_m \end{bmatrix},$$

or by a transposition argument, or (in the case where $\Omega$ is both right and left admissible) by a symmetry argument. □

REMARK 2.12. (2.19) and (2.20) in Theorem 2.10 are both obtained from (2.23) by writing $X - Y = X \cdot I_n - I_n \cdot Y$ and $X - Y = I_n \cdot X - Y \cdot I_n$ respectively. We note that (2.19) and (2.20) are obtained from each other by interchanging $X$ and $Y$. In particular, we have

$$\Delta_R f(Y, X)(X - Y) = \Delta_R f(X, Y)(X - Y).$$

This follows also directly from the intertwining

$$\begin{bmatrix} Y & X-Y \\ 0 & X \end{bmatrix}\begin{bmatrix} I_n & 0 \\ I_n & I_n \end{bmatrix}=\begin{bmatrix} I_n & 0 \\ I_n & I_n \end{bmatrix}\begin{bmatrix} X & X-Y \\ 0 & Y \end{bmatrix},$$

using, if necessary, the similarity envelope $\widetilde{\Omega}$ of the nc set $\Omega$ and the canonical extension $\widetilde{f}$ of the nc function $f$ by comparing the (1,1) block entries in the matrix products. Notice that comparing the (1,2) block entries yields (2.19), and comparing the (2,1) block entries yields (2.20).

An analogous remark applies to (2.21) and (2.22), and the resulting equality

$$\Delta_L f(Y,X)(X-Y)=\Delta_L f(X,Y)(X-Y).$$

REMARK 2.13. Clearly, each of the formulae (2.23) and (2.24) generalizes the condition (2.3).

REMARK 2.14. Let $\mathcal{R}=\mathbb{R}$ or $\mathcal{R}=\mathbb{C}$. Setting $X=Y+tZ$ (with $t\in\mathbb{R}$ or $t\in\mathbb{C}$), we obtain from Theorem 2.10 that

$$f(Y+tZ)-f(Y)=t\Delta_R f(Y,Y+tZ)(Z).$$

Under appropriate continuity conditions, it follows that $\Delta_R f(Y,Y)(Z)$ is the directional derivative of $f$ at $Y$ in the direction $Z$, i.e., $\Delta_R f(Y,Y)$ is the differential of $f$ at $Y$.

## 2.5. Properties of $\Delta_R f(X,Y)$ and $\Delta_L f(X,Y)$ as functions of $X$ and $Y$

A nc function $f$ possesses certain key properties: it respects direct sums, similarities, and intertwinings. We will show now that these properties are inherited, in an appropriate form, by $\Delta_R f(X,Y)(\cdot)$ and by $\Delta_L f(X,Y)(\cdot)$ as functions of $X$ and $Y$.

PROPOSITION 2.15. *Let $f\colon\Omega\to\mathcal{N}_{\rm nc}$ be a nc function on a right admissible nc set $\Omega\subseteq\mathcal{M}_{\rm nc}$. Then:*

(1X)
$$\Delta_R f\left(X'\oplus X'',Y\right)\left(\operatorname{col}[Z',Z'']\right)=\operatorname{col}\left[\Delta_R f\left(X',Y\right)\left(Z'\right),\Delta_R f\left(X'',Y\right)\left(Z''\right)\right]$$
*for $n',n'',m\in\mathbb{N}$, $X'\in\Omega_{n'}$, $X''\in\Omega_{n''}$, $Y\in\Omega_m$, $Z'\in\mathcal{M}^{n'\times m}$, $Z''\in\mathcal{M}^{n''\times m}$;*

(1Y)
$$\Delta_R f\left(X,Y'\oplus Y''\right)\left(\operatorname{row}[Z',Z'']\right)=\operatorname{row}\left[\Delta_R f\left(X,Y'\right)\left(Z'\right),\Delta_R f\left(X,Y''\right)\left(Z''\right)\right]$$
*for $n,m',m''\in\mathbb{N}$, $X\in\Omega_n$, $Y'\in\Omega_{m'}$, $Y''\in\Omega_{m''}$, $Z'\in\mathcal{M}^{n\times m'}$, $Z''\in\mathcal{M}^{n\times m''}$;*

(2X)
$$\Delta_R f\left(TXT^{-1},Y\right)(TZ)=T\,\Delta_R f(X,Y)(Z)$$
*for $n,m\in\mathbb{N}$, $X\in\Omega_n$, $Y\in\Omega_m$, $Z\in\mathcal{M}^{n\times m}$, and an invertible $T\in\mathcal{R}^{n\times n}$ such $TXT^{-1}\in\Omega_n$;*

(2Y)
$$\Delta_R f\left(X,SYS^{-1}\right)\left(ZS^{-1}\right)=\Delta_R f(X,Y)(Z)\,S^{-1}$$
*for $n,m\in\mathbb{N}$, $X\in\Omega_n$, $Y\in\Omega_m$, $Z\in\mathcal{M}^{n\times m}$, and an invertible $S\in\mathcal{R}^{m\times m}$ such that $SYS^{-1}\in\Omega_m$;*

(3X)
$$\text{If}\quad TX=\tilde{X}T\quad\text{then}\quad T\,\Delta_R f(X,Y)(Z)=\Delta_R f(\tilde{X},Y)(TZ)$$
*for $n,\tilde{n},m\in\mathbb{N}$, $X\in\Omega_n$, $\tilde{X}\in\Omega_{\tilde{n}}$, $Y\in\Omega_m$, $Z\in\mathcal{M}^{n\times m}$, and $T\in\mathcal{R}^{\tilde{n}\times n}$;*

(3Y)
$$\text{If}\quad YS=S\tilde{Y}\quad\text{then}\quad \Delta_R f(X,Y)(Z)\,S=\Delta_R f(X,\tilde{Y})(ZS)$$
*for $n,m,\tilde{m}\in\mathbb{N}$, $X\in\Omega_n$, $Y\in\Omega_m$, $\tilde{Y}\in\Omega_{\tilde{m}}$, $Z\in\mathcal{M}^{n\times m}$, and $S\in\mathcal{R}^{m\times\tilde{m}}$.*

PROOF. By (2.11), we may assume, without loss of generality, that $\Omega$ is similarity invariant, i.e., $\widetilde{\Omega} = \Omega$; see Remark 2.5 and Appendix A. Then, by Proposition A.2, for arbitrary $n', n'', m \in \mathbb{N}$, $X' \in \Omega_{n'}$, $X'' \in \Omega_{n''}$, $Y \in \Omega_m$, $Z' \in \mathcal{R}^{n' \times m}$, and $Z'' \in \mathcal{R}^{n'' \times m}$,

$$\begin{bmatrix} X' & Z' \\ 0 & Y \end{bmatrix} \in \Omega_{n'+m}, \quad \begin{bmatrix} X'' & Z'' \\ 0 & Y \end{bmatrix} \in \Omega_{n''+m}, \quad \begin{bmatrix} X' & 0 & Z' \\ 0 & X'' & Z'' \\ 0 & 0 & Y \end{bmatrix} \in \Omega_{n'+n''+m}.$$

We have

$$\begin{bmatrix} I_{n'} & 0 & 0 \\ 0 & 0 & I_m \\ 0 & I_{n''} & 0 \\ 0 & 0 & I_m \end{bmatrix} \begin{bmatrix} X' & 0 & Z' \\ 0 & X'' & Z'' \\ 0 & 0 & Y \end{bmatrix} = \begin{bmatrix} X' & Z' & 0 & 0 \\ 0 & Y & 0 & 0 \\ 0 & 0 & X'' & Z'' \\ 0 & 0 & 0 & Y \end{bmatrix} \begin{bmatrix} I_{n'} & 0 & 0 \\ 0 & 0 & I_m \\ 0 & I_{n''} & 0 \\ 0 & 0 & I_m \end{bmatrix}.$$

By Proposition 2.1,

$$\begin{bmatrix} I_{n'} & 0 & 0 \\ 0 & 0 & I_m \\ 0 & I_{n''} & 0 \\ 0 & 0 & I_m \end{bmatrix} f\left(\begin{bmatrix} X' & 0 & Z' \\ 0 & X'' & Z'' \\ 0 & 0 & Y \end{bmatrix}\right)$$

$$= f\left(\begin{bmatrix} X' & Z' & 0 & 0 \\ 0 & Y & 0 & 0 \\ 0 & 0 & X'' & Z'' \\ 0 & 0 & 0 & Y \end{bmatrix}\right) \begin{bmatrix} I_{n'} & 0 & 0 \\ 0 & 0 & I_m \\ 0 & I_{n''} & 0 \\ 0 & 0 & I_m \end{bmatrix},$$

i.e.,

$$\begin{bmatrix} I_{n'} & 0 & 0 \\ 0 & 0 & I_m \\ 0 & I_{n''} & 0 \\ 0 & 0 & I_m \end{bmatrix} \begin{bmatrix} f(X') & 0 & \Delta_{\mathcal{R}} f(X' \oplus X'', Y)(\operatorname{col}[Z', Z'']) \\ 0 & f(X'') & \\ 0 & 0 & f(Y) \end{bmatrix}$$

$$= \begin{bmatrix} f(X') & \Delta_{\mathcal{R}} f(X', Y)(Z') & 0 & 0 \\ 0 & f(Y) & 0 & 0 \\ 0 & 0 & f(X'') & \Delta_{\mathcal{R}} f(X'', Y)(Z'') \\ 0 & 0 & 0 & f(Y) \end{bmatrix} \begin{bmatrix} I_{n'} & 0 & 0 \\ 0 & 0 & I_m \\ 0 & I_{n''} & 0 \\ 0 & 0 & I_m \end{bmatrix}.$$

Comparing the columns formed by the (1,3) and the (3,3) block entries in the matrix products in the left-hand side and in the right-hand side, we obtain (1X).

The statement (1Y) is proved analogously, using the intertwining

$$\begin{bmatrix} X & Z' & Z'' \\ 0 & Y' & 0 \\ 0 & 0 & Y'' \end{bmatrix} \begin{bmatrix} I_n & 0 & I_n & 0 \\ 0 & I_{m'} & 0 & 0 \\ 0 & 0 & 0 & I_{m''} \end{bmatrix} = \begin{bmatrix} I_n & 0 & I_n & 0 \\ 0 & I_{m'} & 0 & 0 \\ 0 & 0 & 0 & I_{m''} \end{bmatrix} \begin{bmatrix} X & Z' & 0 & 0 \\ 0 & Y' & 0 & 0 \\ 0 & 0 & X & Z'' \\ 0 & 0 & 0 & Y'' \end{bmatrix}.$$

Next, for arbitrary $n, m \in \mathbb{N}$, $X \in \Omega_n$, $Y \in \Omega_m$, $Z \in \mathcal{M}^{n \times m}$, and an invertible $T \in \mathcal{R}^{n \times n}$ we have $\begin{bmatrix} X & Z \\ 0 & Y \end{bmatrix}, \begin{bmatrix} TXT^{-1} & TZ \\ 0 & Y \end{bmatrix} \in \Omega_{n+m}$. Clearly,

$$\begin{bmatrix} TXT^{-1} & TZ \\ 0 & Y \end{bmatrix} = \begin{bmatrix} T & 0 \\ 0 & I_m \end{bmatrix} \begin{bmatrix} X & Z \\ 0 & Y \end{bmatrix} \begin{bmatrix} T & 0 \\ 0 & I_m \end{bmatrix}^{-1}.$$

Then

$$f\left(\begin{bmatrix} TXT^{-1} & TZ \\ 0 & Y \end{bmatrix}\right) = \begin{bmatrix} T & 0 \\ 0 & I_m \end{bmatrix} f\left(\begin{bmatrix} X & Z \\ 0 & Y \end{bmatrix}\right) \begin{bmatrix} T & 0 \\ 0 & I_m \end{bmatrix}^{-1},$$

i.e.,

$$\begin{bmatrix} f\left(TXT^{-1}\right) & \Delta_R f\left(TXT^{-1}, Y\right)(TZ) \\ 0 & f(Y) \end{bmatrix} = \begin{bmatrix} T & 0 \\ 0 & I_m \end{bmatrix} \begin{bmatrix} f(X) & \Delta_R f(X,Y)(Z) \\ 0 & f(Y) \end{bmatrix} \begin{bmatrix} T & 0 \\ 0 & I_m \end{bmatrix}^{-1}.$$

Comparing the (1,2) block entries in the matrix products in the left-hand side and in the right-hand side, we obtain (2X). The statement (2Y) is obtained analogously, using the similarity

$$\begin{bmatrix} X & ZS^{-1} \\ 0 & SYS^{-1} \end{bmatrix} = \begin{bmatrix} I_n & 0 \\ 0 & S \end{bmatrix} \begin{bmatrix} X & Z \\ 0 & Y \end{bmatrix} \begin{bmatrix} I_n & 0 \\ 0 & S \end{bmatrix}^{-1}.$$

The statements (3X) and (3Y) are obtained using the intertwinings

$$\begin{bmatrix} T & 0 \\ 0 & I_m \end{bmatrix} \begin{bmatrix} X & Z \\ 0 & Y \end{bmatrix} = \begin{bmatrix} \tilde{X} & TZ \\ 0 & Y \end{bmatrix} \begin{bmatrix} T & 0 \\ 0 & I_m \end{bmatrix}$$

and

$$\begin{bmatrix} X & Z \\ 0 & Y \end{bmatrix} \begin{bmatrix} I_n & 0 \\ 0 & S \end{bmatrix} = \begin{bmatrix} I_n & 0 \\ 0 & S \end{bmatrix} \begin{bmatrix} X & ZS \\ 0 & \tilde{Y} \end{bmatrix}.$$

□

REMARK 2.16. Much as in Proposition 2.1, the respect of intertwinings ((3X) & (3Y)) is equivalent to the respect of direct sums ((1X) & (1Y)) together with the respect of similarities ((2X) & (2Y)). We will show this equivalence later in Proposition 3.1.

Combining the X and the Y properties in Proposition 2.15, one obtains an equivalent joint formulation.

PROPOSITION 2.17. *Let $f\colon \Omega \to \mathcal{N}_{\rm nc}$ be a nc function on a right admissible nc set $\Omega \subseteq \mathcal{M}_{\rm nc}$. Then:*

$$(2.25)\quad \Delta_R f\left(X' \oplus X'', Y' \oplus Y''\right)\left(\begin{bmatrix} Z^{','} & Z^{','')} \\ Z^{'',}' & Z^{'','')} \end{bmatrix}\right) = \begin{bmatrix} \Delta_R f\left(X', Y'\right)\left(Z^{','}\right) & \Delta_R f\left(X', Y''\right)\left(Z^{','')}\right) \\ \Delta_R f\left(X'', Y'\right)\left(Z^{'','}\right) & \Delta_R f\left(X'', Y''\right)\left(Z^{'',''}\right) \end{bmatrix}$$

*for $n', m' \in \mathbb{N}$, $n'', m'' \in \mathbb{Z}_+$, $X' \in \Omega_{n'}$, $X'' \in \Omega_{n''}$, $Y' \in \Omega_{m'}$, $Y'' \in \Omega_{m''}$, $\begin{bmatrix} Z^{','} & Z^{',''} \\ Z^{'','} & Z^{'',''} \end{bmatrix} \in \mathcal{M}^{(n'+n'')\times(m'+m'')}$, with block entries of appropriate sizes, and if either $n''$ or $m''$ is 0 then the corresponding block entry is void;*

$$(2.26)\qquad \Delta_R f\left(TXT^{-1}, SYS^{-1}\right)\left(TZS^{-1}\right) = T\,\Delta_R f(X,Y)(Z)\,S^{-1}$$

*for $n, m \in \mathbb{N}$, $X \in \Omega_n$, $Y \in \Omega_m$, $Z \in \mathcal{M}^{n\times m}$, and invertible $T \in \mathcal{R}^{n\times n}$, $S \in \mathcal{R}^{m\times m}$ such that $TXT^{-1} \in \Omega_n$, $SYS^{-1} \in \Omega_m$;*

$$(2.27)\qquad TX = \tilde{X}T,\ YS = S\tilde{Y} \Longrightarrow T\,\Delta_R f(X,Y)(Z)\,S = \Delta_R f\left(\tilde{X}, \tilde{Y}\right)(TZS)$$

*for* $n, \tilde{n}, m, \tilde{m} \in \mathbb{N}$, $X \in \Omega_n$, $\tilde{X} \in \Omega_{\tilde{n}}$, $Y \in \Omega_m$, $\tilde{Y} \in \Omega_{\tilde{m}}$, $Z \in \mathcal{M}^{n\times m}$, *and* $T \in \mathcal{R}^{\tilde{n}\times n}$, $S \in \mathcal{R}^{m\times \tilde{m}}$.

In fact, Proposition 2.17 is a special case of Proposition 3.2, which will be proved later.

The "left" counterparts of Propositions 2.15 and 2.17 are as follows. They can either be proved analogously or deduced from the corresponding "right" statements by a transposition argument, or (in the case where the nc set $\Omega$ is both right and left admissible) by a symmetry argument.

PROPOSITION 2.18. *Let* $f\colon \Omega \to \mathcal{N}_{\rm nc}$ *be a nc function on a left admissible nc set* $\Omega \subseteq \mathcal{M}_{\rm nc}$. *Then:*

(1X) $$\Delta_L f(X'\oplus X'', Y)(\mathrm{row}[Z', Z'']) = \mathrm{row}[\Delta_L f(X', Y)(Z'), \Delta_L f(X'', Y)(Z'')]$$

*for* $n', n'', m \in \mathbb{N}$, $X' \in \Omega_{n'}$, $X'' \in \Omega_{n''}$, $Y \in \Omega_m$, $Z' \in \mathcal{M}^{m\times n'}$, $Z'' \in \mathcal{M}^{m\times n''}$;

(1Y) $$\Delta_L f(X, Y'\oplus Y'')(\mathrm{col}[Z', Z'']) = \mathrm{col}[\Delta_L f(X, Y')(Z'), \Delta_L f(X, Y'')(Z'')]$$

*for* $n, m', m'' \in \mathbb{N}$, $X \in \Omega_n$, $Y' \in \Omega_{m'}$, $Y'' \in \Omega_{m''}$, $Z' \in \mathcal{M}^{m'\times n}$, $Z'' \in \mathcal{M}^{m''\times n}$;

(2X) $$\Delta_L f\left(TXT^{-1}, Y\right)\left(ZT^{-1}\right) = \Delta_L f(X, Y)(Z)\, T^{-1}$$

*for* $n, m \in \mathbb{N}$, $X \in \Omega_n$, $Y \in \Omega_m$, $Z \in \mathcal{M}^{m\times n}$, *and an invertible* $T \in \mathcal{R}^{n\times n}$ *such* $TXT^{-1} \in \Omega_n$;

(2Y) $$\Delta_L f\left(X, SYS^{-1}\right)(SZ) = S\, \Delta_L f(X, Y)(Z)$$

*for* $n, m \in \mathbb{N}$, $X \in \Omega_n$, $Y \in \Omega_m$, $Z \in \mathcal{M}^{m\times n}$, *and an invertible* $S \in \mathcal{R}^{m\times m}$ *such that* $SYS^{-1} \in \Omega_m$;

(3X) $$\text{If}\quad XT = T\tilde{X}\quad \text{then}\quad \Delta_L f(X, Y)(Z)\, T = \Delta_L f(\tilde{X}, Y)(ZT)$$

*for* $n, \tilde{n}, m \in \mathbb{N}$, $X \in \Omega_n$, $\tilde{X} \in \Omega_{\tilde{n}}$, $Y \in \Omega_m$, $Z \in \mathcal{M}^{m\times n}$, *and* $T \in \mathcal{R}^{n\times \tilde{n}}$;

(3Y) $$\text{If}\quad SY = \tilde{Y}S\quad \text{then}\quad S\, \Delta_L f(X, Y)(Z) = \Delta_L f(X, \tilde{Y})(SZ)$$

*for* $n, m, \tilde{m} \in \mathbb{N}$, $X \in \Omega_n$, $Y \in \Omega_m$, $\tilde{Y} \in \Omega_{\tilde{m}}$, $Z \in \mathcal{M}^{m\times n}$, *and* $S \in \mathcal{R}^{\tilde{m}\times m}$.

PROPOSITION 2.19. *Let* $f\colon \Omega \to \mathcal{N}_{\rm nc}$ *be a nc function on a left admissible nc set* $\Omega \subseteq \mathcal{M}_{\rm nc}$. *Then:*

$$\Delta_L f(X'\oplus X'', Y'\oplus Y'')\left(\begin{bmatrix} Z^{\prime,\prime} & Z^{\prime,\prime\prime} \\ Z^{\prime\prime,\prime} & Z^{\prime\prime,\prime\prime}\end{bmatrix}\right) = \begin{bmatrix} \Delta_L f(X', Y')(Z^{\prime,\prime}) & \Delta_L f(X'', Y')(Z^{\prime,\prime\prime}) \\ \Delta_L f(X', Y'')(Z^{\prime\prime,\prime}) & \Delta_L f(X'', Y'')(Z^{\prime\prime,\prime\prime})\end{bmatrix} \tag{2.28}$$

*for* $n', m' \in \mathbb{N}$, $n'', m'' \in \mathbb{Z}_+$, $X' \in \Omega_{n'}$, $X'' \in \Omega_{n''}$, $Y' \in \Omega_{m'}$, $Y'' \in \Omega_{m''}$, $\begin{bmatrix} Z^{\prime,\prime} & Z^{\prime,\prime\prime} \\ Z^{\prime\prime,\prime} & Z^{\prime\prime,\prime\prime}\end{bmatrix} \in \mathcal{M}^{(m'+m'')\times(n'+n'')}$, *with block entries of appropriate sizes, and if either* $n''$ *or* $m''$ *is* 0 *then the corresponding block entry is void;*

$$\Delta_L f\left(TXT^{-1}, SYS^{-1}\right)\left(SZT^{-1}\right) = S\, \Delta_L f(X, Y)(Z)\, T^{-1} \tag{2.29}$$

*for* $n, m \in \mathbb{N}$, $X \in \Omega_n$, $Y \in \Omega_m$, $Z \in \mathcal{M}^{m\times n}$, *and invertible* $T \in \mathcal{R}^{n\times n}$, $S \in \mathcal{R}^{m\times m}$ *such that* $TXT^{-1} \in \Omega_n$, $SYS^{-1} \in \Omega_m$;

$$XT = T\tilde{X},\ SY = \tilde{Y}S \Longrightarrow S\, \Delta_L f(X, Y)(Z)\, T = \Delta_L f(\tilde{X}, \tilde{Y})(SZT) \tag{2.30}$$

*for* $n, \tilde{n}, m, \tilde{m} \in \mathbb{N}$, $X \in \Omega_n$, $\tilde{X} \in \Omega_{\tilde{n}}$, $Y \in \Omega_m$, $\tilde{Y} \in \Omega_{\tilde{m}}$, $Z \in \mathcal{M}^{m\times n}$, *and* $T \in \mathcal{R}^{n\times \tilde{n}}$, $S \in \mathcal{R}^{\tilde{m}\times m}$.

## 2.6. Directional nc difference-differential operators

Alongside the full difference-differential operators introduced above, we will also consider their *directional* versions. Let $\Omega \subseteq \mathcal{M}_{\rm nc}$ be a right (resp., left) admissible nc set, and let $f\colon \Omega \to \mathcal{N}_{\rm nc}$ be a nc function. Given $\mu \in \mathcal{M}$, we define for all $n, m \in \mathbb{N}$, $X \in \Omega_n$, $Y \in \Omega_m$ the linear mapping $\Delta_{R,\mu} f(X,Y)\colon \mathcal{R}^{n\times m} \to \mathcal{N}^{n\times m}$ (resp., $\Delta_{L,\mu} f(X,Y)\colon \mathcal{R}^{m\times n} \to \mathcal{N}^{m\times n}$) by

$$\Delta_{R,\mu} f(X,Y)(A) = \Delta_R f(X,Y)(A\mu) \tag{2.31}$$

and, respectively,

$$\Delta_{L,\mu} f(X,Y)(A) = \Delta_L f(X,Y)(A\mu). \tag{2.32}$$

Here for a matrix $A$ over $\mathcal{R}$ and $\mu \in \mathcal{M}$, the product $A\mu$ is a matrix over $\mathcal{M}$ of the same size and $(A\mu)_{ij} = A_{ij}\mu$ for all $i$ and $j$.

In the special case where $\mathcal{M} = \mathcal{R}^d$, we define the *$j$-th right (resp.,left) partial nc difference-differential operator* by $\Delta_{R,j} = \Delta_{R,e_j}$ (resp., $\Delta_{L,j} = \Delta_{L,e_j}$) where $e_j$ is the $j$-th standard basis vector in $\mathcal{R}^d$, $j = 1, \ldots, d$. By linearity, we have

$$\Delta_R f(X,Y)(Z) = \sum_{j=1}^{d} \Delta_{R,j} f(X,Y)(Z_j) \tag{2.33}$$

and

$$\Delta_L f(X,Y)(Z) = \sum_{j=1}^{d} \Delta_{L,j} f(X,Y)(Z_j). \tag{2.34}$$

For the basic rules of nc calculus, 2.3.1, 2.3.2, 2.3.4, and 2.3.5 have exactly the same form for the directional nc difference-differential operators. In the setting of 2.3.3,

$$\Delta_{R,\mu} l(X,Y)(A) = Al(\mu) \quad \text{and} \quad \Delta_{L,\mu} l(X,Y)(A) = Al(\mu);$$

in particular, in the case where $\mathcal{M} = \mathcal{R}^d$ we have

$$\Delta_{R,i} l_j(X,Y)(A) = \delta_{ij} A \quad \text{and} \quad \Delta_{L,i} l_j(X,Y)(A) = \delta_{ij} A.$$

A version of 2.3.6 for the right and left partial nc difference-differential operators in the case where $\mathcal{M} = \mathcal{R}^d$ and $\mathcal{N} = \mathcal{R}^\ell$ is

$$\Delta_{R,i}(g \circ f)(X,Y)(A) = \sum_{j=1}^{\ell} \Delta_{R,j} g(f(X), f(Y))(\Delta_{R,i} f_j(X,Y)(A))$$

and

$$\Delta_{L,i}(g \circ f)(X,Y)(A) = \sum_{j=1}^{\ell} \Delta_{L,j} g(f(X), f(Y))(\Delta_{L,i} f_j(X,Y)(A))$$

for $i = 1, \ldots, d$, where $f(X) = f_1(X)e_1 + \cdots + f_\ell(X)e_\ell$ with the component nc functions $f_j\colon \Omega \to \mathcal{R}_{\rm nc}$.

The first order difference formulae of Theorem 2.10 in the case where $\mathcal{M} = \mathcal{R}^d$ can be written in terms of partial nc difference-differential operators. It follows from (2.19)–(2.20) and (2.33) that

$$f(X) - f(Y) = \sum_{j=1}^{d} \Delta_{R,j} f(X,Y)(X_j - Y_j) = \sum_{j=1}^{d} \Delta_{R,j} f(Y,X)(X_j - Y_j).$$

Similarly, it follows from (2.21)–(2.22) and (2.34) that

$$f(X) - f(Y) = \sum_{j=1}^{d} \Delta_{L,j} f(X,Y)(X_j - Y_j) = \sum_{j=1}^{d} \Delta_{L,j} f(Y,X)(X_j - Y_j).$$

For the properties of $\Delta_{R,\mu} f(X,Y)$ and $\Delta_{L,\mu} f(X,Y)$ as functions of $X$ and $Y$, all the statements of Section 2.5 have exactly the same form.

CHAPTER 3

# Higher order nc functions and their difference-differential calculus

Our next goal is to define higher order differentials of a nc function (more precisely, higher order nc difference-differential operators). As we saw in Chapter 2, for a nc function $f$ on a nc set $\Omega \subseteq \mathcal{M}_{\rm nc}$ to a nc space $\mathcal{N}_{\rm nc}$, $\Delta_R f(X,Y)(\cdot)$ is a function of two arguments $X \in \Omega_n$ and $Y \in \Omega_m$ (for all $n$ and $m$) with values linear mappings $\mathcal{M}^{n\times m} \to \mathcal{N}^{n\times m}$. We also saw that $\Delta_R f(X,Y)(\cdot)$ respects (in an appropriate way) direct sums, similarities, and intertwinings. This motivates the appearance of classes $\mathcal{T}^k = \mathcal{T}^k(\Omega^{(0)},\ldots,\Omega^{(k)};\mathcal{N}_{0,{\rm nc}},\ldots,\mathcal{N}_{k,{\rm nc}})$ of nc functions of order $k$ ($k = 0,1,\ldots$) together with the right nc difference-differential operators

$$\begin{aligned}\Delta_R\colon \mathcal{T}^k(\Omega^{(0)},\ldots,\Omega^{(k)};\mathcal{N}_{0,{\rm nc}},\ldots,\mathcal{N}_{k,{\rm nc}}) \\ \to \mathcal{T}^{k+1}(\Omega^{(0)},\ldots,\Omega^{(k)},\Omega^{(k)};\mathcal{N}_{0,{\rm nc}},\ldots,\mathcal{N}_{k\ {\rm nc}},\mathcal{M}_{k,{\rm nc}}).\end{aligned}$$

A nc function of order $k$ is a function of $k+1$ arguments, in nc sets $\Omega^{(0)} \subseteq \mathcal{M}_{0,{\rm nc}}$, $\ldots$, $\Omega^{(k)} \subseteq \mathcal{M}_{k,{\rm nc}}$, with values certain $k$-linear mappings, which respects (in an appropriate way) direct sums, similarities, and intertwinings. Higher order nc difference-differential operators will be obtained by iterating $\Delta_R$.

(From now on, we will concentrate on right nc difference-differential operators. Of course, one can construct an analogous "left" theory.)

## 3.1. Higher order nc functions

Let $\mathcal{M}_0$, $\ldots$, $\mathcal{M}_k$, $\mathcal{N}_0$, $\ldots$, $\mathcal{N}_k$ be modules over ring $\mathcal{R}$. Let $\Omega^{(j)} \subseteq \mathcal{M}_{j,{\rm nc}}$ be a nc set, $j = 0,\ldots,k$. We consider functions $f$ on $\Omega^{(0)} \times \cdots \times \Omega^{(k)}$ with

$$f(\Omega^{(0)}_{n_0},\ldots,\Omega^{(k)}_{n_k}) \subseteq \hom_{\mathcal{R}}\left({\mathcal{N}_1}^{n_0\times n_1} \otimes \cdots \otimes {\mathcal{N}_k}^{n_{k-1}\times n_k}, {\mathcal{N}_0}^{n_0\times n_k}\right).$$

In other words, for $X^0 \in \Omega^{(0)}_{n_0}$, $\ldots$, $X^k \in \Omega^{(k)}_{n_k}$, $f\left(X^0,\ldots,X^k\right)$ is a $k$-linear mapping over $\mathcal{R}$ from ${\mathcal{N}_1}^{n_0\times n_1} \times \cdots \times {\mathcal{N}_k}^{n_{k-1}\times n_k}$ to ${\mathcal{N}_0}^{n_0\times n_k}$. We will call such a function $f$ a *nc function of order* $k$ if $f$ satisfies the following two conditions:

- $f$ *respects direct sums*: if $n_0,\ldots,n_k \in \mathbb{N}$, $X^0 \in \Omega^{(0)}_{n_0}$, $\ldots$, $X^k \in \Omega^{(k)}_{n_k}$, $Z^1 \in {\mathcal{N}_1}^{n_0\times n_1}$, $\ldots$, $Z^k \in {\mathcal{N}_k}^{n_{k-1}\times n_k}$, then

$$\begin{aligned}(1\mathrm{X}^0)\quad & f({X^0}' \oplus {X^0}'', X^1,\ldots,X^k)(\mathrm{col}\,[{Z^1}',{Z^1}''],Z^2,\ldots,Z^k) \\ &= \mathrm{col}\,[f({X^0}',X^1,\ldots,X^k)({Z^1}',Z^2,\ldots,Z^k), \\ &\qquad f({X^0}'',X^1,\ldots,X^k)({Z^1}'',Z^2,\ldots,Z^k)]\end{aligned}$$

for $n_0', n_0'' \in \mathbb{N}$, $X^{0\,\prime} \in \Omega^{(0)}_{n_0'}$, $X^{0\,\prime\prime} \in \Omega^{(0)}_{n_0''}$, $Z^{1\,\prime} \in \mathcal{N}_1{}^{n_0' \times n_1}$, $Z^{1\,\prime\prime} \in \mathcal{N}_1{}^{n_0'' \times n_1}$, where $Z^2, \ldots, Z^k$ do not show up when $k = 1$;

(1X$^j$)
$$f(X^0, \ldots, X^{j-1}, X^{j\,\prime} \oplus X^{j\,\prime\prime}, X^{j+1}, \ldots, X^k) (Z^1, \ldots, Z^{j-1}, \operatorname{row}[Z^{j\,\prime}, Z^{j\,\prime\prime}], \operatorname{col}[Z^{(j+1)\,\prime}, Z^{(j+1)\,\prime\prime}], Z^{j+2}, \ldots, Z^k) = f(X^0, \ldots, X^{j-1}, X^{j\,\prime}, X^{j+1}, \ldots, X^k) (Z^1, \ldots, Z^{j-1}, Z^{j\,\prime}, Z^{(j+1)\,\prime}, Z^{j+2}, \ldots, Z^k + f(X^0, \ldots, X^{j-1}, X^{j\,\prime\prime}, X^{j+1}, \ldots, X^k) (Z^1, \ldots, Z^{j-1}, Z^{j\,\prime\prime}, Z^{(j+1)\,\prime\prime}, Z^{j+2}, \ldots, Z^k)$$

for every $j \in \{1, \ldots, k-1\}$, and for $n_j', n_j'' \in \mathbb{N}$, $X^{j\,\prime} \in \Omega^{(j)}_{n_j'}$, $X^{j\,\prime\prime} \in \Omega^{(j)}_{n_j''}$, $Z^{j\,\prime} \in \mathcal{N}_j{}^{n_{j-1} \times n_j'}$, $Z^{j\,\prime\prime} \in \mathcal{N}_j{}^{n_{j-1} \times n_j''}$, $Z^{(j+1)\,\prime} \in (\mathcal{N}_{j+1})^{n_j' \times n_{j+1}}$, $Z^{(j+1)\,\prime\prime} \in (\mathcal{N}_{j+1})^{n_j'' \times n_{j+1}}$, where $Z^1$, $\ldots$, $Z^{j-1}$ do not show up for $j = 1$, and $Z^{j+2}, \ldots, Z^k$ do not show up for $j = k-1$;

(1X$^k$)
$$f(X^0, \ldots, X^{k-1}, X^{k\,\prime} \oplus X^{k\,\prime\prime})(Z^1, \ldots, Z^{k-1}, \operatorname{row}[Z^{k\,\prime}, Z^{k\,\prime\prime}]) = \operatorname{row}[f(X^0, \ldots, X^{k-1}, X^{k\,\prime})(Z^1, \ldots, Z^{k-1}, Z^{k\,\prime}, f(X^0, \ldots, X^{k-1}, X^{k\,\prime\prime})(Z^1, \ldots, Z^{k-1}, Z^{k\,\prime\prime})]$$

for $n_k', n_k'' \in \mathbb{N}$, $X^{k\,\prime} \in \Omega^{(k)}_{n_k'}$, $X^{k\,\prime\prime} \in \Omega^{(k)}_{n_k''}$, $Z^{k\,\prime} \in \mathcal{N}_k{}^{n_{k-1} \times n_k'}$, $Z^{k\,\prime\prime} \in \mathcal{N}_k{}^{n_{k-1} \times n_k''}$, where $Z^1, \ldots, Z^{k-1}$ do not show up when $k = 1$.

- $f$ *respects similarities*: if $n_0, \ldots, n_k \in \mathbb{N}$, $X^0 \in \Omega^{(0)}_{n_0}$, $\ldots$, $X^k \in \Omega^{(k)}_{n_k}$, $Z^1 \in \mathcal{N}_1{}^{n_0 \times n_1}$, $\ldots$, $Z^k \in \mathcal{N}_k{}^{n_{k-1} \times n_k}$, then

(2X$^0$)
$$f(S_0 X^0 S_0^{-1}, X^1, \ldots, X^k)(S_0 Z^1, Z^2, \ldots, Z^k) = S_0 f(X^0, \ldots, X^k)(Z^1, \ldots, Z^k)$$

for an invertible $S_0 \in \mathcal{R}^{n_0 \times n_0}$ such that $S_0 X^0 S_0^{-1} \in \Omega^{(0)}_{n_0}$, where $Z^2, \ldots, Z^k$ do not show up when $k = 1$;

(2X$^j$)
$$f(X^0, \ldots, X^{j-1}, S_j X^j S_j^{-1}, X^{j+1}, \ldots, X^k) (Z^1, \ldots, Z^{j-1}, Z^j S_j^{-1}, S_j Z^{j+1}, Z^{j+2}, \ldots, Z^k) = f(X^0, \ldots, X^k)(Z^1, \ldots, Z^k)$$

for every $j \in \{1, \ldots, k-1\}$ and an invertible $S_j \in \mathcal{R}^{n_j \times n_j}$ such that $S_j X^j S_j^{-1} \in \Omega^{(j)}_{n_j}$, where $Z^1, \ldots, Z^{j-1}$ do not show up for $j = 1$, and $Z^{j+2}, \ldots, Z^k$ do not show up for $j = k-1$;

(2X$^k$)
$$f(X^0, \ldots, X^{k-1}, S_k X^k S_k^{-1})(Z^1, \ldots, Z^{k-1}, Z^k S_k^{-1}) = f(X^0, \ldots, X^k)(Z^1, \ldots, Z^k) S_k^{-1}$$

for an invertible $S_k \in \mathcal{R}^{n_k \times n_k}$ such that $S_k X^k S_k^{-1} \in \Omega^{(k)}_{n_k}$, where $Z^1$, $\ldots$, $Z^{k-1}$ do not show up when $k = 1$.

We define *nc functions of order zero* to be simply nc functions as in Section 2.1. We denote by $\mathcal{T}^k = \mathcal{T}^k(\Omega^{(0)}, \ldots, \Omega^{(k)}; \mathcal{N}_{0,\mathrm{nc}}, \ldots, \mathcal{N}_{k,\mathrm{nc}})$ the class of nc functions of order $k$, $k = 0, 1, \ldots$. In the case where $\mathcal{M}_0 = \cdots = \mathcal{M}_k =: \mathcal{M}$ and $\Omega^{(0)} = \cdots = \Omega^{(k)} =: \Omega$, we will write $\mathcal{T}^k = \mathcal{T}^k(\Omega; \mathcal{N}_{0,\mathrm{nc}}, \ldots, \mathcal{N}_{k,\mathrm{nc}})$. We will provide later a slightly different interpretation of nc functions of order $k$ using tensor products rather than multilinear mappings for their values, which shows why the definition of higher order nc functions is natural and generalizes nc functions of order zero.

Similarly to the class $\mathcal{T}^0$ (see Proposition 2.1), the two conditions in the definition of a nc function of order $k$ can be replaced by a single one.

PROPOSITION 3.1. *Let $\Omega^{(j)} \subseteq \mathcal{M}_{j,\mathrm{nc}}$, $j = 0, \ldots, k$, be nc sets. A function $f$ on $\Omega^{(0)} \times \cdots \times \Omega^{(k)}$ with*

$$f(\Omega^{(0)}_{n_0}, \ldots, \Omega^{(k)}_{n_k}) \subseteq \hom_{\mathcal{R}}(\mathcal{N}_1{}^{n_0 \times n_1} \otimes \cdots \otimes \mathcal{N}_k{}^{n_{k-1} \times n_k}, \mathcal{N}_0{}^{n_0 \times n_k})$$

*respects direct sums and similarities, i.e., $f \in \mathcal{T}^k(\Omega^{(0)}, \ldots, \Omega^{(k)}; \mathcal{N}_{0,\mathrm{nc}}, \ldots, \mathcal{N}_{k,\mathrm{nc}})$, if and only if $f$* respects intertwinings*: if $n_0, \ldots, n_k \in \mathbb{N}$, $X^0 \in \Omega^{(0)}_{n_0}$, $\ldots$, $X^k \in \Omega^{(k)}_{n_k}$, $Z^1 \in \mathcal{N}_1{}^{n_0 \times n_1}$, $\ldots$, $Z^k \in \mathcal{N}_k{}^{n_{k-1} \times n_k}$, then:*

(3X$^0$) *if* $T_0 X^0 = \tilde{X}^0 T_0$, *then*

$$T_0 f(X^0, \ldots, X^k)(Z^1, \ldots, Z^k) = f(\tilde{X}^0, X^1, \ldots, X^k)(T_0 Z^1, Z^2, \ldots, Z^k)$$

*for $\tilde{n}_0 \in \mathbb{N}$, $\tilde{X}^0 \in \Omega^{(0)}_{\tilde{n}_0}$, and $T_0 \in \mathcal{R}^{\tilde{n}_0 \times n_0}$, where $Z^2, \ldots, Z^k$ do not show up when $k = 1$;*

(3X$^j$) *if* $T_j X^j = \tilde{X}^j T_j$, *then*

$$\begin{aligned} &f(X^0, \ldots, X^k)(Z^1, \ldots, Z^{j-1}, \tilde{Z}^j T_j, Z^{j+1}, Z^{j+2}, \ldots, Z^k) \\ &= f(X^0, \ldots, X^{j-1}, \tilde{X}^j, X^{j+1}, \ldots, X^k)(Z^1, \ldots, Z^{j-1}, \tilde{Z}^j, T_j Z^{j+1}, Z^{j+2}, \ldots, Z^k) \end{aligned}$$

*for every $j \in \{1, \ldots, k-1\}$, and for $\tilde{n}_j \in \mathbb{N}$, $\tilde{X}^j \in \Omega^{(j)}_{\tilde{n}_j}$, $\tilde{Z}^j \in \mathcal{N}_j{}^{n_{j-1} \times \tilde{n}_j}$, $T_j \in \mathcal{R}^{\tilde{n}_j \times n_j}$, where $Z^1, \ldots, Z^{j-1}$ do not show up for $j = 1$, and $Z^{j+2}, \ldots, Z^k$ do not show up for $j = k-1$;*

(3X$^k$) *if* $X^k T_k = T_k \tilde{X}^k$, *then*

$$f(X^0, \ldots, X^k)(Z^1, \ldots, Z^k) T_k = f(X^0, \ldots, X^{k-1}, \tilde{X}^k)(Z^1, \ldots, Z^{k-1}, Z^k T_k)$$

*for $\tilde{n}_k \in \mathbb{N}$, $\tilde{X}^k \in \Omega^{(k)}_{\tilde{n}_k}$, and $T_k \in \mathcal{R}^{n_k \times \tilde{n}_k}$, where $Z^1, \ldots, Z^{k-1}$ do not show up when $k = 1$.*

*Moreover, conditions* (1X$^0$) *and* (2X$^0$) *in the definition of a nc function of order $k$ together are equivalent to condition* (3X$^0$)*, and similarly,* (1X$^j$) & (2X$^j$) $\Longleftrightarrow$ (3X$^j$) *($j = 1, \ldots, k-1$),* (1X$^k$) & (2X$^k$) $\Longleftrightarrow$ (3X$^k$).

PROOF. We will provide a detailed proof of (1X$^0$) & (2X$^0$) $\Longleftrightarrow$ (3X$^0$). Other equivalences in the proposition can be proved analogously.

Assume first that conditions (1X$^0$) and (2X$^0$) in the definition of a nc function of order $k$ are fulfilled. Let $n_0, \tilde{n}_0, n_1, \ldots, n_k \in \mathbb{N}$, $X^0 \in \Omega^{(0)}_{n_0}$, $\tilde{X}^0 \in \Omega^{(0)}_{\tilde{n}_0}$, $X^1 \in \Omega^{(1)}_{n_1}$, $\ldots$, $X^k \in \Omega^{(k)}_{n_k}$, $Z^1 \in \mathcal{N}_1{}^{n_0 \times n_1}$, $\tilde{Z}^1 \in \mathcal{N}_1{}^{\tilde{n}_0 \times n_1}$, $Z^2 \in \mathcal{N}_2{}^{n_1 \times n_2}$, $\ldots$, $Z^k \in$

${\mathcal{N}_k}^{n_{k-1}\times n_k}$, and let $T_0 \in \mathcal{R}^{\tilde{n}_0\times n_0}$ be such that $T_0X^0 = \tilde{X}^0T_0$. By (1X$^0$), one has

$$\begin{aligned} &f(X^0\oplus\tilde{X}^0, X^1,\ldots,X^k)(\mathrm{col}\,[Z^1,\tilde{Z}^1],Z^2,\ldots,Z^k)\\ &\qquad= \mathrm{col}\,[f(X^0,\ldots,X^k)(Z^1,\ldots,Z^k), f(\tilde{X}^0,X^1,\ldots,X^k)(\tilde{Z}^1,Z^2,\ldots,Z^k)]. \end{aligned}$$

On the other hand, using the identity

$$\begin{bmatrix} X^0 & 0\\ 0 & \tilde{X}^0\end{bmatrix} = \begin{bmatrix} I_{n_0} & 0\\ T_0 & I_{\tilde{n}_0}\end{bmatrix}\begin{bmatrix} X^0 & 0\\ 0 & \tilde{X}^0\end{bmatrix}\begin{bmatrix} I_{n_0} & 0\\ -T_0 & I_{\tilde{n}_0}\end{bmatrix},$$

and conditions (1X$^0$) and (2X$^0$), we obtain

$$\begin{aligned} &f(X^0\oplus\tilde{X}^0, X^1,\ldots,X^k)(\mathrm{col}\,[Z^1,\tilde{Z}^1],Z^2,\ldots,Z^k)\\ &\qquad= f\left(\begin{bmatrix} I_{n_0} & 0\\ T_0 & I_{\tilde{n}_0}\end{bmatrix}\begin{bmatrix} X^0 & 0\\ 0 & \tilde{X}^0\end{bmatrix}\begin{bmatrix} I_{n_0} & 0\\ -T_0 & I_{\tilde{n}_0}\end{bmatrix}, X^1,\ldots,X^k\right)\\ &\qquad\qquad\left(\begin{bmatrix} I_{n_0} & 0\\ T_0 & I_{\tilde{n}_0}\end{bmatrix}\begin{bmatrix} I_{n_0} & 0\\ -T_0 & I_{\tilde{n}_0}\end{bmatrix}\mathrm{col}[Z^1,\tilde{Z}^1],Z^2,\ldots,Z^k\right)\\ &\quad= \begin{bmatrix} I_{n_0} & 0\\ T_0 & I_{\tilde{n}_0}\end{bmatrix} f(X^0\oplus\tilde{X}^0,X^1,\ldots,X^k)(\mathrm{col},[Z^1,-T_0Z^1+\tilde{Z}^1],Z^2,\ldots,Z^k)\\ &= \begin{bmatrix} I_{n_0} & 0\\ T_0 & I_{\tilde{n}_0}\end{bmatrix}\mathrm{col}\,[f(X^0,\ldots,X^k)(Z^1,\ldots,Z^k),\\ &\qquad\qquad f(\tilde{X}^0,X^1,\ldots,X^k)(-T_0Z^1+\tilde{Z}^1,Z^2,\ldots,Z^k)]\\ &= \mathrm{col}\,[f(X^0,\ldots,X^k)(Z^1,\ldots,Z^k),\\ &\qquad T_0f(X^0,\ldots,X^k)(Z^1,\ldots,Z^k)+f(\tilde{X}^0,\ldots,X^k)(-T_0Z^1+\tilde{Z}^1,Z^2,\ldots,Z^k)]. \end{aligned}$$

The comparison of the (2,1) blocks in the two block column expressions for

$$f(X^0\oplus\tilde{X}^0, X^1,\ldots,X^k)(\mathrm{col}\,[Z^1,\tilde{Z}^1],Z^2,\ldots,Z^k)$$

gives

$$\begin{aligned} &f(\tilde{X}^0,X^1,\ldots,X^k)(\tilde{Z}^1,Z^2,\ldots,Z^k)\\ &= T_0f(X^0,\ldots,X^k)(Z^1,\ldots,Z^k)+f(\tilde{X}^0,X^1,\ldots,X^k)(-T_0Z^1+\tilde{Z}^1,Z^2,\ldots,Z^k). \end{aligned}$$

Using linearity in $Z^1$ of $f(X^0,\ldots,X^k)(Z^1,\ldots,Z^k)$, we obtain

$$0 = T_0f(X^0,\ldots,X^k)(Z^1,\ldots,Z^k) - f(\tilde{X}^0,X^1,\ldots,X^k)(T_0Z^1,Z^2,\ldots,Z^k),$$

and (3X$^0$) follows.

Conversely, assume that (3X$^0$) holds. Let $n_0', n_0'', n_1,\ldots,n_k\in\mathbb{N}$, $X^{0\prime}\in\Omega^{(0)}_{n_0'}$, $X^{0\prime\prime}\in\Omega^{(0)}_{n_0''}$, $X^1\in\Omega^{(1)}_{n_1}$, $\ldots$, $X^k\in\Omega^{(k)}_{n_k}$, $Z^{1\prime}\in{\mathcal{N}_1}^{n_0'\times n_1}$, $Z^{1\prime\prime}\in{\mathcal{N}_1}^{n_0''\times n_1}$, $Z^2\in{\mathcal{N}_2}^{n_1\times n_2}$, $\ldots$, $Z^k\in{\mathcal{N}_k}^{n_{k-1}\times n_k}$. The intertwinings

$$\mathrm{col}\,[I_{n_0'},0]X^{0\prime} = (X^{0\prime}\oplus X^{0\prime\prime})\,\mathrm{col}\,[I_{n_0'},0]$$

and

$$\mathrm{col}\,[0,I_{n_0''}]X^{0\prime\prime} = (X^{0\prime}\oplus X^{0\prime\prime})\,\mathrm{col}\,[0,I_{n_0''}]$$

imply, respectively,

$$\begin{aligned} &\mathrm{col}\,[I_{n_0'},0]f(X^{0\prime},X^1,\ldots,X^k)(Z^{1\prime},Z^2,\ldots,Z^k)\\ &\qquad\qquad= f(X^{0\prime}\oplus X^{0\prime\prime},X^1,\ldots,X^k)(\mathrm{col}\,[I_{n_0'},0]Z^{1\prime},Z^2,\ldots,Z^k) \end{aligned}$$

and

$$\operatorname{col}[0, I_{n_0''}]f(X^{0''}, X^1, \dots, X^k)(Z^{1''}, Z^2, \dots, Z^k) \\ = f(X^{0'} \oplus X^{0''}, X^1, \dots, X^k)(\operatorname{col}[0, I_{n_0''}]Z^{1''}, Z^2, \dots, Z^k).$$

Summing up these two equalities, we obtain (1X$^0$) by linearity.

If $n_0 \in \mathbb{N}$, $X^0 \in \Omega^{(0)}_{n_0}$, $Z^1 \in \mathcal{N}_1{}^{n_0 \times n_1}$, and $S_0 \in \mathcal{R}^{n_0 \times n_0}$ is invertible, then, in virtue of (3X$^0$), the intertwining $S_0 X^0 = \left(S_0 X^0 S_0^{-1}\right) S_0$ implies (2X$^0$). □

Similarly to Proposition 2.17, each of the conditions of respecting direct sums, similarities and intertwinings admits an equivalent joint formulation.

PROPOSITION 3.2. (1) *Conditions* (1X$^0$), (1X$^j$) *for* $j = 1, \dots, k-1$, *and* (1X$^k$) *are equivalent to*

$$(3.1)\quad f(X^{0'} \oplus X^{0''}, \dots, X^{k'} \oplus X^{k''}) \left( \begin{bmatrix} Z^{1','} & Z^{1','',} \\ Z^{1'','} & Z^{1'','','} \end{bmatrix}, \dots, \begin{bmatrix} Z^{k','} & Z^{k','',} \\ Z^{k'','} & Z^{k'',''} \end{bmatrix} \right) \\ = \begin{bmatrix} f^{','} & f^{',''} \\ f^{'','} & f^{'',''} \end{bmatrix},$$

*where, for* $\alpha, \beta \in \{','' \}$,

$$(3.2)\quad f^{\alpha,\beta} = \sum_{\alpha_0, \dots, \alpha_k \in \{','' \}:\ \alpha_0 = \alpha, \alpha_k = \beta} f(X^{0\alpha_0}, \dots, X^{k\alpha_k})(Z^{1\alpha_0,\alpha_1}, \dots, Z^{k\alpha_{k-1},\alpha_k}).$$

*Here* $n_j' \in \mathbb{N}$, $n_j'' \in \mathbb{Z}_+$, $X^{j\alpha} \in \Omega^{(j)}_{n_j^\alpha}$ *for* $j = 0, \dots, k$, $\alpha \in \{','' \}$, $Z^{j\alpha,\beta} \in \mathcal{N}_j{}^{n_{j-1}^\alpha \times n_j^\beta}$ *for* $j = 1, \dots, k$, $\alpha, \beta \in \{','' \}$, *the block entry* $f^{\alpha,\beta}$ *is void if either* $n_0^\alpha$ *or* $n_k^\beta$ *is* 0, *and a summand in* (3.2) *is* 0 *if at least one of* $n_j^{\alpha_j}$, $j = 1, \dots, k-1$, *is* 0.

(2) *Conditions* (2X$^0$), (2X$^j$) *for* $j = 1, \dots, k-1$, *and* (2X$^k$) *are equivalent to*

$$(3.3)\quad f(S_0 X^0 S_0^{-1}, \dots, S_k X^k S_k^{-1})(S_0 Z^1 S_1^{-1}, \dots, S_{k-1} Z^k S_k^{-1}) \\ = S_0 f(X^0, \dots, X^k)(Z^1, \dots, Z^k) S_k^{-1}$$

*for* $n_j \in \mathbb{N}$, $X^j \in \Omega^{(j)}_{n_j}$, *and invertible* $S_j \in \mathcal{R}^{n_j \times n_j}$ *such that* $S_j X^j S_j^{-1} \in \Omega^{(j)}_{n_j}$, $j = 0, \dots, k$, *and for* $Z^j \in \mathcal{N}_j{}^{n_{j-1} \times n_j}$, $j = 1, \dots, k$.

(3) *Conditions* (3X$^0$), (3X$^j$) *for* $j = 1, \dots, k-1$, *and* (3X$^k$) *are equivalent to*

(3.4) *If* $T_j X^j = \tilde{X}^j T_j$, $j = 0, \dots, k$, *then*

$$T_0 f(X^0, \dots, X^k)(Z^1 T_1, \dots, Z^k T_k) \\ = f(\tilde{X}^0, \dots, \tilde{X}^k)(T_0 Z^1, \dots, T_{k-1} Z^k) T_k$$

*for* $n_j, \tilde{n}_j \in \mathbb{N}$, $X^j \in \Omega^{(j)}_{n_j}$, $\tilde{X}^j \in \Omega^{(j)}_{\tilde{n}_j}$, $T_j \in \mathcal{R}^{\tilde{n}_j \times n_j}$, $j = 0, \dots, k$, *and for* $Z^j \in \mathcal{N}_j{}^{n_{j-1} \times \tilde{n}_j}$, $j = 1, \dots, k$.

PROOF. (1). Suppose that ($1X^0$)–($1X^k$) hold. Then

$$
\begin{aligned}
&f(X^{0\prime}\oplus X^{0\prime\prime},\ldots,X^{k\prime}\oplus X^{k\prime\prime})\left(\begin{bmatrix} Z^{1\prime,\prime} & Z^{1\prime,\prime\prime}\\ Z^{1\prime\prime,\prime} & Z^{1\prime\prime,\prime\prime}\end{bmatrix},\ldots,\begin{bmatrix} Z^{k\prime,\prime} & Z^{k\prime,\prime\prime}\\ Z^{k\prime\prime,\prime} & Z^{k\prime\prime,\prime\prime}\end{bmatrix}\right)\\
&= \operatorname*{col}_{\alpha_0\in\{\prime,\prime\prime\}}\Bigg[f(X^{0\alpha_0},X^{1\prime}\oplus X^{1\prime\prime},\ldots,X^{k\prime}\oplus X^{k\prime\prime})\\
&\qquad\left(\operatorname{row}[Z^{1\alpha_0,\prime},Z^{1\alpha_0,\prime\prime}],\begin{bmatrix} Z^{2\prime,\prime} & Z^{2\prime,\prime\prime}\\ Z^{2\prime\prime,\prime} & Z^{2\prime\prime,\prime\prime}\end{bmatrix},\ldots,\begin{bmatrix} Z^{k\prime,\prime} & Z^{k\prime,\prime\prime}\\ Z^{k\prime\prime,\prime} & Z^{k\prime\prime,\prime\prime}\end{bmatrix}\right)\Bigg]\\
&= \operatorname*{col}_{\alpha_0\in\{\prime,\prime\prime\}}\Bigg[\sum_{\alpha_1\in\{\prime,\prime\prime\}} f(X^{0\alpha_0},X^{1\alpha_1},X^{2\prime}\oplus X^{2\prime\prime},\ldots,X^{k\prime}\oplus X^{k\prime\prime})\\
&\qquad\left(Z^{1\alpha_0,\alpha_1},\operatorname{row}[Z^{2\alpha_1,\prime},Z^{2\alpha_1,\prime\prime}],\begin{bmatrix} Z^{3\prime,\prime} & Z^{3\prime,\prime\prime}\\ Z^{3\prime\prime,\prime} & Z^{3\prime\prime,\prime\prime}\end{bmatrix},\ldots,\begin{bmatrix} Z^{k\prime,\prime} & Z^{k\prime,\prime\prime}\\ Z^{k\prime\prime,\prime} & Z^{k\prime\prime,\prime\prime}\end{bmatrix}\right)\Bigg]\\
&= \operatorname*{col}_{\alpha_0\in\{\prime,\prime\prime\}}\Bigg[\sum_{\alpha_1,\alpha_2\in\{\prime,\prime\prime\}} f(X^{0\alpha_0},X^{1\alpha_1},X^{2\alpha_2},X^{3\prime}\oplus X^{3\prime\prime},\ldots,X^{k\prime}\oplus X^{k\prime\prime})\\
&\qquad\left(Z^{1\alpha_0,\alpha_1},Z^{2\alpha_1,\alpha_2},\operatorname{row}[Z^{3\alpha_2,\prime},Z^{3\alpha_2,\prime\prime}],\begin{bmatrix} Z^{4\prime,\prime} & Z^{4\prime,\prime\prime}\\ Z^{4\prime\prime,\prime} & Z^{4\prime\prime,\prime\prime}\end{bmatrix},\ldots,\begin{bmatrix} Z^{k\prime,\prime} & Z^{k\prime,\prime\prime}\\ Z^{k\prime\prime,\prime} & Z^{k\prime\prime,\prime\prime}\end{bmatrix}\right)\Bigg]\\
&= \cdots = \operatorname*{col}_{\alpha_0\in\{\prime,\prime\prime\}}\Bigg[\sum_{\alpha_1,\ldots,\alpha_{k-1}\in\{\prime,\prime\prime\}} f(X^{0\alpha_0},\ldots,X^{(k-1)\alpha_{k-1}},X^{k\prime}\oplus X^{k\prime\prime})\\
&\qquad(Z^{1\alpha_0,\alpha_1},\ldots,Z^{(k-1)\alpha_{k-2},\alpha_{k-1}},\operatorname{row}[Z^{k\alpha_{k-1},\prime},Z^{k\alpha_{k-1},\prime\prime}])\Bigg]\\
&= \operatorname*{col}_{\alpha_0\in\{\prime,\prime\prime\}}\Bigg[\sum_{\alpha_1,\ldots,\alpha_{k-1}\in\{\prime,\prime\prime\}} \operatorname*{row}_{\alpha_k\in\{\prime,\prime\prime\}}[f(X^{0\alpha_0},\ldots,X^{k\alpha_k})(Z^{1\alpha_0,\alpha_1},\ldots,Z^{k\alpha_{k-1},\alpha_k})]\Bigg]\\
&= \begin{bmatrix} f^{\prime,\prime} & f^{\prime,\prime\prime}\\ f^{\prime\prime,\prime} & f^{\prime\prime,\prime\prime}\end{bmatrix},
\end{aligned}
$$

where $f^{\alpha\beta}$ are as in (3.2).

Conversely, if (3.1) holds, then each of ($1X^0$)–($1X^k$) is obtained as a special case when all $n_j''$'s but one equal 0.

(2). Suppose that ($2X^0$)–($2X^k$) hold. Then

$$
\begin{aligned}
&f(S_0X^0S_0^{-1},\ldots,S_kX^kS_k^{-1})(S_0Z^1S_1^{-1},\ldots,S_{k-1}Z^kS_k^{-1})\\
&\quad= S_0f(X^0,S_1X^1S_1^{-1},\ldots,S_kX^kS_k^{-1})(Z^1S_1^{-1},S_1Z^2S_2^{-1},\ldots,S_{k-1}Z^kS_k^{-1})\\
&= S_0f(X^0,X^1,S_2X^2S_2^{-1},\ldots,S_kX^kS_k^{-1})(Z^1,Z^2S_2^{-1},S_2Z^3S_3^{-1},\ldots,S_{k-1}Z^kS_k^{-1})\\
&\quad= \cdots = S_0f(X^0,\ldots,X^{k-1},S_kX^kS_k^{-1})(Z^1,\ldots,Z^{k-1},Z^kS_k^{-1})\\
&\quad= S_0f(X^0,\ldots,X^k)(Z^1,\ldots,Z^k)S_k^{-1}.
\end{aligned}
$$

Conversely, if (3.3) holds, then each of ($2X^0$)–($2X^k$) is obtained as a special case when $S_j = I_{n_j}$ for all $j$'s but one.

(3). Suppose that ($3X^0$)–($3X^k$) hold, and $T_jX^j = \tilde{X}^jT_j$, $j = 0,\ldots,k$. Then

$$
\begin{aligned}
&T_0f(X^0,\ldots,X^k)(Z^1T_1,\ldots,Z^kT_k)\\
&\qquad= f(\tilde{X}^0,X^1,\ldots,X^k)(T_0Z^1T_1,Z^2T_2,\ldots,Z^kT_k)\\
&\qquad\quad= f(\tilde{X}^0,\tilde{X}^1,X^2,\ldots,X^k)(T_0Z^1,T_1Z^2T_2,Z^3T_3,\ldots,Z^kT_k)
\end{aligned}
$$

$$= \cdots = f(\tilde{X}^0, \dots, \tilde{X}^{k-1}, X^k)(T_0 Z^1, \dots, T_{k-2} Z^{k-1}, T_{k-1} Z^k T_k)$$
$$= f(\tilde{X}^0, \dots, \tilde{X}^k)(T_0 Z^1, \dots, T_{k-1} Z^k) T_k.$$

Conversely, if (3.4) holds, then each of ($3X^0$)–($3X^k$) is obtained as a special case when $\tilde{n}_j = n_j$ and $T_j = I_{n_j}$ for all $j$'s but one. □

Formulae (3.1)–(3.2) admit a generalization to the case of direct sums of more than two summands as follows. Let

$$X^j = \bigoplus_{\alpha=1}^{m_j} X^{j\alpha}, \quad Z^j = [Z^{j\alpha,\beta}]_{\alpha=1,\dots,m_{j-1},\,\beta=1,\dots,m_j},$$

where $n_j^\alpha \in \mathbb{Z}_+$, $X^{j\alpha} \in \Omega^{(j)}_{n_j^\alpha}$ for $j = 0, \dots, k$, $\alpha = 1, \dots, m_j$; $Z^{j\alpha,\beta} \in \mathcal{N}_j^{n_{j-1}^\alpha \times n_j^\beta}$ for $j = 1, \dots, k$, $\alpha = 1, \dots, m_{j-1}$, $\beta = 1, \dots, m_j$. Then

$$f(X^0, \dots, X^k)(Z^1, \dots, Z^k) = [f^{\alpha,\beta}]_{\alpha=1,\dots,m_0,\,\beta=1,\dots,m_k}, \tag{3.5}$$

where

$$f^{\alpha,\beta} = \sum_{\alpha_j=1,\dots,m_j:\ \alpha_0=\alpha,\alpha_k=\beta} f(X^{0\alpha_0}, \dots, X^{k\alpha_k})(Z^{1\alpha_0,\alpha_1}, \dots, Z^{k\alpha_{k-1},\alpha_k}). \tag{3.6}$$

Here the block entry $f^{\alpha,\beta}$ is void if either $n_0^\alpha$ or $n_k^\beta$ is 0, and a summand in (3.6) is 0 if at least one of $n_j^{\alpha_j}$, $j = 1, \dots, k-1$, is 0. The proof is similar tc the proof of Proposition 3.2(1).

The case where for each $j$ the diagonal blocks $X^{j\alpha}$ are all equal wil. be of special importance (see Chapter 4). Some notations are in order. For $Z^j \in \mathcal{N}_j^{n_{j-1} \times n_j}$, $n_j = m_j s_j$, $j = 1, \dots, k$, we define

$$Z^1 {}_{s_0,s_2}\odot_{s_1} \cdots {}_{s_{k-2},s_k}\odot_{s_{k-1}} Z^k \in \left(\mathcal{N}_1^{s_0 \times s_1} \otimes \cdots \otimes \mathcal{N}_k^{s_{k-1} \times s_k}\right)^{m_0 \times m_k}$$

viewing $Z^j$ as a $m_{j-1} \times m_j$ matrix over $\mathcal{N}_j^{s_{j-1} \times s_j}$ and applying the standard rules of matrix multiplication, i.e.,

$$(Z^1 {}_{s_0,s_2}\odot_{s_1} \cdots {}_{s_{k-2},s_k}\odot_{s_{k-1}} Z^k)^{\alpha,\beta} = \sum_{\alpha_j=1,\dots,m_j:\ \alpha_0=\alpha,\alpha_k=\beta} Z^{1\alpha_0,\alpha_1} \otimes \cdots \otimes Z^{k\alpha_{k-1},\alpha_k}, \tag{3.7}$$

where $Z^j = [Z^{j\alpha,\beta}]_{\alpha=1,\dots,m_{j-1},\,\beta=1,\dots,m_j}$. This is an iteration of a product of matrices over two modules that are endowed with a product operation into a third module, as in 2.3.4. In case two factors have square blocks (necessarily of the same size), we omit the left subscript indices, i.e., we write $Z^1 \odot_s Z^2$ instead of $Z^1 {}_{s,s}\odot_s Z^2$. For $s = 1$ we omit the right subscript, i.e., we write $Z^1 {}_{s_0,s_2}\odot Z^2$ instead of $Z^1 {}_{s_0,s_2}\odot_1 Z^2$, and we write $Z^1 \odot \cdots \odot Z^k$ instead of $Z^1 \odot_1 \cdots \odot_1 Z^k$; see, e.g., [**79**, Page 86] or [**78**, Page 240].

Proposition 3.3. *Let $f \in \mathcal{T}^k(\Omega^{(0)}, \dots, \Omega^{(k)}; \mathcal{N}_{0,\mathrm{nc}}, \dots, \mathcal{N}_{k,\mathrm{nc}})$ and $n_j = m_j s_j$, $j = 0, \dots, k$.*

(1) *For $Y^j \in \Omega^{(j)}_{s_j}$, $X^j = \bigoplus_{\alpha=1}^{m_j} Y^j$, $j = 0, \dots, k$, and for $Z^j \in \mathcal{N}_j^{n_{j-1} \times n_j}$, $j = 1, \dots, k$,*

$$f(X^0, \dots, X^k)(Z^1, \dots, Z^k) = Z^1 {}_{s_0,s_2}\odot_{s_1} \cdots {}_{s_{k-2},s_k}\odot_{s_{k-1}} Z^k f(Y^0, \dots, Y^k). \tag{3.8}$$

*Here the linear mapping*

$$f(Y^0,\dots,Y^k)\colon {\mathcal{N}_1}^{s_0\times s_1}\otimes\cdots\otimes{\mathcal{N}_k}^{s_{k-1}\times s_k}\longrightarrow{\mathcal{N}_0}^{s_0\times s_k}$$

*is acting on the matrix* ${Z^1}_{s_0,s_2}\odot_{s_1}\cdots{}_{s_{k-2},s_k}\odot_{s_{k-1}}Z^k$ *entrywise.*

(2) *For* $Y^j\in\Omega^{(j)}_{s_j}$, $X^j=\bigoplus_{\alpha=1}^{m_j}Y^j$, $j=0,\dots,k-1$, $X^k\in\Omega^{(k)}_{n_k}$, *and for* $Z^j\in{\mathcal{N}_j}^{n_{j-1}\times n_j}$, $j=1,\dots,k$,

(3.9)
$$f(X^0,\dots,X^k)(Z^1,\dots,Z^k)={Z^1}_{s_0,s_2}\odot_{s_1}\cdots{}_{s_{k-2},s_k}\odot_{s_{k-1}}Z^kf(Y^0,\dots,Y^{k-1},X^k).$$

*Here the linear mapping*

$$\begin{aligned}f(Y^0,\dots,Y^{k-1},X^k)\colon {\mathcal{N}_1}^{s_0\times s_1}\otimes\cdots\otimes(\mathcal{N}_{k-1})^{s_{k-2}\times s_{k-1}}\otimes{\mathcal{N}_k}^{s_{k-1}\times m_ks_k}\\ \cong\left({\mathcal{N}_1}^{s_0\times s_1}\otimes\cdots\otimes{\mathcal{N}_k}^{s_{k-1}\times s_k}\right)^{1\times m_k}\longrightarrow{\mathcal{N}_0}^{s_0\times m_ks_k}\cong\left({\mathcal{N}_0}^{s_0\times s_k}\right)^{1\times m_k}\end{aligned}$$

*is acting on the rows of the matrix* ${Z^1}_{s_0,s_2}\odot_{s_1}\cdots{}_{s_{k-2},s_k}\odot_{s_{k-1}}Z^k$; *i.e.,* $f(Y^0,\dots,Y^{k-1},X^k)$ *is a* $m_k\times m_k$ *matrix of linear mappings*

$${\mathcal{N}_1}^{s_0\times s_1}\otimes\cdots\otimes{\mathcal{N}_k}^{s_{k-1}\times s_k}\longrightarrow{\mathcal{N}_0}^{s_0\times s_k}$$

*that multiplies the matrix* ${Z^1}_{s_0,s_2}\odot_{s_1}\cdots{}_{s_{k-2},s_k}\odot_{s_{k-1}}Z^k$ *on the right.*

PROOF. Follows immediately from (3.5)–(3.6). □

REMARK 3.4. We notice that there is a natural mapping

$$\begin{aligned}(3.10)\quad {\mathcal{N}_0}^{n_0\times n_0}\otimes\mathcal{N}_1^{*\,n_1\times n_1}\otimes\cdots\otimes\mathcal{N}_k^{*\,n_k\times n_k}\\ \longrightarrow\hom_{\mathcal{R}}\left({\mathcal{N}_1}^{n_0\times n_1}\otimes\cdots\otimes{\mathcal{N}_k}^{n_{k-1}\times n_k},{\mathcal{N}_0}^{n_0\times n_k}\right)\end{aligned}$$

for $k$ a positive integer, defined on an elementary tensor $Y^0\otimes Y^1\otimes\cdots\otimes Y^k$ by

$$(3.11)\qquad (Z^1,\dots,Z^k)\longmapsto Y^0(Z^1Y^1)\cdots(Z^kY^k),$$

with $Z^jY^j\in\mathcal{R}^{n_{j-1}\times n_j}$ for $Z^j\in{\mathcal{N}_j}^{n_{j-1}\times n_j}$, $j=1,\dots,k$. If the modules $\mathcal{N}_1$, ..., $\mathcal{N}_k$ are free and of finite rank over $\mathcal{R}$ and the module $\mathcal{N}_0$ is free (in particular, when $\mathcal{R}$ is a field and $\mathcal{N}_1$, ..., $\mathcal{N}_k$ are finite-dimensional vector spaces), the natural mapping (3.10)–(3.11) is easily seen to be an isomorphism.

REMARK 3.5. In the case where the natural mapping (3.10)–(3.11) is an isomorphism, we can view a nc function of order $k$ as a function $f$ on $\Omega^{(0)}\times\cdots\times\Omega^{(k)}$ with

$$(3.12)\qquad f(\Omega^{(0)}_{n_0},\dots,\Omega^{(k)}_{n_k})\subseteq{\mathcal{N}_0}^{n_0\times n_0}\otimes\mathcal{N}_1^{*\,n_1\times n_1}\otimes\cdots\otimes\mathcal{N}_k^{*\,n_k\times n_k}.$$

Direct sums and similarities on ${\mathcal{N}_0}^{n_0\times n_0}$, $\mathcal{N}_1^{*\,n_1\times n_1}$, ..., $\mathcal{N}_k^{*\,n_k\times n_k}$ induce $i$-th factor direct sums and similarities on ${\mathcal{N}_0}^{n_0\times n_0}\otimes\mathcal{N}_1^{*\,n_1\times n_1}\otimes\cdots\otimes\mathcal{N}_k^{*\,n_k\times n_k}$ for the corresponding $i=0,\dots,k$. With this interpretation, conditions (1X$^0$)–(1X$^k$) and (2X$^0$)–(2X$^k$) mean that direct sums and similarities in each argument of $f$ result in direct sums and similarities in the corresponding factor in the value of $f$. This explains the form of the conditions of respecting direct sums and similarities in the definition of nc functions of order $k$ and also puts the cases $k=0$ and $k>0$ on an equal footing.

REMARK 3.6. There is also a natural mapping

$$\begin{aligned}(3.13)\quad &\mathcal{T}^k(\Omega^{(0)},\ldots,\Omega^{(k)};\mathcal{N}_{0,\mathrm{nc}},\ldots,\mathcal{N}_{k,\mathrm{nc}})\\ &\quad\otimes\mathcal{T}^\ell(\Omega^{(k+1)},\ldots,\Omega^{(k+\ell+1)};\mathcal{L}_{0,\mathrm{nc}}^*,\mathcal{L}_{1,\mathrm{nc}},\ldots,\mathcal{L}_{\ell,\mathrm{nc}})\\ &\quad\longrightarrow\mathcal{T}^{k+\ell+1}(\Omega^{(0)},\ldots,\Omega^{(k+\ell+1)};\mathcal{N}_{0,\mathrm{nc}},\ldots,\mathcal{N}_{k,\mathrm{nc}},\mathcal{L}_{0,\mathrm{nc}},\ldots,\mathcal{L}_{\ell,\mathrm{nc}})\end{aligned}$$

defined on elementary tensors as

$$f\otimes g\longmapsto h,$$

where

$$(3.14)\qquad h(X^0,\ldots,X^k,U^0,\ldots,U^\ell)=f(X^0,\ldots,X^k)\otimes g(U^0,\ldots,U^\ell),$$

interpreting the values of higher order nc functions as elements of tensor products (see Remark 3.5), or

$$\begin{aligned}(3.15)\quad &h(X^0,\ldots,X^k,U^0,\ldots,U^\ell)(Z^1,\ldots,Z^k,W^0,W^1,\ldots,W^\ell)\\ &\quad=f(X^0,\ldots,X^k)(Z^1,\ldots,Z^k)\Big(W^0g(U^0,\ldots,U^\ell)(W^1,\ldots,W^\ell)\Big),\end{aligned}$$

interpreting the values of higher order nc functions as multilinear mappings. Here $X^0\in\Omega_{n_0}^{(0)}$, ..., $X^k\in\Omega_{n_k}^{(k)}$, $U^0\in\Omega_{m_0}^{(k+1)}$, ..., $U^\ell\in\Omega_{m_\ell}^{(k+\ell+1)}$, $Z^1\in\mathcal{N}_1{}^{n_0\times n_1}$, ..., $Z^k\in\mathcal{N}_k{}^{n_{k-1}\times n_k}$, $W^0\in\mathcal{L}_0{}^{n_k\times m_0}$, $W^1\in\mathcal{L}_1{}^{m_0\times m_1}$, ..., $W^\ell\in\mathcal{L}_\ell{}^{m_{\ell-1}\times m_\ell}$.

Iterating this construction, we obtain a natural mapping

$$\begin{aligned}(3.16)\quad &\mathcal{T}^0(\Omega^{(0)};\mathcal{N}_{0,\mathrm{nc}})\otimes\mathcal{T}^0(\Omega^{(1)};\mathcal{N}_{1,\mathrm{nc}}^*)\otimes\cdots\otimes\mathcal{T}^0(\Omega^{(k)};\mathcal{N}_{k,\mathrm{nc}}^*)\\ &\qquad\longrightarrow\mathcal{T}^k(\Omega^{(0)},\ldots,\Omega^{(k)};\mathcal{N}_{0,\mathrm{nc}},\ldots,\mathcal{N}_{k,\mathrm{nc}}),\end{aligned}$$

defined on elementary tensors as

$$f^0\otimes\cdots\otimes f^k\longmapsto h,$$

where

$$(3.17)\qquad h(X^0,X^1,\ldots,X^k)=f^0(X^0)\otimes f^1(X^1)\otimes\cdots\otimes f^k(X^k),$$

interpreting the values of higher order nc functions as elements of tensor products, or

$$(3.18)\quad h(X^0,X^1,\ldots,X^k)(Z^1,\ldots,Z^k)=f^0(X^0)\Big(Z^1f^1(X^1)\Big)\cdots\Big(Z^kf^k(X^k)\Big),$$

interpreting the values of higher order nc functions as multilinear mappings. Here $X^0\in\Omega_{n_0}^{(0)}$, ..., $X^k\in\Omega_{n_k}^{(k)}$, $Z^1\in\mathcal{N}_1{}^{n_0\times n_1}$, ..., $Z^k\in\mathcal{N}_k{}^{n_{k-1}\times n_k}$.

It is clear that the natural mappings (3.13) & (3.14) (or (3.13) & (3.15)) and, in particular, (3.16) & (3.17) (or (3.16) & (3.18)) are embeddings; we will sometimes abuse the notation writing $h=f\otimes g$ and $h=f^0\otimes\cdots\otimes f^k$ for (3.14) (or (3.15)) and (3.17) (or (3.18)), respectively.

## 3.2. Higher order nc difference-differential operators

Proposition 2.15 tells exactly that, for $f\in\mathcal{T}^0$, one has $\Delta_Rf\in\mathcal{T}^1$. We will extend $\Delta_R$ to an operator from $\mathcal{T}^k$ to $\mathcal{T}^{k+1}$ for all $k$. Similarly to the case $k=0$, it will be done by evaluating a nc function of order $k$ ($>0$) on a $(k+1)$-tuple of square matrices with one of the arguments block upper triangular. We start with the following analogue of Proposition 2.2. It will be shown later, using the tensor product interpretation of the values of higher order nc functions, that this is indeed a natural analogue.

PROPOSITION 3.7. *Let $\mathcal{M}_j$, $\mathcal{N}_j$ be modules, and let $\Omega^{(j)} \subseteq \mathcal{M}_{j,\mathrm{nc}}$ be a nc set, $j = 0, \ldots, k$. Let $f \in \mathcal{T}^k(\Omega^{(0)}, \ldots, \Omega^{(k)}; \mathcal{N}_{0,\mathrm{nc}}, \ldots, \mathcal{N}_{k,\mathrm{nc}})$. Let $X^0 \in \Omega^{(0)}_{n_0}$, $\ldots$, $X^{k-1} \in \Omega^{(k-1)}_{n_{k-1}}$, $X^{k\prime} \in \Omega^{(k)}_{n'_k}$, $X^{k\prime\prime} \in \Omega^{(k)}_{n''_k}$, $Z^1 \in \mathcal{N}_1{}^{n_0 \times n_1}$, $\ldots$, $Z^{k-1} \in (\mathcal{N}_{k-1})^{n_{k-2} \times n_{k-1}}$, $Z^{k\prime} \in \mathcal{N}_k{}^{n_{k-1} \times n'_k}$, $Z^{k\prime\prime} \in \mathcal{N}_k{}^{n_{k-1} \times n''_k}$. Let $Z \in \mathcal{M}_k{}^{n'_k \times n''_k}$ be such that $\begin{bmatrix} X^{k\prime} & Z \\ 0 & X^{k\prime\prime} \end{bmatrix} \in \Omega^{(k)}_{n'_k + n''_k}$. Then*

$$
\begin{aligned}
(3.19) \quad f\left(X^0, \ldots, X^{k-1}, \begin{bmatrix} X^{k\prime} & Z \\ 0 & X^{k\prime\prime} \end{bmatrix}\right) & \left(Z^1, \ldots, Z^{k-1}, \operatorname{row}\left[Z^{k\prime}, Z^{k\prime\prime}\right]\right) \\
= \operatorname{row} \Big[ & f(X^0, \ldots, X^{k-1}, X^{k\prime})(Z^1, \ldots, Z^{k-1}, Z^{k\prime}), \\
& \Delta_R f(X^0, \ldots, X^{k-1}, X^{k\prime}, X^{k\prime\prime})(Z^1, \ldots, Z^{k-1}, Z^{k\prime}, Z) \\
& + f(X^0, \ldots, X^{k-1}, X^{k\prime\prime})(Z^1, \ldots, Z^{k-1}, Z^{k\prime\prime}) \Big].
\end{aligned}
$$

*Here $\Delta_R f = (X^0, \ldots, X^{k-1}, X^{k\prime}, X^{k\prime\prime})(Z^1, \ldots, Z^{k-1}, Z^{k\prime}, Z)$ is determined uniquely by (3.19), is independent of $Z^{k\prime\prime}$, and has the following property. If $r \in \mathcal{R}$ is such that $\begin{bmatrix} X^{k\prime} & rZ \\ 0 & X^{k\prime\prime} \end{bmatrix} \in \Omega^{(k)}_{n'_k + n''_k}$, then*

$$
\begin{aligned}
(3.20) \quad \Delta_R f(X^0, \ldots, X^{k-1}, X^{k\prime}, X^{k\prime\prime}) & (Z^1, \ldots, Z^{k-1}, Z^{k\prime}, rZ) \\
& = r \Delta_R f(X^0, \ldots, X^{k-1}, X^{k\prime}, X^{k\prime\prime})(Z^1, \ldots, Z^{k-1}, Z^{k\prime}, Z).
\end{aligned}
$$

*For $k = 1$, the matrices $Z^1, \ldots, Z^{k-1}$ do not show up in (3.19) and (3.20).*

PROOF. Denote

$$
f\left(X^0, \ldots, X^{k-1}, \begin{bmatrix} X^{k\prime} & Z \\ 0 & X^{k\prime\prime} \end{bmatrix}\right) \left(Z^1, \ldots, Z^{k-1}, \operatorname{row}\left[Z^{k\prime}, Z^{k\prime\prime}\right]\right) =: \operatorname{row}[A, B],
$$

where $A \in \mathcal{N}_0{}^{n_0 \times n'_k}$, $B \in \mathcal{N}_0{}^{n_0 \times n''_k}$. We have

$$
\begin{bmatrix} X^{k\prime} & Z \\ 0 & X^{k\prime\prime} \end{bmatrix} \begin{bmatrix} I_{n'_k} \\ 0 \end{bmatrix} = \begin{bmatrix} I_{n'_k} \\ 0 \end{bmatrix} X^{k\prime}.
$$

By (3X$^k$),

$$
A = f(X^0, \ldots, X^{k-1}, X^{k\prime})(Z^1, \ldots, Z^{k-1}, Z^{k\prime}).
$$

We have also

$$
X^{k\prime\prime} \begin{bmatrix} 0 & I_{n''_k} \end{bmatrix} = \begin{bmatrix} 0 & I_{n''_k} \end{bmatrix} \begin{bmatrix} X^{k\prime} & Z \\ 0 & X^{k\prime\prime} \end{bmatrix}.
$$

By (3X$^k$) and the additivity of

$$
f(X^0, \ldots, X^{k-1}, X^{k\prime})(Z^1, \ldots, Z^{k-1}, Z^k)
$$

as a function of $Z^k$,

$$\begin{aligned}
&\operatorname{row}[0, f(X^0, \ldots, X^{k-1}, X^{k\prime\prime})(Z^1, \ldots, Z^{k-1}, Z^{k\prime\prime})] \\
&\quad = f\left(X^0, \ldots, X^{k-1}, \begin{bmatrix} X^{k\prime} & Z \\ 0 & X^{k\prime\prime} \end{bmatrix}\right)(Z^1, \ldots, Z^{k-1}, \operatorname{row}[0, Z^{k\prime\prime}]) \\
&\quad = \operatorname{row}[A, B] - f\left(X^0, \ldots, X^{k-1}, \begin{bmatrix} X^{k\prime} & Z \\ 0 & X^{k\prime\prime} \end{bmatrix}\right)(Z^1, \ldots, Z^{k-1}, \operatorname{row}[Z^{k\prime}, 0]),
\end{aligned}$$

which implies that

$$\begin{aligned}
B = f\left(X^0, \ldots, X^{k-1}, \begin{bmatrix} X^{k\prime} & Z \\ 0 & X^{k\prime\prime} \end{bmatrix}\right)(Z^1, \ldots, Z^{k-1}, \operatorname{row}[Z^{k\prime}, 0]) \begin{bmatrix} 0 \\ I_{n''_k} \end{bmatrix} \\
+ f(X^0, \ldots, X^{k-1}, X^{k\prime\prime})(Z^1, \ldots, Z^{k-1}, Z^{k\prime\prime}).
\end{aligned}$$

Therefore, we obtain (3.19) with

$$\begin{aligned}
(3.21)\quad &\Delta_R f\left(X^0, \ldots, X^{k-1}, X^{k\prime}, X^{k\prime\prime}\right)\left(Z^1, \ldots, Z^{k-1}, Z^{k\prime}, Z\right) \\
&\quad = f\left(X^0, \ldots, X^{k-1}, \begin{bmatrix} X^{k\prime} & Z \\ 0 & X^{k\prime\prime} \end{bmatrix}\right)\left(Z^1, \ldots, Z^{k-1}, \operatorname{row}\left[Z^{k\prime}, 0\right]\right) \begin{bmatrix} 0 \\ I_{n''_k} \end{bmatrix}.
\end{aligned}$$

We observe that (3.21) implies that for any $r \in \mathcal{R}$ one has

$$\begin{aligned}
(3.22)\quad &\Delta_R f(X^0, \ldots, X^{k-1}, X^{k\prime}, X^{k\prime\prime})(Z^1, \ldots, Z^{k-1}, rZ^{k\prime}, Z) \\
&\qquad = r\Delta_R f(X^0, \ldots, X^{k-1}, X^{k\prime}, X^{k\prime\prime})(Z^1, \ldots, Z^{k-1}, Z^{k\prime}, Z).
\end{aligned}$$

Next, for $r \in \mathcal{R}$ such that $\begin{bmatrix} X^{k\prime} & rZ \\ 0 & X^{k\prime\prime} \end{bmatrix} \in \Omega^{(k)}_{n'_k + n''_k}$ we have

$$\begin{bmatrix} X^{k\prime} & rZ \\ 0 & X^{k\prime\prime} \end{bmatrix} \begin{bmatrix} rI_{n'_k} & 0 \\ 0 & I_{n''_k} \end{bmatrix} = \begin{bmatrix} rI_{n'_k} & 0 \\ 0 & I_{n''_k} \end{bmatrix} \begin{bmatrix} X^{k\prime} & Z \\ 0 & X^{k\prime\prime} \end{bmatrix}.$$

By (3X$^k$),

$$\begin{aligned}
&f\left(X^0, \ldots, X^{k-1}, \begin{bmatrix} X^{k\prime} & rZ \\ 0 & X^{k\prime\prime} \end{bmatrix}\right)(Z^1, \ldots, Z^{k-1}, \operatorname{row}[Z^{k\prime}, Z^{k\prime\prime}]) \begin{bmatrix} rI_{n'_k} & 0 \\ 0 & I_{n''_k} \end{bmatrix} \\
&\quad = f\left(X^0, \ldots, X^{k-1}, \begin{bmatrix} X^{k\prime} & Z \\ 0 & X^{k\prime\prime} \end{bmatrix}\right)(Z^1, \ldots, Z^{k-1}, \operatorname{row}[rZ^{k\prime}, Z^{k\prime\prime}]).
\end{aligned}$$

Multiplying both parts on the right by $\begin{bmatrix} 0 \\ I_{n''_k} \end{bmatrix}$ and subtracting

$$f(X^0, \ldots, X^{k-1}, X^{k\prime\prime})(Z^1, \ldots, Z^{k-1}, Z^{k\prime\prime}),$$

we obtain in virtue of (3.19) that

$$\begin{aligned}
&\Delta_R f(X^0, \ldots, X^{k-1}, X^{k\prime}, X^{k\prime\prime})(Z^1, \ldots, Z^{k-1}, Z^{k\prime}, rZ) \\
&\qquad = \Delta_R f(X^0, \ldots, X^{k-1}, X^{k\prime}, X^{k\prime\prime}) = (Z^1, \ldots, Z^{k-1}, rZ^{k\prime}, Z).
\end{aligned}$$

Then applying (3.22) we obtain (3.20). □

Assume now that the nc set $\Omega^{(k)} \subseteq \mathcal{M}_{k,\mathrm{nc}}$ is right admissible, and let

$$f \in \mathcal{T}^k(\Omega^{(0)}, \ldots, \Omega^{(k)}; \mathcal{N}_{0,\mathrm{nc}}, \ldots, \mathcal{N}_{k,\mathrm{nc}}).$$

Then for every $X^0 \in \Omega^{(0)}_{n_0}$, ..., $X^{k-1} \in \Omega^{(k-1)}_{n_{k-1}}$, $X^{k\prime} \in \Omega^{(k)}_{n'_k}$, $X^{k\prime\prime} \in \Omega^{(k)}_{n''_k}$, $Z^1 \in {\mathcal{N}_1}^{n_0 \times n_1}$, ..., $Z^{k-1} \in (\mathcal{N}_{k-1})^{n_{k-2} \times n_{k-1}}$, $Z^{k\prime} \in {\mathcal{N}_k}^{n_{k-1} \times n'_k}$, $Z \in {\mathcal{M}_k}^{n'_k \times n''_k}$, and for an invertible $r \in \mathcal{R}$ such that $\begin{bmatrix} X^{k\prime} & rZ \\ 0 & X^{k\prime\prime} \end{bmatrix} \in \Omega^{(k)}_{n'_k + n''_k}$, we define first

$$\Delta_R f(X^0, \dots, X^{k-1}, X^{k\prime}, X^{k\prime\prime})(Z^1, \dots, Z^{k-1}, Z^{k\prime}, rZ)$$

by (3.19), with $Z$ replaced by $rZ$ (for an arbitrary choice of $Z^{k\prime\prime} \in {\mathcal{N}_k}^{n_{k-1} \times n''_k}$), and then define

$$\begin{aligned} (3.23) \quad & \Delta_R f(X^0, \dots, X^{k-1}, X^{k\prime}, X^{k\prime\prime})(Z^1, \dots, Z^{k-1}, Z^{k\prime}, Z) \\ & \qquad = r^{-1} \Delta_R f(X^0, \dots, X^{k-1}, X^{k\prime}, X^{k\prime\prime})(Z^1, \dots, Z^{k-1}, Z^{k\prime}, rZ). \end{aligned}$$

By Proposition 3.7, the right-hand side of (3.23) is independent of $r$, hence the left-hand side is defined unambiguously. Using Proposition A.5, we can establish a higher order version of (2.11),

$$\begin{aligned} (3.24) \quad & \Delta_R f(X^0, \dots, X^{k-1}, X^{k\prime}, X^{k\prime\prime})(Z^1, \dots, Z^{k-1}, Z^{k\prime}, Z) \\ & \qquad = \Delta_R \widetilde{f}(X^0, \dots, X^{k-1}, X^{k\prime}, X^{k\prime\prime})(Z^1, \dots, Z^{k-1}, Z^{k\prime}, Z). \end{aligned}$$

We have the following analogue of Proposition 2.4, with essentially the same proof.

PROPOSITION 3.8. *Let $f \in \mathcal{T}^k(\Omega^{(0)}, \dots, \Omega^{(k)}; \mathcal{N}_{0,\mathrm{nc}}, \dots, \mathcal{N}_{k,\mathrm{nc}})$, with $\Omega^{(k)} \subseteq \mathcal{M}_{k,\mathrm{nc}}$ a right admissible nc set. The mapping*

$$Z \longmapsto \Delta_R f(X^0, \dots, X^{k-1}, X^{k\prime}, X^{k\prime\prime})(Z^1, \dots, Z^{k-1}, Z^{k\prime}, Z)$$

*is homogeneous as an operator from ${\mathcal{M}_k}^{n'_k \times n''_k}$ to ${\mathcal{N}_0}^{n_0 \times n''_k}$, i.e.,* (3.20) *holds for every $X^0 \in \Omega^{(0)}_{n_0}$, ..., $X^{k-1} \in \Omega^{(k-1)}_{n_{k-1}}$, $X^{k\prime} \in \Omega^{(k)}_{n'_k}$, $X^{k\prime\prime} \in \Omega^{(k)}_{n''_k}$, $Z^1 \in {\mathcal{N}_1}^{n_0 \times n_1}$, ..., $Z^{k-1} \in (\mathcal{N}_{k-1})^{n_{k-2} \times n_{k-1}}$, $Z^{k\prime} \in {\mathcal{N}_k}^{n_{k-1} \times n'_k}$, $Z \in {\mathcal{M}_k}^{n'_k \times n''_k}$, and every $r \in \mathcal{R}$.*

We proceed with establishing an analogue of Proposition 2.6.

PROPOSITION 3.9. *Let $f \in \mathcal{T}^k(\Omega^{(0)}, \dots, \Omega^{(k)}; \mathcal{N}_{0,\mathrm{nc}}, \dots, \mathcal{N}_{k,\mathrm{nc}})$, with $\Omega^{(k)} \subseteq \mathcal{M}_{k,\mathrm{nc}}$ a right admissible nc set. The mapping*

$$Z \longmapsto \Delta_R f(X^0, \dots, X^{k-1}, X^{k\prime}, X^{k\prime\prime})(Z^1, \dots, Z^{k-1}, Z^{k\prime}, Z)$$

*is additive as an operator from ${\mathcal{M}_k}^{n'_k \times n''_k}$ to ${\mathcal{N}_0}^{n_0 \times n''_k}$, i.e.,*

$$\begin{aligned} (3.25) \quad & \Delta_R f(X^0, \dots, X^{k-1}, X^{k\prime}, X^{k\prime\prime})(Z^1, \dots, Z^{k-1}, Z^{k\prime}, Z_I + Z_{II}) \\ & \quad = \Delta_R f(X^0, \dots, X^{k-1}, X^{k\prime}, X^{k\prime\prime})(Z^1, \dots, Z^{k-1}, Z^{k\prime}, Z_I) \\ & \qquad + \Delta_R f(X^0, \dots, X^{k-1}, X^{k\prime}, X^{k\prime\prime})(Z^1, \dots, Z^{k-1}, Z^{k\prime}, Z_{II}) \end{aligned}$$

*for every $X^0 \in \Omega^{(0)}_{n_0}$, ..., $X^{k-1} \in \Omega^{(k-1)}_{n_{k-1}}$, $X^{k\prime} \in \Omega^{(k)}_{n'_k}$, $X^{k\prime\prime} \in \Omega^{(k)}_{n''_k}$, $Z^1 \in {\mathcal{N}_1}^{n_0 \times n_1}$, ..., $Z^{k-1} \in (\mathcal{N}_{k-1})^{n_{k-2} \times n_{k-1}}$, $Z^{k\prime} \in {\mathcal{N}_k}^{n_{k-1} \times n'_k}$, and $Z_I, Z_{II} \in {\mathcal{M}_k}^{n'_k \times n''_k}$.*

PROOF. Let $X^0 \in \Omega^{(0)}_{n_0}$, ..., $X^{k-1} \in \Omega^{(k-1)}_{n_{k-1}}$, $X^{k\prime} \in \Omega^{(k)}_{n'_k}$, $X^{k\prime\prime} \in \Omega^{(k)}_{n''_k}$, $Z^1 \in {\mathcal{N}_1}^{n_0 \times n_1}$, ..., $Z^{k-1} \in (\mathcal{N}_{k-1})^{n_{k-2} \times n_{k-1}}$, $Z^{k\prime} \in {\mathcal{N}_k}^{n_{k-1} \times n'_k}$, and $Z_I, Z_{II} \in {\mathcal{M}_k}^{n'_k \times n''_k}$ be arbitrary. By (3.24), we may assume, without loss of generality, that $\Omega^{(k)}$ is similarity invariant, i.e., $\widetilde{\Omega}^{(k)} = \Omega^{(k)}$; see Remark 2.5 and Appendix A. Then by Proposition A.2,

$$\begin{bmatrix} X^{k\prime} & Z_I \\ 0 & X^{k\prime\prime} \end{bmatrix} \in \Omega^{(k)}_{n'_k+n''_k}, \quad \begin{bmatrix} X^{k\prime} & Z_{II} \\ 0 & X^{k\prime\prime} \end{bmatrix} \in \Omega^{(k)}_{n'_k+n''_k}, \quad \begin{bmatrix} X^{k\prime} & (Z_I+Z_{II}) \\ 0 & X^{k\prime\prime} \end{bmatrix} \in \Omega^{(k)}_{n'_k+n''_k},$$

and

$$\begin{bmatrix} X^{k\prime} & 0 & Z_I \\ 0 & X^{k\prime} & Z_{II} \\ 0 & 0 & X^{k\prime\prime} \end{bmatrix} \in \Omega^{(k)}_{2n'_k+n''_k}$$

(for the latter, we use the fact that if $X^{k\prime} \in \Omega^{(k)}_{n'_k}$ then $\begin{bmatrix} X^{k\prime} & 0 \\ 0 & X^{k\prime} \end{bmatrix} \in \Omega^{(k)}_{2n'_k}$). We have

$$\begin{bmatrix} X^{k\prime} & Z_I \\ 0 & X^{k\prime\prime} \end{bmatrix} \begin{bmatrix} I_{n'_k} & 0 & 0 \\ 0 & 0 & I_{n''_k} \end{bmatrix} = \begin{bmatrix} I_{n'_k} & 0 & 0 \\ 0 & 0 & I_{n''_k} \end{bmatrix} \begin{bmatrix} X^{k\prime} & 0 & Z_I \\ 0 & X^{k\prime} & Z_{II} \\ 0 & 0 & X^{k\prime\prime} \end{bmatrix}.$$

By (3X$^k$),

$$\begin{aligned} &f\left(X^0,\ldots,X^{k-1},\begin{bmatrix} X^{k\prime} & Z_I \\ 0 & X^{k\prime\prime} \end{bmatrix}\right)(Z^1,\ldots,Z^{k-1},\operatorname{row}[Z^{k\prime},0])\begin{bmatrix} I_{n'_k} & 0 & 0 \\ 0 & 0 & I_{n''_k} \end{bmatrix} \\ &\quad = f\left(X^0,\ldots,X^{k-1},\begin{bmatrix} X^{k\prime} & 0 & Z_I \\ 0 & X^{k\prime} & Z_{II} \\ 0 & 0 & X^{k\prime\prime} \end{bmatrix}\right)(Z^1,\ldots,Z^{k-1},\operatorname{row}[Z^{k\prime},0,0]). \end{aligned}$$

Multiplying both parts by $\operatorname{col}[0,0,I_{n''_k}]$ and taking into account (3.21), we obtain

$$\begin{aligned} (3.26)\quad &\Delta_R f(X^0,\ldots,X^{k-1},X^{k\prime},X^{k\prime\prime})(Z^1,\ldots,Z^{k-1},Z^{k\prime},Z_I) \\ &= \Delta_R f(X^0,\ldots,X^{k-1},X^{k\prime}\oplus X^{k\prime},X^{k\prime\prime})(Z^1,\ldots,Z^{k-1},\operatorname{row}[Z^{k\prime},0],\operatorname{col}[Z_I,Z_{II}]). \end{aligned}$$

Similarly, using the intertwining relation

$$\begin{bmatrix} X^{k\prime} & Z_{II} \\ 0 & X^{k\prime\prime} \end{bmatrix} \begin{bmatrix} 0 & I_{n'_k} & 0 \\ 0 & 0 & I_{n''_k} \end{bmatrix} = \begin{bmatrix} 0 & I_{n'_k} & 0 \\ 0 & 0 & I_{n''_k} \end{bmatrix} \begin{bmatrix} X^{k\prime} & 0 & Z_I \\ 0 & X^{k\prime} & Z_{II} \\ 0 & 0 & X^{k\prime\prime} \end{bmatrix},$$

we obtain

$$\begin{aligned} (3.27)\quad &\Delta_R f(X^0,\ldots,X^{k-1},X^{k\prime},X^{k\prime\prime})(Z^1,\ldots,Z^{k-1},Z^{k\prime},Z_{II}) \\ &= \Delta_R f(X^0,\ldots,X^{k-1},X^{k\prime}\oplus X^{k\prime},X^{k\prime\prime})(Z^1,\ldots,Z^{k-1},\operatorname{row}[0,Z^{k\prime}],\operatorname{col}[Z_I,Z_{II}]). \end{aligned}$$

Next, using the intertwining relation

$$\begin{bmatrix} X^{k\prime} & Z_I+Z_{II} \\ 0 & X^{k\prime\prime} \end{bmatrix} \begin{bmatrix} I_{n'_k} & I_{n'_k} & 0 \\ 0 & 0 & I_{n''_k} \end{bmatrix} = \begin{bmatrix} I_{n'_k} & I_{n'_k} & 0 \\ 0 & 0 & I_{n''_k} \end{bmatrix} \begin{bmatrix} X^{k\prime} & 0 & Z_I \\ 0 & X^{k\prime} & Z_{II} \\ 0 & 0 & X^{k\prime\prime} \end{bmatrix},$$

we obtain

$$\begin{aligned} (3.28)\quad &\Delta_R f(X^0,\ldots,X^{k-1},X^{k\prime},X^{k\prime\prime})(Z^1,\ldots,Z^{k-1},Z^{k\prime},Z_I+Z_{II}) \\ &= \Delta_R f(X^0,\ldots,X^{k-1},X^{k\prime}\oplus X^{k\prime},X^{k\prime\prime})(Z^1,\ldots,Z^{k-1},\operatorname{row}[Z^{k\prime},Z^{k\prime}],\operatorname{col}[Z_I,Z_{II}]). \end{aligned}$$

We observe that (3.21) implies the additivity of

$$\Delta_R f(X^0,\ldots,X^{k-1},X^{k\prime},X^{k\prime\prime})(Z^1,\ldots,Z^{k-1},Z^{k\prime},Z)$$

as a function of $Z^{k\prime}$. Therefore, the sum of the right-hand sides of (3.26) and (3.27) equals the right-hand side of (3.28), and (3.25) follows. □

THEOREM 3.10. *Let* $f \in \mathcal{T}^k(\Omega^{(0)}, \ldots, \Omega^{(k)}; \mathcal{N}_{0,\mathrm{nc}}, \ldots, \mathcal{N}_{k,\mathrm{nc}})$, *where* $\Omega^{(k)} \subseteq \mathcal{M}_{k,\mathrm{nc}}$ *is a right admissible nc set. For every* $X^0 \in \Omega^{(0)}_{n_0}$, $\ldots$, $X^k \in \Omega^{(k)}_{n_k}$, $X^{k+1} \in \Omega^{(k)}_{n_{k+1}}$, *the mapping*

$$(Z^1, \ldots, Z^{k+1}) \longmapsto \Delta_R f(X^0, \ldots, X^{k+1})(Z^1, \ldots, Z^{k+1})$$

*from* $\mathcal{N}_1{}^{n_0 \times n_1} \times \cdots \times \mathcal{N}_k{}^{n_{k-1} \times n_k} \times \mathcal{M}_k{}^{n_k \times n_{k+1}}$ *to* $\mathcal{N}_0{}^{n_0 \times n_{k+1}}$ *is* $(k+1)$*-linear over* $\mathcal{R}$. *Furthermore,* $\Delta_R f \in \mathcal{T}^{k+1}(\Omega^{(0)}, \ldots, \Omega^{(k)}, \Omega^{(k)}; \mathcal{N}_{0,\mathrm{nc}}, \ldots, \mathcal{N}_{k,\mathrm{nc}}, \mathcal{M}_{k,\mathrm{nc}})$.

PROOF. The fact that $\Delta_R f\left(X^0, \ldots, X^{k+1}\right)\left(Z^1, \ldots, Z^{k+1}\right)$ is linear as a function of $Z^j$, for every $j = 1, \ldots, k$, follows from (3.21). The linearity in $Z^{k+1}$ follows from Propositions 3.8 and 3.9. We also observe that (3.21) implies that the function $\Delta_R f\left(X^0, \ldots, X^{k+1}\right)\left(Z^1, \ldots, Z^{k+1}\right)$ respects direct sums and similarities in variables $X^0, \ldots, X^{k-1}$. Thus, it only remains to show that the function $\Delta_R f\left(X^0, \ldots, X^{k+1}\right)\left(Z^1, \ldots, Z^{k+1}\right)$ respects direct sums and similarities, or equivalently (by Proposition 3.1), respects intertwinings, in variables $X^k$ and $X^{k+1}$.

Let $X^0 \in \Omega^{(0)}_{n_0}$, $\ldots$, $X^k \in \Omega^{(k)}_{n_k}$, $X^{k+1} \in \Omega^{(k)}_{n_{k+1}}$, $\widetilde{X}^k \in \Omega^{(k)}_{\widetilde{n}_k}$, $\widetilde{X}^{k+1} \in \Omega^{(k)}_{\widetilde{n}_{k+1}}$, $Z^1 \in \mathcal{N}_1{}^{n_0 \times n_1}$, $\ldots$, $Z^{k-1} \in (\mathcal{N}_{k-1})^{n_{k-2} \times n_{k-1}}$, $Z^k \in \mathcal{N}_k{}^{n_{k-1} \times \widetilde{n}_k}$, and $Z^{k+1} \in \mathcal{M}_k{}^{n_k \times \widetilde{n}_{k+1}}$ be arbitrary, and let $T_k \in \mathcal{R}^{\widetilde{n}_k \times n_k}$ and $T_{k+1} \in \mathcal{R}^{\widetilde{n}_{k+1} \times n_{k+1}}$ satisfy the identities $T_k X^k = \widetilde{X}^k T_k$ and $T_{k+1} X^{k+1} = \widetilde{X}^{k+1} T_{k+1}$. By (3.24), we may assume, without loss of generality, that $\Omega^{(k)}$ is similarity invariant, i.e., $\widetilde{\Omega}^{(k)} = \Omega^{(k)}$; see Remark 2.5 and Appendix A. Then by Proposition A.2,

$$\begin{bmatrix} X^k & Z^{k+1} T_{k+1} \\ 0 & X^{k+1} \end{bmatrix} \in \Omega^{(k)}_{n_k + n_{k+1}}, \quad \begin{bmatrix} \widetilde{X}^{k+1} & T_k Z^{k+1} \\ 0 & \widetilde{X}^{k+1} \end{bmatrix} \in \Omega^{(k)}_{\widetilde{n}_k + \widetilde{n}_{k+1}}.$$

We have

$$\begin{bmatrix} T_k & 0 \\ 0 & T_{k+1} \end{bmatrix} \begin{bmatrix} X^k & Z^{k+1} T_{k+1} \\ 0 & X^{k+1} \end{bmatrix} = \begin{bmatrix} \widetilde{X}^{k+1} & T_k Z^{k+1} \\ 0 & \widetilde{X}^{k+1} \end{bmatrix} \begin{bmatrix} T_k & 0 \\ 0 & T_{k+1} \end{bmatrix}.$$

Using (3.21) and applying (3X$^k$) to $f$, we obtain

$$\begin{aligned}
&\Delta_R f(X^0, \ldots, X^{k-1}, \widetilde{X}^k, \widetilde{X}^{k+1})(Z^1, \ldots, Z^k, T_k Z^{k+1}) T_{k+1} \\
&= f\left(X^0, \ldots, X^{k-1}, \begin{bmatrix} \widetilde{X}^k & T_k Z^{k+1} \\ 0 & \widetilde{X}^{k+1} \end{bmatrix}\right) (Z^1, \ldots, Z^{k-1}, \operatorname{row}[Z^k, 0]) \begin{bmatrix} 0 \\ I_{n_{k+1}} \end{bmatrix} T_{k+1} \\
&= f\left(X^0, \ldots, X^{k-1}, \begin{bmatrix} \widetilde{X}^k & T_k Z^{k+1} \\ 0 & \widetilde{X}^{k+1} \end{bmatrix}\right) \left(Z^1, \ldots, Z^{k-1}, \operatorname{row}[Z^k, 0]\right) \begin{bmatrix} T_k & 0 \\ 0 & T_{k+1} \end{bmatrix} \\
&\qquad \times \begin{bmatrix} 0 \\ I_{n_{k+1}} \end{bmatrix} \\
&= f\left(X^0, \ldots, X^{k-1}, \begin{bmatrix} X^k & Z^{k+1} T_{k+1} \\ 0 & X^{k+1} \end{bmatrix}\right) \left(Z^1, \ldots, Z^{k-1}, \operatorname{row}[Z^k, 0] \begin{bmatrix} T_k & 0 \\ 0 & T_{k+1} \end{bmatrix}\right) \\
&\qquad \times \begin{bmatrix} 0 \\ I_{n_{k+1}} \end{bmatrix} \\
&\qquad = f(X^0, \ldots, X^{k+1})(Z^1, \ldots, Z^{k-1}, Z^k T_k, Z^{k+1} T_{k+1}),
\end{aligned}$$

which is a special case of (3X$^k$) applied to $\Delta_R f$. □

We have therefore the right nc difference-differential operator

$$\Delta_R\colon \mathcal{T}^k(\Omega^{(0)},\ldots,\Omega^{(k)};\mathcal{N}_{0,\mathrm{nc}},\ldots,\mathcal{N}_{k,\mathrm{nc}}) \\ \longrightarrow \mathcal{T}^{k+1}(\Omega^{(0)},\ldots,\Omega^{(k)},\Omega^{(k)};\mathcal{N}_{0,\mathrm{nc}},\ldots,\mathcal{N}_{k,\mathrm{nc}},\mathcal{M}_{k,\mathrm{nc}}),$$

for $k=0,1,\ldots$. Iterating this operator $\ell$ times, we obtain the *$\ell$-th order right nc difference-differential operator*

$$\Delta_R^\ell\colon \mathcal{T}^k(\Omega^{(0)},\ldots,\Omega^{(k)};\mathcal{N}_{0,\mathrm{nc}},\ldots,\mathcal{N}_{k,\mathrm{nc}}) \\ \longrightarrow \mathcal{T}^{k+\ell}(\Omega^{(0)},\ldots,\Omega^{(k)},\underbrace{\Omega^{(k)},\ldots,\Omega^{(k)}}_{\ell\text{ times}};\mathcal{N}_{0,\mathrm{nc}},\ldots,\mathcal{N}_{k,\mathrm{nc}},\underbrace{\mathcal{M}_{k,\mathrm{nc}},\ldots,\mathcal{M}_{k,\mathrm{nc}}}_{\ell\text{ times}}),$$

for $k=0,1,\ldots$. According to our definition, $\Delta_R^\ell f$ is calculated iteratively by evaluating nc functions of increasing orders on $2\times 2$ block upper triangular matrices at each step. It turns out that $\Delta_R^\ell f$ can also be calculated in a single step by evaluating $f$ on $(\ell+1)\times(\ell+1)$ block upper bidiagonal matrices. We formulate the result separately for the case $k=0$ and for the case $k>0$.

THEOREM 3.11. *Let $f\in\mathcal{T}^0(\Omega;\mathcal{N}_{\mathrm{nc}})$, with $\Omega\subseteq\mathcal{M}_{\mathrm{nc}}$ a right admissible nc set. For every $X^0\in\Omega_{n_0}$, ..., $X^\ell\in\Omega_{n_\ell}$, $Z^1\in\mathcal{M}^{n_0\times n_1}$, ..., $Z^\ell\in\mathcal{M}^{n_{\ell-1}\times n_\ell}$ such that*

$$\begin{bmatrix} X^0 & Z^1 & 0 & \cdots & 0\\ 0 & X^1 & \ddots & \ddots & \vdots\\ \vdots & \ddots & \ddots & \ddots & 0\\ \vdots & & \ddots & X^{\ell-1} & Z^\ell\\ 0 & \cdots & \cdots & 0 & X^\ell \end{bmatrix}\in\Omega_{n_0+\cdots+n_\ell},$$

*one has*

$$f\left(\begin{bmatrix} X^0 & Z^1 & 0 & \cdots & 0\\ 0 & X^1 & \ddots & \ddots & \vdots\\ \vdots & \ddots & \ddots & \ddots & 0\\ \vdots & & \ddots & X^{\ell-1} & Z^\ell\\ 0 & \cdots & \cdots & 0 & X^\ell \end{bmatrix}\right) \\ =\begin{bmatrix} f(X^0) & \Delta_R f(X^0,X^1)(Z^1) & \cdots & \cdots & \Delta_R^\ell f(X^0,\ldots,X^\ell)(Z^1,\ldots,Z^\ell)\\ 0 & f(X^1) & \ddots & & \Delta_R^{\ell-1} f(X^1,\ldots,X^\ell)(Z^2,\ldots,Z^\ell)\\ \vdots & \ddots & \ddots & \ddots & \vdots\\ \vdots & & \ddots & f(X^{\ell-1}) & \Delta_R f(X^{\ell-1},X^\ell)(Z^\ell)\\ 0 & \cdots & \cdots & 0 & f(X^\ell) \end{bmatrix}. \tag{3.29}$$

THEOREM 3.12. *Let $f\in\mathcal{T}^k(\Omega^{(0)},\ldots,\Omega^{(k)};\mathcal{N}_{0,\mathrm{nc}},\ldots,\mathcal{N}_{k,\mathrm{nc}})$, where $\Omega^{(k)}\subseteq\mathcal{M}_{k,\mathrm{nc}}$ is a right admissible nc set, $k>0$. For every $X^0\in\Omega^{(0)}_{n_0}$, ..., $X^{k-1}\in\Omega^{(k-1)}_{n_{k-1}}$, $X^{k,0}\in\Omega^{(k)}_{n_{k,0}}$, ..., $X^{k,\ell}\in\Omega^{(k)}_{n_{k,\ell}}$, $Z^{k+1,1}\in\mathcal{M}_k{}^{n_{k,0}\times n_{k,1}}$, ..., $Z^{k+1,\ell}\in$*

${\mathcal{M}_k}^{n_{k,\ell-1}\times n_{k,\ell}}$ *such that*

$$\begin{bmatrix} X^{k,0} & Z^{k+1,1} & 0 & \cdots & 0 \\ 0 & X^{k,1} & \ddots & \ddots & \vdots \\ \vdots & \ddots & \ddots & \ddots & 0 \\ \vdots & & \ddots & X^{k,\ell-1} & Z^{k+1,\ell} \\ 0 & \cdots & \cdots & 0 & X^{k,\ell} \end{bmatrix} \in \Omega^{(k)}_{n_{k,0}+\cdots+n_{k,\ell}},$$

*one has*

$$\begin{aligned} (3.30)\quad & f\left( X^0,\ldots,X^{k-1}, \begin{bmatrix} X^{k,0} & Z^{k+1,1} & 0 & \cdots & 0 \\ 0 & X^{k,1} & \ddots & \ddots & \vdots \\ \vdots & \ddots & \ddots & \ddots & 0 \\ \vdots & & \ddots & X^{k,\ell-1} & Z^{k+1,\ell} \\ 0 & \cdots & \cdots & 0 & X^{k,\ell} \end{bmatrix}\right) \\ & \qquad (Z^1,\ldots,Z^{k-1},\operatorname{row}[Z^{k,0},\ldots,Z^{k,\ell}]) \\ & = \operatorname*{row}_{0\le j\le \ell}\Big[\sum_{i=0}^{j} \Delta_R^{j-i} f(X^0,\ldots,X^{k-1},X^{k,i},\ldots,X^{k,j}) \\ & \qquad\qquad (Z^1,\ldots,Z^{k-1},Z^{k,i},Z^{k+1,i+1},\ldots,Z^{k+1,j})\Big] \end{aligned}$$

*for all* $Z^1 \in {\mathcal{N}_1}^{n_0\times n_1}$, ..., $Z^{k-1} \in (\mathcal{N}_{k-1})^{n_{k-2}\times n_{k-1}}$, $Z^{k,0} \in {\mathcal{N}_k}^{n_{k-1}\times n_{k,0}}$, ..., $Z^{k,\ell} \in {\mathcal{N}_k}^{n_{k-1}\times n_{k,\ell}}$.

REMARK 3.13. In the special case where $Z^{k,1} = 0$, ..., $Z^{k,\ell} = 0$, the right-hand side of (3.30) becomes

$$\begin{aligned} \operatorname{row}\big[ & f(X^0,\ldots,X^{k-1},X^{k,0})(Z^1,\ldots,Z^{k-1},Z^{k,0}), \\ & \Delta_R f(X^0,\ldots,X^{k-1},X^{k,0},X^{k,1})(Z^1,\ldots,Z^{k-1},Z^{k,0},Z^{k+1,1}),\ldots, \\ & \Delta_R^{\ell} f(X^0,\ldots,X^{k-1},X^{k,0},\ldots,X^{k,\ell})(Z^1,\ldots,Z^{k-1},Z^{k,0},Z^{k+1,1},\ldots,Z^{k+1,\ell})\big]. \end{aligned}$$

For the proofs of Theorems 3.11 and 3.12, we will need the following lemma.

LEMMA 3.14. *Let* $\Omega \subseteq \mathcal{M}_{\rm nc}$ *be similarity invariant, i.e.,* $\widetilde{\Omega} = \Omega$; *see Appendix* A. *Then for all* $n_0$, ..., $n_\ell$, $X^0 \in \Omega_{n_0}$, ..., $X^\ell \in \Omega_{n_\ell}$, $Z^1 \in \mathcal{M}^{n_0\times n_1}$, ..., $Z^\ell \in \mathcal{M}^{n_{\ell-1}\times n_\ell}$ *one has*

$$\begin{bmatrix} X^{j} & Z^{j+1} & 0 & \cdots & 0 \\ 0 & X^{j+1} & \ddots & \ddots & \vdots \\ \vdots & \ddots & \ddots & \ddots & 0 \\ \vdots & & \ddots & X^{\ell-1} & Z^{\ell} \\ 0 & \cdots & \cdots & 0 & X^{\ell} \end{bmatrix} \in \Omega_{n_j+\cdots+n_\ell}, \quad j=0,\ldots,\ell-1.$$

PROOF. Induction on $\ell$, using Proposition A.2. □

PROOF OF THEOREM 3.11. By (3.24), we may assume, without loss of generality, that $\Omega$ is similarity invariant, i.e., $\widetilde{\Omega} = \Omega$; see Remark 2.5 and Appendix A. We will prove the statement by induction on $\ell$. For $\ell = 0$, the two sides of (3.29) are identical. For $\ell = 1$, (3.29) coincides with (2.5) up to notation. Suppose now that the statement is true for $\ell = m$, with some $m \in \mathbb{N}$. Let $X^0 \in \Omega_{n_0}$, $\ldots$, $X^{m+1} \in \Omega_{n_{m+1}}$, $Z^1 \in \mathcal{M}^{n_0 \times n_1}$, $\ldots$, $Z^{m+1} \in \mathcal{M}^{n_m \times n_{m+1}}$. Using induction hypothesis, Lemma 3.14, and (3.19), we obtain

$$f\left(\begin{bmatrix} X^0 & Z^1 & 0 & \cdots & \cdots & 0 \\ 0 & X^1 & \ddots & \ddots & & \vdots \\ \vdots & \ddots & \ddots & \ddots & \ddots & \vdots \\ \vdots & & \ddots & X^{m-1} & Z^m & 0 \\ 0 & \cdots & \cdots & 0 & X^m & Z^{m+1} \\ 0 & \cdots & \cdots & \cdots & 0 & X^{m+1} \end{bmatrix}\right)$$

$$= \begin{bmatrix} f(X^0) & \cdots & \cdots & \Delta_R^m f\left(X^0, \ldots, X^{m-1}, \begin{bmatrix} X^m & Z^{m+1} \\ 0 & X^{m+1} \end{bmatrix}\right)(Z^1, \ldots, Z^{m-1}, \mathrm{row}[Z^m, 0]) \\ 0 & \ddots & & \vdots \\ \vdots & \ddots & f(X^{m-1}) & \vdots \\ 0 & \cdots & 0 & f\left(\begin{bmatrix} X^m & Z^{m+1} \\ 0 & X^{m+1} \end{bmatrix}\right) \end{bmatrix}$$

$$= \begin{bmatrix} f(X^0) & \Delta_R f(X^0, X^1)(Z^1) & \cdots & \Delta_R^{m+1} f\left(X^0, \ldots, X^{m+1}\right)(Z^1, \ldots, Z^{m+1}) \\ 0 & \ddots & \ddots & \vdots \\ \vdots & \ddots & f(X^m) & \Delta_R f(X^m, X^{m+1})(Z^{m+1}) \\ 0 & \cdots & 0 & f(X^{m+1}) \end{bmatrix}$$

Thus, we have (3.29) for $\ell = m + 1$. $\square$

PROOF OF THEOREM 3.12. By (3.24), we may assume, without loss of generality, that $\Omega^{(k)}$ is similarity invariant, i.e., $\widetilde{\Omega}^{(k)} = \Omega^{(k)}$; see Remark 2.5 and Appendix A. We will prove the statement by induction on $\ell$. For $\ell = 0$, the two sides of (3.30) are identical. For $\ell = 1$, (3.30) coincides with (3.19) up to notation. Suppose now that the statement is true for all $k > 0$ and $\ell = m$, with some $m \in \mathbb{N}$. Let $k > 0$ be fixed. Let $X^0 \in \Omega^{(0)}_{n_0}$, $\ldots$, $X^{k-1} \in \Omega^{(k-1)}_{n_{k-1}}$, $X^{k,0} \in \Omega^{(k)}_{n_{k,0}}$, $\ldots$, $X^{k,m+1} \in \Omega^{(k)}_{n_{k,m+1}}$, $Z^{k+1,1} \in \mathcal{M}_k{}^{n_{k,0} \times n_{k,1}}$, $\ldots$, $Z^{k+1,m+1} \in \mathcal{M}_k{}^{n_{k,m} \times n_{k,m+1}}$, $Z^1 \in \mathcal{N}_1{}^{n_0 \times n_1}$, $\ldots$, $Z^{k-1} \in (\mathcal{N}_{k-1})^{n_{k-2} \times n_{k-1}}$, $Z^{k,0} \in \mathcal{N}_k{}^{n_{k-1} \times n_{k,0}}$, $\ldots$, $Z^{k,m+1} \in \mathcal{N}_k{}^{n_{k-1} \times n_{k,m+1}}$. By Lemma 3.14,

$$\begin{bmatrix} X^{k,j} & Z^{k+1,j+1} & 0 & \cdots & 0 \\ 0 & X^{k,j+1} & \ddots & \ddots & \vdots \\ \vdots & \ddots & \ddots & \ddots & 0 \\ \vdots & & \ddots & X^{k,m} & Z^{k+1,m+1} \\ 0 & \cdots & \cdots & 0 & X^{k,m+1} \end{bmatrix} \in \Omega^{(k)}_{n_{k,j}+\cdots+n_{k,m+1}}$$

for each $j = 0, \dots, m$. By (3.19), we have

$$f\left(X^0, \dots, X^{k-1}, \begin{bmatrix} X^{k,0} & Z^{k+1,1} & 0 & \cdots & 0 \\ 0 & X^{k,1} & \ddots & \ddots & \vdots \\ \vdots & \ddots & \ddots & \ddots & 0 \\ \vdots & & \ddots & X^{k,m} & Z^{k+1,m+1} \\ 0 & \cdots & \cdots & 0 & X^{k,m+1} \end{bmatrix}\right)$$

$$(Z^1, \dots, Z^{k-1}, \operatorname{row}[Z^{k,0}, \dots, Z^{k,m+1}])$$

$$= \operatorname{row}\Big[f(X^0, \dots, X^{k-1}, X^{k,0})(Z^1, \dots, Z^{k-1}, Z^{k,0}),$$

$$\Delta_R f\left(X^0, \dots, X^{k-1}, \begin{bmatrix} X^{k,1} & Z^{k+1,2} & 0 & \cdots & 0 \\ 0 & X^{k,2} & \ddots & \ddots & \vdots \\ \vdots & \ddots & \ddots & \ddots & 0 \\ \vdots & & \ddots & X^{k,m} & Z^{k+1,m+1} \\ 0 & \cdots & \cdots & 0 & X^{k,m+1} \end{bmatrix}\right)$$

$$(Z^1, \dots, Z^{k-1}, Z^{k,0}, \operatorname{row}[Z^{k+1,1}, 0, \dots, 0])$$

$$+ f\left(X^0, \dots, X^{k-1}, \begin{bmatrix} X^{k,1} & Z^{k+1,2} & 0 & \cdots & 0 \\ 0 & X^{k,2} & \ddots & \ddots & \vdots \\ \vdots & \ddots & \ddots & \ddots & 0 \\ \vdots & & \ddots & X^{k,m} & Z^{k+1,m+1} \\ 0 & \cdots & \cdots & 0 & X^{k,m+1} \end{bmatrix}\right)$$

$$(Z^1, \dots, Z^{k-1}, \operatorname{row}[Z^{k,1}, \dots, Z^{k,m+1}])\Big].$$

Applying the induction assumption to the right-hand side of the latter (see also Remark 3.13), we obtain

$$f\left(X^0, \dots, X^{k-1}, \begin{bmatrix} X^{k,0} & Z^{k+1,1} & 0 & \cdots & 0 \\ 0 & X^{k,1} & \ddots & \ddots & \vdots \\ \vdots & \ddots & \ddots & \ddots & 0 \\ \vdots & & \ddots & X^{k,m} & Z^{k+1,m+1} \\ 0 & \cdots & \cdots & 0 & X^{k,m+1} \end{bmatrix}\right)$$

$$(Z^1, \dots, Z^{k-1}, \operatorname{row}[Z^{k,0}, \dots, Z^{k,m+1}])$$

$$= \operatorname{row}\Big[f(X^0, \dots, X^{k-1}, X^{k,0})(Z^1, \dots, Z^{k-1}, Z^{k,0}),$$

$$\Delta_R f(X^0, \dots, X^{k-1}, X^{k,0}, X^{k,1})(Z^1, \dots, Z^{k-1}, Z^{k,0}, Z^{k+1,1})$$

$$+ f(X^0, \dots, X^{k-1}, X^{k,1})(Z^1, \dots, Z^{k-1}, Z^{k,1}),$$

$$\Delta_R^2 f(X^0, \dots, X^{k-1}, X^{k,0}, X^{k,1}, X^{k,2})(Z^1, \dots, Z^{k-1}, Z^{k,0}, Z^{k+1,1}, Z^{k+1,2})$$

$$+ \Delta_R f(X^0, \dots, X^{k-1}, X^{k,1}, X^{k,2})(Z^1, \dots, Z^{k-1}, Z^{k,1}, Z^{k+1,2})$$

$$+ f(X^0, \dots, X^{k-1}, X^{k,2})(Z^1, \dots, Z^{k-1}, Z^{k,2}), \dots,$$

$$\Delta_R^{m+1}(X^0,\dots,X^{k-1},X^{k,0},\dots,X^{k,m+1})$$
$$(Z^1,\dots,Z^{k-1},Z^{k,0},Z^{k+1,1},\dots,Z^{k+1,m+1})$$
$$+\sum_{i=0}^{m}\Delta_R^{m-i}f(X^0,\dots,X^{k-1},X^{k,i+1},\dots,X^{k,m+1})$$
$$(Z^1,\dots,Z^{k-1},Z^{k,i+1},Z^{k+1,i+2},\dots,Z^{k+1,m+1})\Big]$$

$$=\operatorname*{row}_{0\le j\le m+1}\Big[\sum_{i=0}^{j}\Delta_R^{j-i}f(X^0,\dots,X^{k-1},X^{k,i},\dots,X^{k,j})$$
$$(Z^1,\dots,Z^{k-1},Z^{k,i},Z^{k+1,i+1},\dots,Z^{k+1,j})\Big].$$

Thus the statement is true also for $\ell=m+1$. □

REMARK 3.15. It follows from Theorem 3.11 (and Lemma 3.14) that if $\Omega'$ a right admissible nc subset of $\Omega$ and $X^0,\ \dots,\ X^\ell\in\Omega'$, then $\Delta_R^\ell f(X^0,\dots,X^\ell)$ is determined by $f|_{\Omega'}$.

REMARK 3.16. Using the tensor product interpretation of the values of a nc function $f$ of order $k$ (see Remark 3.5), we can view the definition of $\Delta_R f$ as follows. Let $n_k=n_k'+n_k''$. Then we have the $2\times 2$ block matrix decomposition of $n_k\times n_k$ matrices over $\mathcal{N}_k^*$:

$$\mathcal{N}_k^{*\,n_k\times n_k}=\mathcal{N}_k^{*\,n_k'\times n_k'}\oplus\mathcal{N}_k^{*\,n_k'\times n_k''}\oplus\mathcal{N}_k^{*\,n_k''\times n_k'}\oplus\mathcal{N}_k^{*\,n_k''\times n_k''}.$$

This decomposition induces the corresponding decomposition of

$$\mathcal{N}_0^{\,n_0\times n_0}\otimes\mathcal{N}_1^{*\,n_1\times n_1}\otimes\cdots\otimes\mathcal{N}_k^{*\,n_k\times n_k},$$

and

$$f\left(X^0,\dots,X^{k-1},\begin{bmatrix}X^{k\prime} & Z\\ 0 & X^{k\prime\prime}\end{bmatrix}\right)=f(X^0,\dots,X^{k-1},X^{k\prime})$$
$$+\Delta_R f(X^0,\dots,X^{k-1},X^{k\prime},X^{k\prime\prime})(\cdot,\dots,\cdot,Z)+f(X^0,\dots,X^{k-1},X^{k\prime\prime}),$$

where

$$f(X^0,\dots,X^{k-1},X^{k\prime})\in\mathcal{N}_0^{\,n_0\times n_0}\otimes\mathcal{N}_1^{*\,n_1\times n_1}\otimes\cdots\otimes(\mathcal{N}_{k-1}^*)^{n_{k-1}\times n_{k-1}}\otimes\mathcal{N}_k^{*\,n_k'\times n_k'},$$

$$\Delta_R f(X^0,\dots,X^{k-1},X^{k\prime},X^{k\prime\prime})(\cdot,\dots,\cdot,Z)$$
$$\in\mathcal{N}_0^{\,n_0\times n_0}\otimes\mathcal{N}_1^{*\,n_1\times n_1}\otimes\cdots\otimes(\mathcal{N}_{k-1}^*)^{n_{k-1}\times n_{k-1}}\otimes\mathcal{N}_k^{*\,n_k'\times n_k''},$$

$$f(X^0,\dots,X^{k-1},X^{k\prime\prime})\in\mathcal{N}_0^{\,n_0\times n_0}\otimes\mathcal{N}_1^{*\,n_1\times n_1}\otimes\cdots\otimes(\mathcal{N}_{k-1}^*)^{n_{k-1}\times n_{k-1}}\otimes\mathcal{N}_k^{*\,n_k''\times n_k''}$$

(this is equivalent to (3.19) when using the multilinear mapping interpretation of the values of $f$). Furthermore, the mapping

$$\mathcal{M}_k^{\,n_k'\times n_k''}\longrightarrow\mathcal{N}_0^{\,n_0\times n_0}\otimes\mathcal{N}_1^{*\,n_1\times n_1}\otimes\cdots\otimes(\mathcal{N}_{k-1}^*)^{n_{k-1}\times n_{k-1}}\otimes\mathcal{N}_k^{*\,n_k'\times n_k''},\tag{3.31}$$

$$Z\longmapsto\Delta_R f(X^0,\dots,X^{k-1},X^{k\prime},X^{k\prime\prime})(\cdot,\dots,\cdot,Z)\tag{3.32}$$

is linear (this is equivalent to Propositions 3.8 and 3.9). We have a special case of the natural isomorphism (3.10)–(3.11),

$$\mathcal{N}_k^{*\,n_k'\times n_k'}\otimes\mathcal{M}_k^{*\,n_k''\times n_k''}\xrightarrow{\sim}\hom_{\mathcal{R}}\left(\mathcal{M}_k^{\,n_k'\times n_k''},\mathcal{N}_k^{*\,n_k'\times n_k''}\right)$$

(we assume here that the module $\mathcal{M}_k$ is free and of finite rank, in addition to our previous assumptions that the modules $\mathcal{N}_1, \ldots, \mathcal{N}_k$ are free and of finite rank and that the module $\mathcal{N}_0$ is free). Tensoring with ${\mathcal{N}_0}^{n_0\times n_0} \otimes {\mathcal{N}_1^*}^{n_1\times n_1} \otimes \cdots \otimes (\mathcal{N}_{k-1}^*)^{n_{k-1}\times n_{k-1}}$ yields a natural isomorphism

$$\begin{aligned}&{\mathcal{N}_0}^{n_0\times n_0} \otimes {\mathcal{N}_1^*}^{n_1\times n_1} \cdots \otimes (\mathcal{N}_{k-1}^*)^{n_{k-1}\times n_{k-1}} \otimes {\mathcal{N}_k^*}^{n_k'\times n_k'} \otimes {\mathcal{M}_k^*}^{n_k''\times n_k''}\\ &\xrightarrow{\sim} \hom_{\mathcal{R}}\Big({\mathcal{M}_k}^{n_k'\times n_k''}, {\mathcal{N}_0}^{n_0\times n_0} \otimes {\mathcal{N}_1^*}^{n_1\times n_1} \otimes \cdots \otimes (\mathcal{N}_{k-1}^*)^{n_{k-1}\times n_{k-1}} \otimes {\mathcal{N}_k^*}^{n_k'\times n_k''}\Big).\end{aligned}$$

It follows from (3.31)–(3.32) that

$$\begin{aligned}&\Delta_R f(X^0, \ldots, X^{k-1}, X^{k\prime}, X^{k\prime\prime})\\ &\qquad \in {\mathcal{N}_0}^{n_0\times n_0} \otimes {\mathcal{N}_1^*}^{n_1\times n_1} \otimes \cdots \otimes (\mathcal{N}_{k-1}^*)^{n_{k-1}\times n_{k-1}} \otimes {\mathcal{N}_k^*}^{n_k'\times n_k'} \otimes {\mathcal{M}_k^*}^{n_k''\times n_k''},\end{aligned}$$

which is equivalent to the first part of Theorem 3.10.

REMARK 3.17. It follows from the definition of $\Delta_R$ that its action on a tensor product of (possibly, higher order) nc functions (see Remark 3.6) boils down to its action on the last factor:

$$\Delta_R(f \otimes g) = f \otimes \Delta_R g, \tag{3.33}$$

where

$$f \in \mathcal{T}^k(\Omega^{(0)}, \ldots, \Omega^{(k)}; \mathcal{N}_{0,\mathrm{nc}}, \ldots, \mathcal{N}_{k,\mathrm{nc}}),$$
$$g \in \mathcal{T}^\ell(\Omega^{(k+1)}, \ldots, \Omega^{(k+\ell)}; \mathcal{L}_{0,\mathrm{nc}}^*, \mathcal{L}_{1,\mathrm{nc}}, \ldots, \mathcal{L}_{\ell,\mathrm{nc}}),$$

and

$$\Delta_R(f^0 \otimes \cdots \otimes f^{k-1} \otimes f^k) = f^0 \otimes \cdots \otimes f^{k-1} \otimes \Delta_R f^k, \tag{3.34}$$

where $f^0 \in \mathcal{T}^0(\Omega^{(0)}; \mathcal{N}_{0,\mathrm{nc}})$, $f^1 \in \mathcal{T}^0(\Omega^{(1)}; \mathcal{N}_{1,\mathrm{nc}}^*)$, $\ldots$, $f^k \in \mathcal{T}^0(\Omega^{(k)}; \mathcal{N}_{k,\mathrm{nc}}^*)$.

REMARK 3.18. More generally, we can define a difference-differential operator ${}_j\Delta_R$ by evaluating an nc function of order $k$ on a $(k+1)$-tuple of square matrices with the $(j+1)$-th argument upper triangular, $j = 0, \ldots, k$. Hence, $\Delta_R = {}_k\Delta_R$. To define ${}_0\Delta_R f$ for $f \in \mathcal{T}^k(\Omega^{(0)}, \ldots, \Omega^{(k)}; \mathcal{N}_{0,\mathrm{nc}}, \ldots, \mathcal{N}_{k,\mathrm{nc}})$, where $\Omega^{(0)} \subseteq \mathcal{M}_{0,\mathrm{nc}}$ is a right admissible nc set, let $X^{0\prime} \in \Omega^{(0)}_{n_0'}$, $X^{0\prime\prime} \in \Omega^{(0)}_{n_0''}$, $X^1 \in \Omega^{(1)}_{n_1}$, $\ldots$, $X^k \in \Omega^{(k)}_{n_k}$, $Z^{1\prime} \in {\mathcal{N}_1}^{n_0'\times n_1}$, $Z^{1\prime\prime} \in {\mathcal{N}_1}^{n_0''\times n_1}$, $Z^2 \in {\mathcal{N}_2}^{n_1\times n_2}$ $\ldots$, $Z^k \in {\mathcal{N}_k}^{n_{k-1}\times n_k}$. Let $Z \in {\mathcal{M}_0}^{n_0'\times n_0''}$ be such that $\begin{bmatrix} X^{0\prime} & Z \\ 0 & X^{0\prime\prime}\end{bmatrix} \in \Omega^{(0)}_{n_0'+n_0''}$. Then, similarly to Proposition 3.7,

$$\begin{aligned}&f\left(\begin{bmatrix} X^{0\prime} & Z \\ 0 & X^{0\prime\prime}\end{bmatrix}, X^1, \ldots, X^k\right)(\mathrm{col}\,[Z^{1\prime}, Z^{1\prime\prime}], Z^2, \ldots, Z^k)\\ &= \mathrm{col}\Big[f(X^{0\prime}, X^1, \ldots, X^k)(Z^{1\prime}, Z^2, \ldots, Z^k)\\ &\qquad + {}_0\Delta_R f(X^{0\prime}, X^{0\prime\prime}, X^1, \ldots, X^k)(Z, Z^{1\prime\prime}, Z^2, \ldots, Z^k),\\ &\qquad\qquad f(X^{0\prime\prime}, X^1, \ldots, X^k)(Z^{1\prime\prime}, Z^2, \ldots, Z^k)\Big].\end{aligned} \tag{3.35}$$

Here ${}_0\Delta_R f(X^{0\prime}, X^{0\prime\prime}, X^1, \ldots, X^k)(Z, Z^{1\prime\prime}, Z^2, \ldots, Z^k)$ is determined uniquely by (3.35), is independent of $Z^{1\prime}$ and is homogeneous in $Z$ whenever $\begin{bmatrix} X^{0\prime} & Z \\ 0 & X^{0\prime\prime}\end{bmatrix} \in \Omega^{(0)}_{n_0'+n_0''}$. We can define ${}_0\Delta_R f(X^{0\prime}, X^{0\prime\prime}, X^1, \ldots, X^k)(Z, Z^{1\prime\prime}, Z^2, \ldots, Z^k)$ for all $Z \in$

${\mathcal{M}_0}^{n_0'\times n_0''}$ by homogeneity analogously to (3.23). Similarly to Proposition 3.9, it is linear in $Z$; similarly to Theorem 3.10,

$$ {}_0\Delta_R f \in \mathcal{T}^{k+1}(\Omega^{(0)},\Omega^{(0)},\Omega^{(1)},\ldots,\Omega^{(k)};\mathcal{N}_{0,\mathrm{nc}},\mathcal{M}_{0,\mathrm{nc}},\mathcal{N}_{1,\mathrm{nc}},\ldots,\mathcal{N}_{k,\mathrm{nc}}). $$

To define ${}_j\Delta_R f$, $1\le j\le k-1$, we assume now that $\Omega^{(j)}$ is a right admissible nc set. Let $X^0\in\Omega^{(0)}_{n_0}$, …, $X^{j-1}\in\Omega^{(j-1)}_{n_{j-1}}$, $X^{j\prime}\in\Omega^{(j)}_{n_j'}$, $X^{j\prime\prime}\in\Omega^{(j)}_{n_j''}$, $X^{j+1}\in\Omega^{(j+1)}_{n_{j+1}}$, …, $X^k\in\Omega^{(k)}_{n_k}$, $Z^1\in{\mathcal{N}_1}^{n_0\times n_1}$, …, $Z^{j-1}\in(\mathcal{N}_{j-1})^{n_{j-2}\times n_{j-1}}$, $Z^{j\prime}\in{\mathcal{N}_j}^{n_{j-1}\times n_j'}$, $Z^{j\prime\prime}\in{\mathcal{N}_j}^{n_{j-1}\times n_j''}$, $Z^{j+1\prime}\in(\mathcal{N}_{j+1})^{n_j'\times n_{j+1}}$, $Z^{j+1\prime\prime}\in(\mathcal{N}_{j+1})^{n_j''\times n_{j+1}}$, $Z^{j+2}\in(\mathcal{N}_{j+2})^{n_{j+1}\times n_{j+2}}$, …, $Z^k\in{\mathcal{N}_k}^{n_{k-1}\times n_k}$. Let $Z\in{\mathcal{M}_j}^{n_j'\times n_j''}$ be such that $\begin{bmatrix} X^{j\prime} & Z\\ 0 & X^{j\prime\prime}\end{bmatrix}\in\Omega^{(j)}_{n_j'+n_j''}$. Then, similarly to Proposition 3.7,

$$ \begin{aligned} (3.36)\quad & f\left(X^0,\ldots,X^{j-1},\begin{bmatrix} X^{j\prime} & Z\\ 0 & X^{j\prime\prime}\end{bmatrix},X^{j+1},\ldots,X^k\right)\\ &\qquad (Z^1,\ldots,Z^{j-1},\mathrm{row}\,[Z^{j\prime},Z^{j\prime\prime}],\mathrm{col}\,[Z^{j+1\prime},Z^{j+1\prime\prime}],Z^{j+2},\ldots,Z^k)\\ &= f(X^0,\ldots,X^{j-1},X^{j\prime},X^{j+1},\ldots,X^k)(Z^1,\ldots,Z^{j-1},Z^{j\prime},Z^{j+1\prime},Z^{j+2},\ldots,Z^k)\\ &\quad + {}_j\Delta_R f(X^0,\ldots,X^{j-1},X^{j\prime},X^{j\prime\prime},X^{j+1},\ldots,X^k)\\ &\qquad (Z^1,\ldots,Z^{j-1},Z^{j\prime},Z,Z^{j+1\prime\prime},Z^{j+2},\ldots,Z^k)\\ &\quad + f(X^0,\ldots,X^{j-1},X^{j\prime\prime},X^{j+1},\ldots,X^k)(Z^1,\ldots,Z^{j-1},Z^{j\prime\prime},Z^{j+1\prime\prime},Z^{j+2},\ldots,Z^k). \end{aligned} $$

Here

$$ {}_j\Delta_R f(X^0,\ldots,X^{j-1},X^{j\prime},X^{j\prime\prime},X^{j+1},\ldots,X^k)(Z^1,\ldots,Z^{j-1},Z^{j\prime},Z,Z^{j+1\prime\prime},Z^{j+2},\ldots,Z^k) $$

is determined uniquely by (3.36), is independent of $Z^{j\prime\prime}$ and $Z^{j+1\prime}$ and is homogeneous in $Z$ whenever $\begin{bmatrix} X^{j\prime} & Z\\ 0 & X^{j\prime\prime}\end{bmatrix}\in\Omega^{(j)}_{n_j'+n_j''}$. We can define

$$ {}_j\Delta_R f(X^0,\ldots,X^{j-1},X^{j\prime},X^{j\prime\prime},X^{j+1},\ldots,X^k)(Z^1,\ldots,Z^{j-1},Z^{j\prime},Z,Z^{j+1\prime\prime},Z^{j+2},\ldots,Z^k) $$

for all $Z\in{\mathcal{M}_j}^{n_j'\times n_j''}$ by homogeneity analogously to (3.23). Similarly to Proposition 3.9, it is linear in $Z$; similarly to Theorem 3.10,

$$ {}_j\Delta_R f\in\mathcal{T}^{k+1}(\Omega^{(0)},\ldots,\Omega^{(j-1)},\Omega^{(j)},\Omega^{(j)},\Omega^{(j+1)},\ldots,\Omega^{(k)};\mathcal{N}_{0,\mathrm{nc}},\ldots,\mathcal{N}_{j,\mathrm{nc}},\mathcal{M}_{j,\mathrm{nc}},\mathcal{N}_{j+1,\mathrm{nc}},\ldots,\mathcal{N}_{k,\mathrm{nc}}). $$

Using Proposition A.5, we can establish a $j$-analogue of (3.24),

$$ \begin{aligned} (3.37)\quad & {}_j\Delta_R f(X^0,\ldots,X^{j-1},X^{j\prime},X^{j\prime\prime},X^{j+1},\ldots,X^k)\\ &\qquad (Z^1,\ldots,Z^{j-1},Z^{j\prime},Z,Z^{j+1\prime\prime},Z^{j+2},\ldots,Z^k)\\ &= {}_j\Delta_R \widetilde{f}(X^0,\ldots,X^{j-1},X^{j\prime},X^{j\prime\prime},X^{j+1},\ldots,X^k)\\ &\qquad (Z^1,\ldots,Z^{j-1},Z^{j\prime},Z,Z^{j+1\prime\prime},Z^{j+2},\ldots,Z^k) \end{aligned} $$

One can also formulate the definition of ${}_j\Delta_R f$, $0 \le j \le k$, equivalently, using the tensor product interpretation of the values of the nc function $f$ of order $k$, similarly to Remark 3.16.

Analogously to Remark 3.17, it follows from the definition of ${}_j\Delta_R$ that its action on a tensor product of (possibly, higher order) nc functions boils down to its action on one of the factors:

$$ {}_j\Delta_R(f \otimes g) = \begin{cases} {}_j\Delta_R(f) \otimes g, & 0 \le j \le k, \\ f \otimes {}_{j-k-1}\Delta_R(g), & k+1 \le j \le k+\ell+1, \end{cases} \tag{3.38} $$

where

$$ f \in \mathcal{T}^k(\Omega^{(0)}, \ldots, \Omega^{(k)}; \mathcal{N}_{0,\mathrm{nc}}, \ldots, \mathcal{N}_{k,\mathrm{nc}}), $$
$$ g \in \mathcal{T}^\ell(\Omega^{(k+1)}, \ldots, \Omega^{(k+\ell)}; \mathcal{L}^*_{0,\mathrm{nc}}, \ldots, \mathcal{L}_{\ell,\mathrm{nc}}), $$

and

$$ {}_j\Delta_R(f^0 \otimes \cdots \otimes f^{j-1} \otimes f^j \otimes f^{j+1} \otimes \cdots \otimes f^k) = (f^0 \otimes \cdots \otimes f^{j-1} \otimes \Delta_R f^j \otimes f^{j+1} \otimes \cdots \otimes f^k). \tag{3.39} $$

## 3.3. First order difference formulae for higher order nc functions

We establish now the analogue of Theorem 2.10 for higher order nc functions. An iterative use of this result will lead us in Chapter 4 to a nc analogue of Brook Taylor expansion.

THEOREM 3.19. *Let $f \in \mathcal{T}^k(\Omega^{(0)}, \ldots, \Omega^{(k)}; \mathcal{N}_{0,\mathrm{nc}}, \ldots, \mathcal{N}_{k,\mathrm{nc}})$, where $\Omega^{(k)} \subseteq \mathcal{M}_{k,\mathrm{nc}}$ is a right admissible nc set. Then for all $n_0, \ldots, n_k \in \mathbb{N}$, and arbitrary $X^0 \in \Omega^{(0)}_{n_0}$, ..., $X^{k-1} \in \Omega^{(k-1)}_{n_{k-1}}$, $X, Y \in \Omega^{(k)}_{n_k}$, $Z^1 \in \mathcal{N}_1{}^{n_0 \times n_1}$, ..., $Z^k \in \mathcal{N}_k{}^{n_{k-1} \times n_k}$,*

$$ \begin{aligned} f(X^0, \ldots, X^{k-1}, X)(Z^1, \ldots, Z^k) - f(X^0, \ldots, X^{k-1}, Y)(Z^1, \ldots, Z^k) \\ = \Delta_R f(X^0, \ldots, X^{k-1}, Y, X)(Z^1, \ldots, Z^k, X - Y) \end{aligned} \tag{3.40} $$

*and*

$$ \begin{aligned} f(X^0, \ldots, X^{k-1}, X)(Z^1, \ldots, Z^k) - f(X^0, \ldots, X^{k-1}, Y)(Z^1, \ldots, Z^k) \\ = \Delta_R f(X^0, \ldots, X^{k-1}, X, Y)(Z^1, \ldots, Z^k, X - Y). \end{aligned} \tag{3.41} $$

As with Theorem 2.10, it is a consequence of the following more general result.

THEOREM 3.20. *Let $f \in \mathcal{T}^k(\Omega^{(0)}, \ldots, \Omega^{(k)}; \mathcal{N}_{0,\mathrm{nc}}, \ldots, \mathcal{N}_{k,\mathrm{nc}})$, where $\Omega^{(k)} \subseteq \mathcal{M}_{k,\mathrm{nc}}$ is a right admissible nc set. Then for all $n_0, \ldots, n_{k-1} \in \mathbb{N}$ and arbitrary $X^0 \in \Omega^{(0)}_{n_0}$, ..., $X^{k-1} \in \Omega^{(k-1)}_{n_{k-1}}$, $X \in \Omega^{(k)}_n$, $Y \in \Omega^{(k)}_m$, $Z^1 \in \mathcal{N}_1{}^{n_0 \times n_1}$, ..., $Z^{k-1} \in (\mathcal{N}_{k-1})^{n_{k-2} \times n_{k-1}}$, $Z^k \in \mathcal{N}_k{}^{n_{k-1} \times m}$, $S \in \mathcal{R}^{m \times n}$,*

$$ \begin{aligned} f(X^0, \ldots, X^{k-1}, X)(Z^1, \ldots, Z^k S) - f(X^0, \ldots, X^{k-1}, Y)(Z^1, \ldots, Z^k)S \\ = \Delta_R f(X^0, \ldots, X^{k-1}, Y, X)(Z^1, \ldots, Z^k, SX - YS). \end{aligned} \tag{3.42} $$

PROOF. By (3.24), we may assume, without loss of generality, that $\Omega^{(k)}$ is similarity invariant, i.e., $\widetilde{\Omega}^{(k)} = \Omega^{(k)}$; see Remark 2.5 and Appendix A. By Proposition A.2,

$$ \begin{bmatrix} Y & SX - YS \\ 0 & X \end{bmatrix} \in \Omega^{(k)}_{m+n}. $$

We have

$$\begin{bmatrix} Y & SX - YS \\ 0 & X \end{bmatrix} \begin{bmatrix} S \\ I_n \end{bmatrix} = \begin{bmatrix} S \\ I_n \end{bmatrix} X.$$

By ($3X^k$),

$$f\left(X^0, \dots, X^{k-1}, \begin{bmatrix} Y & SX - YS \\ 0 & X \end{bmatrix}\right) (Z^1, \dots, Z^{k-1}, \operatorname{row}[Z^k, 0]) \begin{bmatrix} S \\ I_n \end{bmatrix}$$
$$= f(X^0, \dots, X^{k-1}, X)(Z^1, \dots, Z^{k-1}, Z^k S).$$

By Proposition 3.7, the left-hand side is equal to

$$\begin{aligned}
&\operatorname{row}[f(X^0, \dots, X^{k-1}, Y)(Z^1, \dots, Z^k),\\
&\qquad \Delta_R f(X^0, \dots, X^{k-1}, Y, X)(Z^1, \dots, Z^k, SX - YS)] \operatorname{col}[S, I_n]\\
&= f(X^0, \dots, X^{k-1}, Y)(Z^1, \dots, Z^k) S\\
&\qquad + \Delta_R f(X^0, \dots, X^{k-1}, Y, X)(Z^1, \dots, Z^k, SX - YS),
\end{aligned}$$

hence (3.42) follows. □

REMARK 3.21. Let $\mathcal{R} = \mathbb{R}$ or $\mathcal{R} = \mathbb{C}$. Setting $X = Y + tZ$ (with $t \in \mathbb{R}$ or $t \in \mathbb{C}$) in Theorem 3.19, we obtain

$$\begin{aligned}
&f(X^0, \dots, X^{k-1}, Y + tZ)(Z^1, \dots, Z^k) - f(X^0, \dots, X^{k-1}, Y)(Z^1, \dots, Z^k)\\
&\qquad = t\Delta_R f(X^0, \dots, X^{k-1}, Y, Y + tZ)(Z^1, \dots, Z^k, Z).
\end{aligned}$$

Under appropriate continuity conditions, it follows that

$$\Delta_R f(X^0, \dots, X^{k-1}, Y, Y)(Z^1, \dots, Z^k, Z)$$

is the directional derivative of

$$f(X^0, \dots, X^{k-1}, \cdot)(Z^1, \dots, Z^k)$$

at $Y$ in the direction $Z$, i.e.,

$$\Delta_R f(X^0, \dots, X^{k-1}, Y, Y)(Z^1, \dots, Z^k, \cdot)$$

is the differential of

$$f(X^0, \dots, X^{k-1}, \cdot)(Z^1, \dots, Z^k)$$

at $Y$.

REMARK 3.22. Using the difference-differential operators ${}_j\Delta_R$ (see Remark 3.18), we can establish first order difference formulae for higher order nc functions (cf. Theorem 3.19) with respect to the $j$-th variable: for $j = 0, \dots, k$,

$$\begin{aligned}
(3.43)\quad &f(X^0, \dots, X^{j-1}, X, X^{j+1}, \dots, X^k)(Z^1, \dots, Z^k)\\
&\qquad - f(X^0, \dots, X^{j-1}, Y, X^{j+1}, \dots, X^k)(Z^1, \dots, Z^k)\\
&= {}_j\Delta_R f(X^0, \dots, X^{j-1}, Y, X, X^{j+1}, \dots, X^k)(Z^1, \dots, Z^j, X - Y, Z^{j+1}, \dots, Z^k)
\end{aligned}$$

and

$$\begin{aligned}
(3.44)\quad &f(X^0, \dots, X^{j-1}, X, X^{j+1}, \dots, X^k)(Z^1, \dots, Z^k)\\
&\qquad - f(X^0, \dots, X^{j-1}, Y, X^{j+1}, \dots, X^k)(Z^1, \dots, Z^k)\\
&= {}_j\Delta_R f(X^0, \dots, X^{j-1}, X, Y, X^{j+1}, \dots, X^k)(Z^1, \dots, Z^j, X - Y, Z^{j+1}, \dots, Z^k).
\end{aligned}$$

More generally (cf. 3.20), we have

$$Sf(X,X^1,\dots,X^k)(Z^1,\dots,Z^k)-f(Y,X^1,\dots,X^k)(SZ^1,Z^2,\dots,Z^k) \\ = {}_0\Delta_R f(Y,X,X^1,\dots,X^k)(SX-YS,Z^1,\dots,Z^k) \tag{3.45}$$

and, for $0<j<k$,

$$f(X^0,\dots,X^{j-1},X,X^{j+1},\dots,X^k)(Z^1,\dots,Z^{j-1},Z^jS,Z^{j+1},\dots,Z^k) \\ -f(X^0,\dots,X^{j-1},Y,X^{j+1},\dots,X^k)(Z^1,\dots,Z^j,SZ^{j+1},Z^{j+2},\dots,Z^k) \\ ={}_j\Delta_R f(X^0,\dots,X^{j-1},Y,X,X^{j+1},\dots,X^k)(Z^1,\dots,Z^j,SX-YS,Z^{j+1},\dots,Z^k). \tag{3.46}$$

The proof is similar to the proof of Theorem 3.20, except that we use ($3\mathrm{X}^0$) together with (3.35) (in the case of (3.45)) or ($3\mathrm{X}^j$) together with (3.36) (in the case of (3.46)) instead of using ($3\mathrm{X}^k$) together with (3.19). We can also use the similarity invariant envelope $\widetilde{\Omega}^{(j)}$ of the corresponding right admissible nc set $\Omega^{(j)}$, and the equality (3.37).

REMARK 3.23. Clearly, the formulae (3.45), (3.46), and (3.42) generalize ($3\mathrm{X}^0$), ($3\mathrm{X}^j$), and ($3\mathrm{X}^k$), respectively.

## 3.4. NC integrability

We derive now necessary conditions for a higher order nc function to be integrable, i.e., to be the result of applying a higher order difference-differential operator to a nc function (or to a higher order nc function of a lower order).

THEOREM 3.24. *Let $f\in\mathcal{T}^k(\Omega;\mathcal{N}_{\rm nc},\mathcal{M}_{\rm nc},\dots,\mathcal{M}_{\rm nc})$, with $\Omega\subseteq\mathcal{M}_{\rm nc}$ a right admissible nc set. If there exists a nc function $g\in\mathcal{T}^0(\Omega;\mathcal{N}_{\rm nc})$ such that $f=\Delta_R^k g$, then*

$${}_i\Delta_R f={}_j\Delta_R f,\qquad i,j=0,\dots,k. \tag{3.47}$$

COROLLARY 3.25. *For every $g\in\mathcal{T}^0(\Omega;\mathcal{N}_{\rm nc})$, $\ell\in\mathbb{N}$, and $j_s\in\mathbb{N}$ with $0\le j_s\le s$, $s=1,\dots,\ell$, one has*

$${}_{j_\ell}\Delta_R\cdots{}_{j_1}\Delta_R g=\Delta_R^\ell g. \tag{3.48}$$

PROOF OF THEOREM 3.24. It clearly suffices to show that ${}_j\Delta_R\Delta_R^k g=\Delta_R^{k+1}g$ for $0\le j<k$. We prove this for $0<j<k$ (the proof for the case where $j=0$ is analogous, using (3.35) instead of (3.36)). By (3.37), we may assume, without loss of generality, that $\Omega$ is similarity invariant, i.e., $\widetilde{\Omega}=\Omega$; see Remark 2.5 and Appendix A. Let $X^0\in\Omega_{n_0}$, …, $X^{j-1}\in\Omega_{n_{j-1}}$, $X^{j\prime}\in\Omega_{n'_j}$, $X^{j\prime\prime}\in\Omega_{n''_j}$, $X^{j+1}\in\Omega_{n_{j+1}}$, …, $X^k\in\Omega_{n_k}$, $Z^1\in\mathcal{M}^{n_0\times n_1}$, …, $Z^{j-1}\in\mathcal{M}^{n_{j-2}\times n_{j-1}}$, $Z^{j\prime}\in\mathcal{M}^{n_{j-1}\times n'_j}$, $Z\in\mathcal{M}^{n'_j\times n''_j}$, $Z^{j+1\prime\prime}\in\mathcal{M}^{n''_j\times n_{j+1}}$, $Z^{j+2}\in\mathcal{M}^{n_{j+1}\times n_{j+2}}$, …, $Z^k\in\mathcal{M}^{n_{k-1}\times n_k}$. By Proposition A.2, $\begin{bmatrix}X^{j\prime}&Z\\0&X^{j\prime\prime}\end{bmatrix}\in\Omega_{n'_j+n''_j}$. According to (3.36),

$${}_j\Delta_R\Delta_R^k g(X^0,\dots,X^{j-1},X^{j\prime},X^{j\prime\prime},X^{j+1},\dots,X^k) \\ (Z^1,\dots,Z^{j-1},Z^{j\prime},Z,Z^{j+1\prime\prime},Z^{j+2},\dots,Z^k) \\ =\Delta_R^k g(X^0,\dots,X^{j-1},\begin{bmatrix}X^{j\prime}&Z\\0&X^{j\prime\prime}\end{bmatrix},X^{j+1},\dots,X^k) \\ (Z^1,\dots,Z^{j-1},\operatorname{row}[Z^{j\prime},0],\operatorname{col}[0,Z^{j+1\prime\prime}],Z^{j+2},\dots,Z^k). \tag{3.49}$$

By (3.29), this is equal to the $(1, k+1)$ block entry of the $(k+1) \times (k+1)$ block matrix

$$f\left(\begin{bmatrix} X^0 & Z^1 & 0 & \cdots & \cdots & \cdots & \cdots & \cdots & 0 \\ 0 & X^1 & \ddots & \ddots & & & & & \vdots \\ \vdots & \ddots & \ddots & Z^{j-1} & \ddots & & & & \vdots \\ \vdots & & \ddots & X^{j-1} & \begin{bmatrix} Z^{j\prime} & 0 \end{bmatrix} & \ddots & & & \vdots \\ \vdots & & & \ddots & \begin{bmatrix} X^{j\prime} & Z \\ 0 & X^{j\prime\prime} \end{bmatrix} & \begin{bmatrix} 0 \\ Z^{j+1\prime\prime} \end{bmatrix} & \ddots & & \vdots \\ \vdots & & & & \ddots & X^{j+1} & Z^{j+2} & \ddots & \vdots \\ \vdots & & & & & \ddots & \ddots & \ddots & 0 \\ \vdots & & & & & & \ddots & X^{k-1} & Z^k \\ 0 & \cdots & \cdots & \cdots & \cdots & \cdots & \cdots & 0 & X^k \end{bmatrix}\right),$$

where the evaluation of $f$ at the block bidiagonal matrix above is well defined, since by Lemma 3.14 that matrix belongs to $\Omega_{n_0+\cdots+n_{j-1}+n_j'+n_j''+n_{j+1}+\cdots+n_k}$. We consider $\begin{bmatrix} Z^{j\prime} & 0 \end{bmatrix}$, $\begin{bmatrix} X^{j\prime} & Z \\ 0 & X^{j\prime\prime} \end{bmatrix}$, and $\begin{bmatrix} 0 \\ Z^{j+1\prime\prime} \end{bmatrix}$ as single blocks. If we view this block matrix as a $(k+2) \times (k+2)$ block matrix, removing the brackets from those blocks, then the expression in (3.49) becomes the $(1, k+2)$ block entry, which by (3.29) is equal to

$$\Delta_R^{k+1} g(X^0, \dots, X^{j-1}, X^{j\prime}, X^{j\prime\prime}, X^{j+1}, \dots, X^k)$$
$$(Z^1, \dots, Z^{j-1}, Z^{j\prime}, Z, Z^{j+1\prime\prime}, Z^{j+2}, \dots, Z^k).$$

□

## 3.5. Higher order directional nc difference-differential operators

Similarly to Section 2.6, we will also consider a directional version of higher order difference-differential operators. It is introduced by iterating directional nc difference-differential operators on higher order nc functions.

Let $\Omega^{(j)} \subseteq \mathcal{M}_{j,\text{nc}}$, $j = 0, \dots, k$, be nc sets, and let $\Omega^{(k)} \subseteq \mathcal{M}_{k,\text{nc}}$ be right admissible. Let $f \in \mathcal{T}^k(\Omega^{(0)}, \dots, \Omega^{(k)}; \mathcal{N}_{0,\text{nc}}, \dots, \mathcal{N}_{k,\text{nc}})$. Given $\mu \in \mathcal{M}_k$, we define for all $n_0, \dots, n_{k+1} \in \mathbb{N}$, $X^0 \in \Omega^{(0)}_{n_0}, \dots, X^k \in \Omega^{(k)}$, $X^{k+1} \in \Omega^{(k)}_{n_{k+1}}$ the $(k+1)$-linear mapping

$$\Delta_{R,\mu} f(X^0, \dots, X^{k+1}) \colon \mathcal{N}_1^{n_0 \times n_1} \times \cdots \times \mathcal{N}_k^{n_{k-1} \times n_k} \times \mathcal{R}^{n_k \times n_{k+1}} \longrightarrow \mathcal{N}_0^{n_0 \times n_{k+1}}$$

by

$$\text{(3.50)} \quad \Delta_{R,\mu} f(X^0, \dots, X^{k+1})(Z^1, \dots, Z^k, A)$$
$$= \Delta_R f(X^0, \dots, X^{k+1})(Z^1, \dots, Z^k, A\mu).$$

Furthermore, $\Delta_{R,\mu} f \in \mathcal{T}^{k+1}(\Omega^{(0)}, \dots, \Omega^{(k)}, \Omega^{(k)}; \mathcal{N}_{0,\text{nc}}, \dots, \mathcal{N}_{k,\text{nc}}, \mathcal{R}_{\text{nc}})$.

In the special case of $\mathcal{M}_k = \mathcal{R}^d$, we define the *$j$-th right partial nc difference-differential operator* by $\Delta_{R,j} = \Delta_{R,e_j}$. By linearity, we have

$$\begin{aligned}(3.51)\quad \Delta_R f(X^0, \ldots, X^{k+1})&(Z^1, \ldots, Z^{k+1}) \\ &= \sum_{j=1}^{d} \Delta_{R,j} f(X^0, \ldots, X^{k+1})(Z^1, \ldots, Z^k, Z_j^{k+1}).\end{aligned}$$

It follows that the first order difference formulae of Theorem 3.19 can be written in this case in terms of partial nc difference-differential operators:

$$\begin{aligned}&f(X^0, \ldots, X^{k-1}, X)(Z^1, \ldots, Z^k) - f(X^0, \ldots, X^{k-1}, Y)(Z^1, \ldots, Z^k) \\ &= \sum_{j=1}^{d} \Delta_{R,j} f(X^0, \ldots, X^{k-1}, Y, X)(Z^1, \ldots, Z^k, X_j - Y_j) \\ &= \sum_{j=1}^{d} \Delta_{R,j} f(X^0, \ldots, X^{k-1}, X, Y)(Z^1, \ldots, Z^k, X_j - Y_j).\end{aligned}$$

Given $\mu^1, \ldots, \mu^\ell \in \mathcal{M}_k$, we define the corresponding *$\ell$-th order directional nc difference-differential operators*

$$\begin{aligned}\Delta^\ell_{R,\mu^1,\ldots,\mu^\ell} &= \Delta_{R,\mu^\ell} \cdots \Delta_{R,\mu^1} \colon\ \mathcal{T}^k(\Omega^{(0)}, \ldots, \Omega^{(k)}; \mathcal{N}_{0,\mathrm{nc}}, \ldots, \mathcal{N}_{k,\mathrm{nc}}) \\ &\longrightarrow \mathcal{T}^{k+\ell}(\Omega^{(0)}, \ldots, \Omega^{(k)}, \underbrace{\Omega^{(k)}, \ldots, \Omega^{(k)}}_{\ell \text{ times}}; \mathcal{N}_{0,\mathrm{nc}}, \ldots, \mathcal{N}_{k,\mathrm{nc}}, \underbrace{\mathcal{R}_{\mathrm{nc}}, \ldots, \mathcal{R}_{\mathrm{nc}}}_{\ell \text{ times}}).\end{aligned}$$

More explicitly, using (3.21) and induction on $\ell$, one obtains

$$\begin{aligned}\Delta^\ell_{R,\mu^1,\ldots,\mu^\ell} f(X^0, \ldots, X^{k+\ell})&(Z^1, \ldots, Z^k, A^1, \ldots, A^\ell) \\ &= \Delta^\ell_R f(X^0, \ldots, X^{k+\ell})(Z^1, \ldots, Z^k, A^1\mu^1, \ldots, A^\ell\mu^\ell).\end{aligned}$$

In the tensor product interpretation of the values of nc functions (see Remark 3.5), we have

$$\begin{aligned}&\Delta^\ell_R f(X^0, \ldots, X^{k+\ell}) \\ &\in \mathcal{N}_0^{n_0 \times n_0} \otimes \mathcal{N}_1^{*n_1 \times n_1} \otimes \cdots \otimes \mathcal{N}_k^{*n_k \times n_k} \otimes \mathcal{M}_k^{*n_{k+1} \times n_{k+1}} \otimes \cdots \otimes \mathcal{M}_k^{*n_{k+\ell} \times n_{k+\ell}},\end{aligned}$$

$$\begin{aligned}&\Delta^\ell_{R,\mu^1,\ldots,\mu^\ell} f(X^0, \ldots, X^{k+\ell}) \\ &\quad \in \mathcal{N}_0^{n_0 \times n_0} \otimes \mathcal{N}_1^{*n_1 \times n_1} \otimes \cdots \otimes \mathcal{N}_k^{*n_k \times n_k} \otimes \mathcal{R}^{n_{k+1} \times n_{k+1}} \otimes \cdots \otimes \mathcal{R}^{n_{k+\ell} \times n_{k+\ell}},\end{aligned}$$

and

$$\begin{aligned}&\Delta^\ell_{R,\mu^1,\ldots,\mu^\ell} f(X^0, \ldots, X^{k+\ell}) \\ &\qquad = (I_{n_0} \otimes \cdots \otimes I_{n_k} \otimes I_{n_{k+1}}\mu^1 \otimes \cdots \otimes I_{n_{k+\ell}}\mu^\ell)\Delta^\ell_R f(X^0, \ldots, X^{k+\ell}).\end{aligned}$$

Setting $Z^1 = A^1\mu^1, \dots, Z^\ell = A^\ell\mu^\ell$ in Theorem 3.11, we can rewrite (3.29) as

$$(3.52)\quad f\left(\begin{bmatrix} X^0 & A^1\mu^1 & 0 & \cdots & 0 \\ 0 & X^1 & \ddots & \ddots & \vdots \\ \vdots & \ddots & \ddots & \ddots & 0 \\ \vdots & & \ddots & X^{\ell-1} & A^\ell\mu^\ell \\ 0 & \cdots & \cdots & 0 & X^\ell \end{bmatrix}\right)$$

$$= \begin{bmatrix} f(X^0) & \Delta_{R,\mu^1} f(X^0, X^1)(A^1) & \cdots & \cdots & \Delta^\ell_{R,\mu^1,\dots,\mu^\ell} f(X^0, \dots, X^\ell)(A^1, \dots, A^\ell) \\ 0 & f(X^1) & \ddots & & \vdots \\ \vdots & \ddots & \ddots & \ddots & \vdots \\ \vdots & & \ddots & \ddots & \Delta_{R,\mu^\ell} f(X^{\ell-1}, X^\ell)(A^\ell) \\ 0 & \cdots & \cdots & 0 & f(X^\ell) \end{bmatrix}.$$

There is a similar version of Theorem 3.12 in terms of higher order directional nc difference-differential operators.

In the special case where $\mathcal{M} = \mathcal{R}^d$, we define the *$\ell$-th order partial nc difference-differential operator* corresponding to a word $w = g_{i_1} \cdots g_{i_\ell} \in \mathcal{G}_d$ by

$$\Delta_R^w = \Delta^\ell_{R,e_{i_1},\dots,e_{i_\ell}} = \Delta_{R,i_\ell} \cdots \Delta_{R,i_1};$$

see Section 1.1 for the definition of the free monoid $\mathcal{G}_d$ and nc multipowers. (Notice that we can interpret $\Delta_R^w$ as a nc multipower if we abuse the notation by letting $\Delta_R$ denote the $d$-tuple $(\Delta_{R,1}, \dots, \Delta_{R,d})$ of first order partial difference-differential operators as well as the first order full difference-differential operator.)

By multilinearity we have, for $f \in \mathcal{T}^k$,

$$(3.53)\quad \Delta_R^\ell f(X^0, \dots, X^{k+\ell})(Z^1, \dots, Z^{k+\ell})$$
$$= \sum_{w=g_{i_1}\cdots g_{i_\ell}} \Delta_R^{w^\top} f(X^0, \dots, X^{k+\ell})(Z^1, \dots, Z^k, Z^{k+1}_{i_1}, \dots, Z^{k+\ell}_{i_\ell}),$$

where the summation is over all the words of length $\ell$ (notice the transposition of words: for $w = g_{i_1} \cdots g_{i_\ell}$ we use the notation $w^\top = g_{i_\ell} \cdots g_{i_1}$).

Setting $Z^1 = A^1 e_{i_1}, \dots, Z^\ell = A^\ell e_{i_\ell}$ in Theorem 3.11, i.e., choosing $\mu^1 = e_{i_1}, \dots, \mu^\ell = e_{i_\ell}$ in (3.52), we obtain

$$(3.54)\quad f\left(\begin{bmatrix} X^0 & A^1 e_{i_1} & 0 & \cdots & 0 \\ 0 & X^1 & \ddots & \ddots & \vdots \\ \vdots & \ddots & \ddots & \ddots & 0 \\ \vdots & & \ddots & X^{\ell-1} & A^\ell e_{i_\ell} \\ 0 & \cdots & \cdots & 0 & X^\ell \end{bmatrix}\right)$$

$$= \begin{bmatrix} f(X^0) & \Delta_R^{g_{i_1}} f(X^0, X^1)(A^1) & \cdots & \cdots & \Delta_R^{g_{i_\ell}\cdots g_{i_1}} f(X^0, \dots, X^\ell)(A^1, \dots, A^\ell) \\ 0 & f(X^1) & \ddots & & \vdots \\ \vdots & \ddots & \ddots & \ddots & \vdots \\ \vdots & & \ddots & \ddots & \Delta_R^{g_{i_\ell}} f(X^{\ell-1}, X^\ell)(A^\ell) \\ 0 & \cdots & \cdots & 0 & f(X^\ell) \end{bmatrix}.$$

CHAPTER 4

# The Taylor–Taylor formula

We use now the calculus of higher order nc difference-differential operators to derive a nc analogue of the Brook Taylor expansion, which we call the TT (Taylor–Taylor) expansion in honour of Brook Taylor and Joseph L. Taylor.

THEOREM 4.1. *Let $f \in \mathcal{T}^0(\Omega; \mathcal{N}_{\rm nc})$ with $\Omega \subseteq \mathcal{M}_{\rm nc}$ a right admissible nc set, $n \in \mathbb{N}$, and $Y \in \Omega_n$. Then for each $N \in \mathbb{N}$ and arbitrary $X \in \Omega_n$,*

$$
\begin{aligned}
(4.1)\quad f(X) = \sum_{\ell=0}^{N} \Delta_R^\ell f(\underbrace{Y, \ldots, Y}_{\ell+1 \text{ times}})(\underbrace{X - Y, \ldots, X - Y}_{\ell \text{ times}}) \\
+ \Delta_R^{N+1} f(\underbrace{Y, \ldots, Y}_{N+1 \text{ times}}, X)(\underbrace{X - Y, \ldots, X - Y}_{N+1 \text{ times}}).
\end{aligned}
$$

PROOF. Using Theorems 2.10 and 3.19, we obtain

$$
\begin{aligned}
f(X) &= f(Y) + \Delta_R f(Y, X)(X - Y) \\
&= f(Y) + \Delta_R f(Y, Y)(X - Y) + \Delta_R^2 f(Y, Y, X)(X - Y, X - Y) \\
&= \cdots \\
&= f(Y) + \Delta_R f(Y, Y)(X - Y) + \cdots + \Delta_R^N f(\underbrace{Y, \ldots, Y}_{N+1 \text{ times}})(\underbrace{X - Y, \ldots, X - Y}_{N \text{ times}}) \\
&\quad + \Delta_R^{N+1} f(\underbrace{Y, \ldots, Y}_{N+1 \text{ times}}, X)(\underbrace{X - Y, \ldots, X - Y}_{N+1 \text{ times}}).
\end{aligned}
$$

□

It is possible to obtain a version of a power expansion centered at a matrix $Y \in \Omega_s$ valid in matrix dimensions which are multiples of $s$. For $X \in \mathcal{M}^{n \times n}$ and $n = sm$ we denote

$$X^{\odot_s \ell} := \underbrace{X \odot_s \cdots \odot_s X}_{\ell \text{ times}},$$

where we use the notation introduced prior to Proposition 3.3. Notice that $X^{\odot_s \ell}$ is the $\ell$-th power of $X$ viewed as a $m \times m$ matrix over $\mathbf{T}(\mathcal{M}^{s \times s})$. Here

$$\mathbf{T}(\mathcal{L}) := \bigoplus_{\ell=0}^{\infty} \mathcal{L}^{\otimes \ell}$$

denotes the tensor algebra over a module $\mathcal{L}$ over $\mathcal{R}$. We also write $X^{\odot \ell}$ instead of $X^{\odot_1 \ell}$, thus omitting the subscript in case $s = 1$.

THEOREM 4.2. *Let $f \in \mathcal{T}^0(\Omega;\mathcal{N}_{\rm nc})$ with $\Omega \subseteq \mathcal{M}_{\rm nc}$ a right admissible nc set, and $Y \in \Omega_s$. Then for each $N \in \mathbb{N}$ and arbitrary $m \in \mathbb{N}$ and $X \in \Omega_{ms}$,*

$$f(X) = \sum_{\ell=0}^{N} \Big(X - \bigoplus_{\alpha=1}^{m} Y\Big)^{\odot_s \ell} \Delta_R^{\ell} f(\underbrace{Y,\dots,Y}_{\ell+1 \text{ times}}) + \Big(X - \bigoplus_{\alpha=1}^{m} Y\Big)^{\odot_s N+1} \Delta_R^{N+1} f(\underbrace{Y,\dots,Y}_{N+1 \text{ times}}, X). \tag{4.2}$$

*In particular, if $\mu \in \Omega_1$ then for each $N \in \mathbb{N}$ and arbitrary $m \in \mathbb{N}$ and $X \in \Omega_m$,*

$$f(X) = \sum_{\ell=0}^{N} (X - I_m\mu)^{\odot \ell} \Delta_R^{\ell} f(\underbrace{\mu,\dots,\mu}_{\ell+1 \text{ times}}) + (X - I_m\mu)^{\odot N+1} \Delta_R^{N+1} f(\underbrace{\mu,\dots,\mu}_{N+1 \text{ times}}, X). \tag{4.3}$$

PROOF. The first statement is immediate from Theorem 4.1 and Proposition 3.3. The second statement is a special case of the first one when $s = 1$. □

REMARK 4.3. It follows from (3.45), (3.46), (3.42), and (3.48) that the multilinear mappings $f_\ell := \Delta_R^\ell f(Y,\dots,Y)\colon (\mathcal{M}^{s\times s})^\ell \to \mathcal{N}^{s\times s}$, satisfy

$$Sf_0 - f_0 S = f_1(SY - YS), \tag{4.4}$$

and for $\ell = 1, \dots,$

$$Sf_\ell(Z^1,\dots,Z^\ell) - f_\ell(SZ^1, Z^2,\dots,Z^\ell) = f_{\ell+1}(SY - YS, Z^1,\dots,Z^\ell), \tag{4.5}$$

$$f_\ell(Z^1,\dots,Z^{j-1}, Z^j S, Z^{j+1},\dots,Z^\ell) - f_\ell(Z^1,\dots,Z^j, SZ^{j+1}, Z^{j+2},\dots,Z^\ell) = f_{\ell+1}(Z^1,\dots,Z^j, SY - YS, Z^{j+1},\dots,Z^\ell), \tag{4.6}$$

$$f_\ell(Z^1,\dots,Z^{\ell-1}, Z^\ell S) - f_\ell(Z^1,\dots,Z^\ell)S = f_{\ell+1}(Z^1,\dots,Z^\ell, SY - YS), \tag{4.7}$$

for every $S \in \mathcal{R}^{s\times s}$.

We next specialize Theorem 4.2 to the case where $\mathcal{M} = \mathcal{R}^d$, using higher order partial nc difference-differential operators. For $X \in (\mathcal{R}^{n\times n})^d$, $n = sm$, and $w = g_{i_1} \cdots g_{i_\ell}$ we denote

$$X^{\odot_s w} := X_{i_1} \odot_s \cdots \odot_s X_{i_\ell}. \tag{4.8}$$

Notice that $X^{\odot_s w}$ is the $w$-th nc power of $X$ viewed as a $d$-tuple of $m \times m$ matrices over $\mathbf{T}(\mathcal{R}^{s\times s})$. We also write $X^{\odot w}$ instead of $X^{\odot_1 w}$, thus omitting the subscript in case $s = 1$. Notice that

$$Z^{\odot_s \ell} = \sum_{w = g_1 \cdots g_{i_\ell}} (Z^{\odot_s w}) e_{i_1} \otimes \cdots \otimes e_{i_\ell}, \tag{4.9}$$

for every $\ell, m \in \mathbb{N}$, and $Z \in (\mathcal{R}^{sm\times sm})^d$, where $e_i$, $i = 1,\dots,d$, is the standard basis for $\mathcal{R}^d$. One also has

$$\Delta_R^\ell f(Y,\dots,Y)(Z^1,\dots,Z^\ell) = \sum_{w = g_{i_1}\cdots g_{i_\ell}} \Delta_R^{w^\top} f(Y,\dots,Y)(Z^1_{i_1},\dots,Z^\ell_{i_\ell}) \tag{4.10}$$

(see (3.53) with $k = 0$) and, similarly,

$$Z^{\odot_s \ell}\Delta_R^\ell f(Y,\dots,Y) = \sum_{w=g_1\cdots g_{i_\ell}} Z^{\odot_s w}\Delta_R^{w^\top} f(Y,\dots,Y) \tag{4.11}$$

for a nc function $f$ on a nc set $\Omega \subseteq (\mathcal{R}^d)_{\rm nc}$ with values in $\mathcal{N}_{\rm nc}$, and every $Y \in \Omega_s$, $Z^1, \dots, Z^\ell \in (\mathcal{R}^{s\times s})^d$, and $Z \in (\mathcal{R}^{sm\times sm})^d$. Theorem 4.2 implies the following.

COROLLARY 4.4. *Let $f \in \mathcal{T}^0(\Omega;\mathcal{N}_{\rm nc})$ with $\Omega \subseteq (\mathcal{R}^d)_{\rm nc}$ a right admissible nc set, $m \in \mathbb{N}$, and $Y \in \Omega_s$. Then for each $N \in \mathbb{N}$ and arbitrary $X \in \Omega_{sm}$,*

$$\begin{aligned} f(X) = \sum_{\ell=0}^N \sum_{|w|=\ell} \Big(X - \bigoplus_{\alpha=1}^m Y\Big)^{\odot_s w} \Delta_R^{w^\top} f(\underbrace{Y,\dots,Y}_{\ell+1 \text{ times}}) \\ + \sum_{|w|=N+1} \Big(X - \bigoplus_{\alpha=1}^m Y\Big)^{\odot_s w} \Delta_R^{w^\top} f(\underbrace{Y,\dots,Y}_{N+1 \text{ times}}, X). \end{aligned} \tag{4.12}$$

*In particular, for $s = 1$ and $\mu \in \Omega_1$, we obtain a genuine nc power expansion*

$$\begin{aligned} f(X) = \sum_{\ell=0}^N \sum_{|w|=\ell} (X - I_m\mu)^w \Delta_R^{w^\top} f(\underbrace{\mu,\dots,\mu}_{\ell+1 \text{ times}}) \\ + \sum_{|w|=N+1} (X - I_m\mu)^w \Delta_R^{w^\top} f(\underbrace{\mu,\dots,\mu}_{N+1 \text{ times}}, X). \end{aligned} \tag{4.13}$$

*Here we identify the $\ell$-linear mapping $\Delta_R^{w^\top} f(\mu,\dots,\mu)\colon \mathcal{R}\times\cdots\times\mathcal{R} \to \mathcal{N}$ with the vector $\Delta_R^{w^\top} f(\mu,\dots,\mu)(1,\dots,1) \in \mathcal{N}$, and the $m\times m$ matrix $\Delta_R^{w^\top} f(\mu,\dots,\mu,X)$ of $(N+1)$-linear mappings $\mathcal{R}\times\cdots\times\mathcal{R}\to\mathcal{N}$ (see formula (3.9) in Proposition 3.3) with the $m\times m$ matrix over $\mathcal{N}$ whose $i$-th row is equal to $\Delta_R^{w^\top} f(\mu,\dots,\mu,X)(1,\dots,1,e_i^\top)$ where $e_i$ is the $i$-th standard basis vector of $\mathcal{R}^m$.*

REMARK 4.5. Specializing Remark 4.3 to the case of $\mathcal{M} = \mathcal{R}^d$ and using (4.10), we obtain that the multilinear forms $f_w = \Delta_R^{w^\top} f(Y,\dots,Y)$, $w \in \mathcal{G}_d$, satisfy the conditions

$$Sf_\emptyset - f_\emptyset S = \sum_{k=1}^d f_{g_k}(SY_k - Y_kS), \tag{4.14}$$

and for $w = g_{i_1}\cdots g_{i_\ell} \neq \emptyset$,

$$Sf_w(A^1,\dots,A^\ell) - f_w(SA^1,A^2,\dots,A^\ell) = \sum_{k=1}^d f_{g_kw}(SY_k - Y_kS, A^1,\dots,A^\ell), \tag{4.15}$$

$$\begin{aligned} f_w(A^1,\dots,A^{j-1},A^jS,A^{j+1},\dots,A^\ell) - f_w(A^1,\dots,A^j,SA^{j+1},A^{j+2},\dots,A^\ell) \\ = \sum_{k=1}^d f_{g_{i_1}\cdots g_{i_j}g_kg_{i_{j+1}}\cdots g_{i_\ell}}(A^1,\dots,A^j,SY_k - Y_kS,A^{j+1},\dots,A^\ell), \end{aligned} \tag{4.16}$$

$$(4.17)\quad f_w(A^1,\dots,A^{\ell-1},A^\ell S)-f_w(A^1,\dots,A^\ell)S = \sum_{k=1}^{d} f_{wg_k}(A^1,\dots,A^\ell,SY_k-Y_kS),$$

for all $S,A^1,\dots,A^\ell\in\mathcal{R}^{s\times s}$.

We proceed to rewrite the TT formula (4.12) in a more concrete fashion. Using the tensor product interpretation for the values of higher order nc functions (see Remark 3.5 — we assume that the module $\mathcal{N}$ is free),

$$\Delta_R^{w^\top} f(\underbrace{Y,\dots,Y}_{\ell+1\text{ times}})\in\mathcal{N}^{s\times s}\otimes\underbrace{\mathcal{R}^{s\times s}\otimes\cdots\otimes\mathcal{R}^{s\times s}}_{\ell\text{ times}}$$

and therefore

$$(4.18)\qquad \Delta_R^{w^\top} f(\underbrace{Y,\dots,Y}_{\ell+1\text{ times}}) = A_{w,(0)}\otimes A_{w,(1)}\otimes\cdots\otimes A_{w,(\ell)}.$$

Here we use an analogue of the *sumless Sweedler notation* familiar in the theory of coalgebras [**60**],

$$(4.19)\qquad A_{w,(0)}\otimes A_{w,(1)}\otimes\cdots\otimes A_{w,(\ell)} = \sum_j A_{w,(0)}^{(j)}\otimes A_{w,(1)}^{(j)}\otimes\cdots\otimes A_{w,(\ell)}^{(j)},$$

with a finite sum (having at most $(s^2)^\ell$ terms) and $A_{w,(0)}^{(j)}\in\mathcal{N}^{s\times s}$, $A_{w,(1)}^{(j)}$, $\dots$, $A_{w,(\ell)}^{(j)}\in\mathcal{R}^{s\times s}$. According to (3.11),

$$(4.20)\quad \Delta_R^{w^\top} f(\underbrace{Y,\dots,Y}_{\ell+1\text{ times}})(X_{i_1}-Y_{i_1},\dots,X_{i_\ell}-Y_{i_\ell}) = (A_{w,(0)}\otimes A_{w,(1)}\otimes\cdots\otimes A_{w,(\ell)})\star(X-Y)^{[w]},$$

where we introduce the notation

$$(4.21)\quad (C_{w,(0)}\otimes C_{w,(1)}\otimes\cdots\otimes C_{w,(\ell)})\star Z^{[w]} = \sum_j C_{w,(0)}^{(j)}\left(Z_{i_1}C_{w,(1)}^{(j)}\right)\cdots\left(Z_{i_\ell}C_{w,(\ell)}^{(j)}\right),$$

where $C_{w,(0)}\otimes C_{w,(1)}\otimes\cdots\otimes C_{w,(\ell)}\in\mathcal{N}^{s\times s}\otimes\underbrace{\mathcal{R}^{s\times s}\otimes\cdots\otimes\mathcal{R}^{s\times s}}_{\ell\text{ times}}$ in the sumless Sweedler notation, and $Z\in(\mathcal{R}^{n\times n})^d$. Recall that for $Z\in(\mathcal{R}^{sm\times sm})^d$, $w=g_{i_1}\cdots g_{i_\ell}\in\boldsymbol{\mathcal{G}}_d$, and $1\le i,k\le m$ we have

$$(Z^{\odot_s w})_{ik} = \sum_{1\le j_1,\dots,j_{\ell-1}\le m} (Z_{i_1})_{ij_1}\otimes(Z_{i_2})_{j_1j_2}\otimes\cdots\otimes(Z_{i_{\ell-1}})_{j_{\ell-2}j_{\ell-1}}\otimes(Z_{i_\ell})_{j_{\ell-1}k}.$$

Therefore,

$$\Big(Z^{\odot_s w}\Delta_R^{w^\top} f(\underbrace{Y,\dots,Y}_{\ell+1\ \text{times}})\Big)_{ik}$$
$$= \sum_{1\le j_1,\dots,j_{\ell-1}\le m} \Delta_R^{w^\top} f(\underbrace{Y,\dots,Y}_{\ell+1\ \text{times}})\Big((Z_{i_1})_{ii_1},(Z_{i_2})_{j_1j_2},\dots,(Z_{i_{\ell-1}})_{j_{\ell-2}j_{\ell-1}},(Z_{i_\ell})_{j_{\ell-1}k}\Big)$$
$$= \sum_j \sum_{1\le j_1,\dots,j_{\ell-1}\le m} A_{w,(0)}^{(j)}\Big((Z_{i_1})_{ii_1}A_{w,(1)}^{(j)}\Big)\Big((Z_{i_2})_{j_1j_2}A_{w,(2)}^{(j)}\Big)$$
$$\cdots\Big((Z_{i_{\ell-1}})_{j_{\ell-2}j_{\ell-1}}A_{w,(\ell-1)}^{(j)}\Big)\Big((Z_{i_\ell})_{j_{\ell-1}k}A_{w,(\ell)}^{(j)}\Big)$$
$$= \left(\left(\Big(\bigoplus_{\alpha=1}^m A_{w,(0)}\Big)\otimes\cdots\otimes\Big(\bigoplus_{\alpha=1}^m A_{w,(\ell)}\Big)\right)\star Z^{[w]}\right)_{ik}$$

for every $Z\in(\mathcal{R}^{sm\times sm})^d$, so that

$$(4.22)\quad \Big(X-\bigoplus_{\alpha=1}^m Y\Big)^{\odot_s w}\Delta_R^{w^\top} f(\underbrace{Y,\dots,Y}_{\ell+1\ \text{times}})$$
$$= \left(\Big(\bigoplus_{\alpha=1}^m A_{w,(0)}\Big)\otimes\cdots\otimes\Big(\bigoplus_{\alpha=1}^m A_{w,(\ell)}\Big)\right)\star\Big(X-\bigoplus_{\alpha=1}^m Y\Big)^{[w]}$$

for every $X\in(\mathcal{R}^{sm\times sm})^d$.

Analogously, for the remainder of the TT formula (4.12) (the sum over words of length $N+1$) we have

$$\Delta_R^{w^\top} f(\underbrace{Y,\dots,Y}_{N+1\ \text{times}},X)\in\mathcal{N}^{n\times n}\otimes\underbrace{\mathcal{R}^{n\times n}\otimes\cdots\otimes\mathcal{R}^{n\times n}}_{N+1\ \text{times}}$$

and therefore

$$(4.23)\qquad \Delta_R^{w^\top} f(\underbrace{Y,\dots,Y}_{N+1\ \text{times}},X)=B_{w,(0)}\otimes B_{w,(1)}\otimes\cdots\otimes B_{w,(N+1)}$$

and

$$(4.24)\quad \Big(X-\bigoplus_{\alpha=1}^m Y\Big)^{\odot_s w}\Delta_R^{w^\top} f(\underbrace{Y,\dots,Y}_{N+1\ \text{times}},X)$$
$$= \left(\Big(\bigoplus_{\alpha=1}^m B_{w,(0)}\Big)\otimes\cdots\otimes\Big(\bigoplus_{\alpha=1}^m B_{w,(N)}\Big)\otimes B_{w,(N+1)}\right)\star\Big(X-\bigoplus_{\alpha=1}^m Y\Big)^{[w]}$$

for every $X\in(\mathcal{R}^{sm\times sm})^d$. Notice that while the "TT coefficients" $A_{w,(0)}\otimes A_{w,(1)}\otimes\cdots\otimes A_{w,(\ell)}$ depend only on the center point $Y$, the "remainder coefficients" $B_{w,(0)}\otimes B_{w,(1)}\otimes\cdots\otimes B_{w,(N+1)}$ depend on both $Y$ and $X$.

Thus, we obtain the TT formula in the form of "nc pseudo-power" expansion, which is a generalization of nc power expansion (4.13) for the case of $s=1$ to the case of general s.

THEOREM 4.6. *Let $f \in \mathcal{T}^0(\Omega;\mathcal{N}_{\rm nc})$ with $\Omega \subseteq (\mathcal{R}^d)_{\rm nc}$ a right admissible nc set, and $Y \in \Omega_s$. Then for each $N \in \mathbb{N}$, an arbitrary $m \in \mathbb{N}$, and $X \in \Omega_{sm}$,*

$$f(X) = \sum_{\ell=0}^{N} \sum_{|w|=\ell} \left( \Big( \bigoplus_{\alpha=1}^{m} A_{w,(0)} \Big) \otimes \cdots \otimes \Big( \bigoplus_{\alpha=1}^{m} A_{w,(\ell)} \Big) \right) \star \Big( X - \bigoplus_{\alpha=1}^{m} Y \Big)^{[w]} \tag{4.25}$$

$$+ \sum_{|w|=N+1} \left( \Big( \bigoplus_{\alpha=1}^{m} B_{w,(0)} \Big) \otimes \cdots \otimes \Big( \bigoplus_{\alpha=1}^{m} B_{w,(N)} \Big) \otimes B_{w,(N+1)} \right) \star \Big( X - \bigoplus_{\alpha=1}^{m} Y \Big)^{[w]}.$$

*Here $A_{w,(0)} \otimes \cdots \otimes A_{w,(\ell)} = \Delta_R^{w^\top} f(\underbrace{Y,\ldots,Y}_{\ell+1 \text{ times}})$ and $B_{w,(0)} \otimes \cdots \otimes B_{w,(N)} \otimes B_{w,(N+1)} = \Delta_R^{w^\top} f(\underbrace{Y,\ldots,Y}_{N+1 \text{ times}}, X)$ with the tensor product interpretation for the values of $\Delta_R^{w^\top} f$ (Remark* 3.5*), the sumless Sweedler notation* (4.19)*, and the pseudo-power notation* (4.21)*.*

CHAPTER 5

# NC functions on nilpotent matrices

As the first application of the TT formula, we will now describe nc functions on nilpotent matrices as sums of their TT series. Conversely, we will see that the sum of nc power series evaluated on nilpotent matrices defines a nc function, provided the coefficients of the series satisfy the conditions listed in Remarks 4.3 and 4.5.

## 5.1. From nc functions to nc power series

The TT formula (4.3) allows us to describe nc functions on nilpotent matrices over $\mathcal{M}$ as sums of nc power series. Let $n, \kappa \in \mathbb{N}$. Define the set $\mathrm{Nilp}(\mathcal{M}; n, \kappa)$ of $n \times n$ matrices $X$ over $\mathcal{M}$ which are *nilpotent of rank at most* $\kappa$, i.e., $X^{\odot \ell} = 0$ for all $\ell \geq \kappa$. Define also the set $\mathrm{Nilp}(\mathcal{M}; n) = \bigcup_{\kappa=1}^{\infty} \mathrm{Nilp}(\mathcal{M}; n, \kappa)$ of nilpotent $n \times n$ matrices over $\mathcal{M}$ and the set $\mathrm{Nilp}(\mathcal{M}) = \coprod_{n=1}^{\infty} \mathrm{Nilp}(\mathcal{M}; n)$ of nilpotent matrices over $\mathcal{M}$. Notice that $\Omega = \mathrm{Nilp}(\mathcal{M}) \subseteq \mathcal{M}_{\mathrm{nc}}$ is a right (as well as left) admissible nc set, and $\Omega_n = \mathrm{Nilp}(\mathcal{M}; n)$, $n = 1, 2, \ldots$.

In the special case of $\mathcal{M} = \mathcal{R}^d$, the set $\mathrm{Nilp}_d(\mathcal{R}; n, \kappa) := \mathrm{Nilp}(\mathcal{R}^d; n, \kappa)$ consists of $d$-tuples $X$ of $n \times n$ matrices over $\mathcal{R}$ which are *jointly nilpotent of rank at most* $\kappa$, i.e., $X^w = 0$ for all $w \in \mathcal{G}_d$ such that $|w| \geq \kappa$; see (4.9). We also have the set $\mathrm{Nilp}_d(\mathcal{R}; n) := \mathrm{Nilp}(\mathcal{R}^d; n)$ of jointly nilpotent $d$-tuples of $n \times n$ matrices over $\mathcal{R}$ and the set $\mathrm{Nilp}_d(\mathcal{R}) := \mathrm{Nilp}(\mathcal{R}^d)$ of jointly nilpotent $d$-tuples of matrices over $\mathcal{R}$.

REMARK 5.1. In the case where $\mathcal{R} = \mathbb{K}$ is a field, a matrix $X \in \mathrm{Nilp}(\mathcal{M}; n)$ is similar to a strictly upper triangular matrix. Indeed, consider a finite-dimensional subspace of $\mathcal{M}$ spanned by the entries of $X$ and choose its basis $v_1, \ldots, v_d$; we can then write $X = A_1 v_1 + \cdots + A_d v_d$, where $(A_1, \ldots, A_d) \in \mathrm{Nilp}_d(\mathbb{K}; n)$. The Lie subalgebra of $\mathbb{K}^{n \times n}$ generated by $A_1, \ldots, A_d$ consists of nilpotent matrices. By Engel's theorem (see, e.g., [**63**, Sections 3.2, 3.3]), the matrices $A_1, \ldots, A_d$ are jointly similar to strictly upper triangular matrices. Hence, $X$ is similar to a strictly upper triangular matrix.

THEOREM 5.2. *Let $f \in \mathcal{T}^0(\mathrm{Nilp}(\mathcal{M}); \mathcal{N}_{\mathrm{nc}})$. Then for all $X \in \mathrm{Nilp}(\mathcal{M})$*

$$f(X) = \sum_{\ell=0}^{\infty} X^{\odot \ell} \Delta_R^{\ell} f(\underbrace{0, \ldots, 0}_{\ell+1 \text{ times}}), \tag{5.1}$$

*where the sum has finitely many nonzero terms.*

PROOF. Let $X \in \mathrm{Nilp}(\mathcal{M}; n, \kappa)$ for some positive integers $n$ and $\kappa$. Then the TT formula (4.3) centered at $\mu = 0$ for $N > \kappa$ becomes

$$f(X) = \sum_{\ell=0}^{\kappa-1} X^{\odot \ell} \Delta_R^{\ell} f(\underbrace{0, \ldots, 0}_{\ell+1 \text{ times}}).$$

∎

REMARK 5.3. It follows from (5.1) that $f(X) - I_n f(0) \in \mathrm{Nilp}(\mathcal{N}; n)$ for $X \in \mathrm{Nilp}(\mathcal{M}; n)$. Alternatively (cf. Remark 5.1), if $X$ is strictly upper triangular, then $f(X)$ is upper triangular with all diagonal entries equal to $f(0)$.

In the special case of $\mathcal{M} = \mathcal{R}^d$, we can use (4.13) instead of (4.3) to obtain:

THEOREM 5.4. *Let $f \in \mathcal{T}^0(\mathrm{Nilp}_d(\mathcal{R}); \mathcal{N}_{\mathrm{nc}})$. Then for all $X \in \mathrm{Nilp}_d(\mathcal{R})$*

$$f(X) = \sum_{w \in \mathcal{G}_d} X^w \Delta_R^{w^\top} f(\underbrace{0, \ldots, 0}_{|w|+1 \text{ times}}), \tag{5.2}$$

*where the sum has finitely many nonzero terms.*

We proceed now to generalize Theorems 5.2 and 5.4 to the case of an arbitrary center. For $Y \in \mathcal{M}^{s \times s}$, we define the set $\mathrm{Nilp}(\mathcal{M}, Y; sm, \kappa)$ of $sm \times sm$ matrices $X$ over $\mathcal{M}$ which are *nilpotent about $Y$ of rank at most $\kappa$*, i.e., $(X - \bigoplus_{\alpha=1}^m Y)^{\odot_s \ell} = 0$ for all $\ell \geq \kappa$. We also define the set $\mathrm{Nilp}(\mathcal{M}, Y; sm) = \bigcup_{\kappa=1}^\infty \mathrm{Nilp}(\mathcal{M}, Y; sm, \kappa)$ and the set $\mathrm{Nilp}(\mathcal{M}, Y) = \coprod_{m=1}^\infty \mathrm{Nilp}(\mathcal{M}, Y; sm)$. Notice that $\mathrm{Nilp}(\mathcal{M}, Y)$ is a right (as well as left) admissible nc set.

In the special case of $\mathcal{M} = \mathcal{R}^d$, the set $\mathrm{Nilp}_d(\mathcal{R}, Y; sm, \kappa) := \mathrm{Nilp}(\mathcal{R}^d; sm, \kappa)$ consists of $d$-tuples $X$ of $sm \times sm$ matrices over $\mathcal{R}$ which are *jointly nilpotent about $Y$ of rank at most $\kappa$*, i.e., $(X - \bigoplus_{\alpha=1}^m Y)^{\odot_s w} = 0$ for all $w \in \mathcal{G}_d$ such that $|w| \geq \kappa$. We also have the sets $\mathrm{Nilp}_d(\mathcal{R}, Y; sm) := \mathrm{Nilp}(\mathcal{R}^d, Y; sm)$ and $\mathrm{Nilp}_d(\mathcal{R}, Y) := \mathrm{Nilp}(\mathcal{R}^d, Y)$.

REMARK 5.5. Clearly, we have $X \in \mathrm{Nilp}(\mathcal{M}, Y; sm, \kappa)$ if and only if $X - \bigoplus_{\alpha=1}^m Y \in \mathrm{Nilp}(\mathcal{M}^{s \times s}; m, \kappa)$. It follows from Remark 5.1 that when $\mathcal{R} = \mathbb{K}$ is a field, $X \in \mathrm{Nilp}(\mathcal{M}, Y; sm)$ is similar, with a similarity matrix of the form $S \otimes I_s \in \mathbb{K}^{sm \times sm}$, to an $s \times s$-block upper triangular matrix with all diagonal blocks equal to $Y$.

We have the following analogue of Theorems 5.2, with exactly the same proof.

THEOREM 5.6. *Let $Y \in \mathcal{M}^{s \times s}$ and $f \in \mathcal{T}^0(\mathrm{Nilp}(\mathcal{M}, Y); \mathcal{N}_{\mathrm{nc}})$. Then for all $X \in \mathrm{Nilp}(\mathcal{M}, Y; sm)$*

$$f(X) = \sum_{\ell=0}^\infty \left(X - \bigoplus_{\alpha=1}^m Y\right)^{\odot_s \ell} \Delta_R^\ell f(\underbrace{Y, \ldots, Y}_{\ell+1 \text{ times}}), \tag{5.3}$$

*where the sum has finitely many nonzero terms.*

REMARK 5.7. It follows from (5.3) that we have $f(X) \in \mathrm{Nilp}(\mathcal{N}, f(Y))$ for $X \in \mathrm{Nilp}(\mathcal{M}, Y)$. Alternatively (cf. Remark 5.5), if $X$ is $s \times s$-block upper triangular with all diagonal blocks equal to $Y$, then $f(X)$ is $s \times s$-block upper triangular with all diagonal entries equal to $f(Y)$.

In the special case of $\mathcal{M} = \mathcal{R}^d$, we have the following analogue of Theorem 5.4.

THEOREM 5.8. *Let $Y \in (\mathcal{R}^{s\times s})^d$ and $f \in \mathcal{T}^0(\mathrm{Nilp}_d(\mathcal{R}, Y); \mathcal{N}_{\mathrm{nc}})$. Then for all $X \in \mathrm{Nilp}_d(\mathcal{R}, Y; sm)$*

$$f(X) = \sum_{w\in\mathcal{G}_d} \Big(X - \bigoplus_{\alpha=1}^{m} Y\Big)^{\odot_s w} \Delta_R^{w^\top} f(\underbrace{Y, \ldots, Y}_{|w|+1 \text{ times}}) \tag{5.4}$$

$$= \sum_{\ell=0}^{N} \sum_{|w|=\ell} \left(\Big(\bigoplus_{\alpha=1}^{m} A_{w,(0)}\Big) \otimes \cdots \otimes \Big(\bigoplus_{\alpha=1}^{m} A_{w,(\ell)}\Big)\right) \star \Big(X - \bigoplus_{\alpha=1}^{m} Y\Big)^{[w]},$$

*where $A_{w,(0)} \otimes \cdots \otimes A_{w,(\ell)} = \Delta_R^{w^\top} f(\underbrace{Y, \ldots, Y}_{\ell+1 \text{ times}})$ with the tensor product interpretation for the values of $\Delta_R^{w^\top} f$ (Remark 3.5), the sumless Sweedler notation (4.19), and the pseudo-power notation (4.21), and the sum has finitely many nonzero terms.*

We establish now the uniqueness of nc power expansions for nc functions on nilpotent matrices.

THEOREM 5.9. *Let $Y \in \mathcal{M}^{s\times s}$, $f \in \mathcal{T}^0(\mathrm{Nilp}(\mathcal{M}, Y); \mathcal{N}_{\mathrm{nc}})$, and assume that*

$$f(X) = \sum_{\ell=0}^{\infty} \Big(X - \bigoplus_{\alpha=1}^{m} Y\Big)^{\odot_s \ell} f_\ell$$

*for all $m \in \mathbb{N}$, $X \in \mathrm{Nilp}(\mathcal{M}, Y; sm)$, and some $\ell$-linear mappings $f_\ell \colon (\mathcal{M}^{s\times s})^\ell \to \mathcal{N}^{s\times s}$, $\ell = 0, 1, \ldots$ (where the sum has finitely many nonzero terms). Then $f_\ell = \Delta_R^\ell f(Y, \ldots, Y)$ for all $\ell$.*

PROOF. By Theorem 5.6, it suffices to show that the coefficients $f_\ell$ are uniquely determined by the nc function $f$. For any fixed $\ell$ and any $Z^1, \ldots, Z^\ell \in \mathcal{M}^{s\times s}$,

$$f\left(\begin{bmatrix} Y & Z^1 & 0 & \cdots & 0 \\ \vdots & \ddots & \ddots & \ddots & \vdots \\ \vdots & & \ddots & \ddots & 0 \\ \vdots & & & \ddots & Z^\ell \\ 0 & \cdots & \cdots & \cdots & Y \end{bmatrix}\right) = \begin{bmatrix} f_0 & f_1(Z^1) & \cdots & \cdots & f_\ell(Z^1, \ldots, Z^\ell) \\ 0 & f_0 & \ddots & & \vdots \\ \vdots & \ddots & \ddots & \ddots & \vdots \\ \vdots & & \ddots & \ddots & f_1(Z^\ell) \\ 0 & \cdots & \cdots & 0 & f_0 \end{bmatrix}.$$

The $(1, \ell + 1)$ entry of the matrix on the right-hand side, i.e., the value of $f_\ell$ on arbitrary $\ell$ matrices in $\mathcal{M}^{s\times s}$ is uniquely determined by $f$. (To argue somewhat differently, we see directly that $f_\ell = \Delta_R^\ell f(Y, \ldots, Y)$ because of (3.29).) $\square$

There is also a version of Theorem 5.9 in the special case of $\mathcal{M} = \mathcal{R}^d$ for the coefficients of nc power series along $\mathcal{G}_d$, as in (5.4).

THEOREM 5.10. *Let $Y \in \mathcal{M}^{s\times s}$, $f \in \mathcal{T}^0(\mathrm{Nilp}_d(\mathcal{R}, Y); \mathcal{N}_{\mathrm{nc}})$, and assume that*

$$f(X) = \sum_{w\in\mathcal{G}_d} \Big(X - \bigoplus_{\alpha=1}^{m} Y\Big)^{\odot_s w} f_w$$

*for all $m \in \mathbb{N}$, $X \in \mathrm{Nilp}_d(\mathcal{R}, Y; sm)$, and some $\ell$-linear mappings $f_w \colon (\mathcal{R}^{s\times s})^{|w|} \to \mathcal{N}^{s\times s}$, $w \in \mathcal{G}_d$ (where the sum has finitely many nonzero terms). Then $f_w = \Delta_R^{w^\top} f(Y, \ldots, Y)$ for all $w$.*

PROOF. We define $\ell$-linear mappings $f_\ell\colon \left((\mathcal{R}^{s\times s})^d\right)^\ell \to \mathcal{N}^{s\times s}$ by

$$f_\ell(Z^1,\ldots,Z^\ell) = \sum_{w=g_{i_1}\cdots g_{i_\ell}\in\mathcal{G}_d} f_w(Z^1_{i_1},\ldots,Z^\ell_{i_\ell}), \tag{5.5}$$

so that for every $m\in\mathbb{N}$ and $Z\in(\mathcal{R}^{sm\times sm})^d$,

$$Z^{\odot_s \ell} f_\ell = \sum_{|w|=\ell} Z^{\odot_s w} f_w. \tag{5.6}$$

Clearly,

$$f_w(A^1,\ldots,A^\ell) = f_\ell(A^1 e_{i_1},\ldots,A^\ell e_{i_\ell}) \quad \text{for every } w = g_{i_1}\cdots g_{i_\ell}. \tag{5.7}$$

The result now follows from Theorem 5.9 and the definition of $\Delta_R^{w^\top} f$. □

REMARK 5.11. A simple corollary of Theorem 5.6 or 5.8 is a fairly explicit description of nc functions from $\mathbb{K}_{\rm nc}$ to $\mathcal{W}_{\rm nc}$ for a field $\mathbb{K}$ and a vector space $\mathcal{W}$ over $\mathbb{K}$. More precisely, let $\Lambda_n$ be the set of all $X\in\mathbb{K}^{n\times n}$ such that all the roots of the characteristic polynomial of $X$ are in $\mathbb{K}$, let $\Lambda = \coprod_{n=1}^\infty \Lambda_n$ (clearly, $\Lambda\subseteq\mathbb{K}_{\rm nc}$ is a nc set), and let $f\colon\Lambda\to\mathcal{W}_{\rm nc}$ be a nc function. For any $X\in\Lambda_n$, there exists an invertible matrix $T\in\mathbb{K}^{n\times n}$ so that

$$X = T^{-1}\left(\bigoplus_\lambda X_\lambda\right)T, \tag{5.8}$$

where the summation is over all the distinct eigenvalues $\lambda$ of $X$ with algebraic multiplicities $n_\lambda$ and $X_\lambda\in \operatorname{Nilp}_1(\mathbb{K},\lambda;n_\lambda)$ (if we want we can take $X_\lambda$ to be a direct sum of Jordan cells with eigenvalue $\lambda$). It follows that

$$f(X) = T^{-1}\left(\bigoplus_\lambda \sum_{\ell=0}^{n_\lambda}(X_\lambda - I_{n_\lambda}\lambda)^\ell \Delta_R^\ell f(\lambda,\ldots,\lambda)\right)T. \tag{5.9}$$

Conversely, given any functions $f_\ell\colon\mathbb{K}\to\mathcal{W}$, $\ell=0,1,\ldots$, we can define for any matrix $X\in\Lambda_n$ as above

$$f(X) = T^{-1}\left(\bigoplus_\lambda \sum_{\ell=0}^{n_\lambda}(X_\lambda - I_{n_\lambda}\lambda)^\ell f_\ell(\lambda)\right)T. \tag{5.10}$$

It is seen immediately that the right hand side is well defined independently of the representation (5.8) of $X$, and that if $XS = SY$ then $f(X)S = Sf(Y)$; therefore (5.10) defines a nc function $f\colon\Lambda\to\mathcal{W}_{\rm nc}$, and it follows from Theorem 5.9 or 5.10 that $\Delta_R^\ell f(\lambda,\ldots,\lambda) = f_\ell(\lambda)$.

## 5.2. From nc power series to nc functions

In this chapter, we consider the sum of power series of the form

$$\sum_{\ell=0}^\infty \left(X - \bigoplus_{\alpha=1}^m Y\right)^{\odot_s \ell} f_\ell \tag{5.11}$$

on nilpotent matrices $X$ about $Y$ (so that the sum has finitely many nonzero terms). Here $Y\in\mathcal{M}^{s\times s}$ is a given center; $X\in\operatorname{Nilp}(\mathcal{M},Y;sm)$, $m=1,2,\ldots$; $f_\ell\colon(\mathcal{M}^{s\times s})^\ell\to\mathcal{N}^{s\times s}$, $\ell=0,1,\ldots$, is a given sequence of $\ell$-linear mappings, i.e., a linear mapping $f\colon\mathbf{T}(\mathcal{M}^{s\times s})\to\mathcal{N}^{s\times s}$; and conditions (4.4)–(4.7) are satisfied.

Notice that the latter conditions are trivially satisfied for any sequence $f_\ell$ in the case of $s = 1$.

The necessity of conditions (4.4)–(4.7) on the coefficients $f_\ell$ for the sum of the series to be a nc function on $\mathrm{Nilp}(\mathcal{M}, Y)$ follows from Theorem 5.6 and Remark 4.3. It turns out that these conditions are also sufficient; this will essentially follow from the following lemma.

LEMMA 5.12. *Let $\mathcal{M}$ and $\mathcal{N}$ be modules over a ring $\mathcal{R}$. Let $Y \in \mathcal{M}^{s\times s}$, and let $f_\ell \colon (\mathcal{M}^{s\times s})^\ell \to \mathcal{N}^{s\times s}$, $\ell = 0, 1, \ldots$, be a sequence of $\ell$-linear mappings satisfying (4.4)–(4.7). Let $X \in \mathcal{M}^{sm\times sm}$, $\widetilde{X} \in \mathcal{M}^{s\widetilde{m}\times s\widetilde{m}}$, and $S \in \mathcal{R}^{s\widetilde{m}\times sm}$ satisfy $SX = \widetilde{X}S$. Then*

$$S\left(\sum_{\ell=0}^{N}\left(X-\bigoplus_{\nu=1}^{m}Y\right)^{\odot_s \ell}f_\ell\right)-\left(\sum_{\ell=0}^{N}\left(\widetilde{X}-\bigoplus_{\nu=1}^{\widetilde{m}}Y\right)^{\odot_s \ell}f_\ell\right)S \tag{5.12}$$
$$=\sum_{k=0}^{N}\left(\left(\widetilde{X}-\bigoplus_{\nu=1}^{\widetilde{m}}Y\right)^{\odot_s k}\odot_s\left(S\bigoplus_{\nu=1}^{m}Y-\bigoplus_{\nu=1}^{\widetilde{m}}YS\right)\odot_s\left(X-\bigoplus_{\nu=1}^{m}Y\right)^{\odot_s(N-k)}\right)f_{N+1}.$$

PROOF. In what follows, we view $X$, $\widetilde{X}$, and $S$ as block matrices of sizes $m\times m$, $\widetilde{m}\times\widetilde{m}$, and $\widetilde{m}\times m$, respectively, with $s\times s$ blocks. Since $SX = \widetilde{X}S$, we have

$$\sum_{\beta=1}^{m}S_{\alpha\beta}X_{\beta\gamma}=\sum_{\widetilde{\beta}=1}^{\widetilde{m}}\widetilde{X}_{\alpha\widetilde{\beta}}S_{\widetilde{\beta}\gamma},\quad \alpha=1,\ldots,\widetilde{m},\ \gamma=1,\ldots,m.$$

Denote

$$L=S\left(\sum_{\ell=0}^{N}\left(X-\bigoplus_{\nu=1}^{m}Y\right)^{\odot_s \ell}f_\ell\right),\qquad \widetilde{L}=\left(\sum_{\ell=0}^{N}\left(\widetilde{X}-\bigoplus_{\nu=1}^{\widetilde{m}}Y\right)^{\odot_s \ell}f_\ell\right)S.$$

Then

$$\begin{aligned}L_{\alpha\gamma}&=\sum_{\beta_0=1}^{m}S_{\alpha\beta_0}\sum_{\ell=0}^{N}\left[\left(X-\bigoplus_{\nu=1}^{m}Y\right)^{\odot_s \ell}f_\ell\right]_{\beta_0\gamma}\\&=S_{\alpha\gamma}f_0+\sum_{\ell=1}^{N}\sum_{\beta_0,\beta_1,\ldots,\beta_{\ell-1}=1}^{m}S_{\alpha\beta_0}f_\ell(X_{\beta_0\beta_1}-\delta_{\beta_0\beta_1}Y,\ldots,X_{\beta_{\ell-2}\beta_{\ell-1}}-\delta_{\beta_{\ell-2}\beta_{\ell-1}}Y,\\&\qquad\qquad X_{\beta_{\ell-1}\gamma}-\delta_{\beta_{\ell-1}\gamma}Y),\end{aligned}$$

$$\begin{aligned}\widetilde{L}_{\alpha\gamma}&=\sum_{\widetilde{\beta}_0=1}^{\widetilde{m}}\sum_{\ell=0}^{N}\left[\left(\widetilde{X}-\bigoplus_{\nu=1}^{\widetilde{m}}Y\right)^{\odot_s \ell}f_\ell\right]_{\alpha\widetilde{\beta}_0}S_{\widetilde{\beta}_0\gamma}\\&=f_0S_{\alpha\gamma}+\sum_{\ell=1}^{N}\sum_{\widetilde{\beta}_0,\widetilde{\beta}_1,\ldots,\widetilde{\beta}_{\ell-1}=1}^{\widetilde{m}}f_\ell(\widetilde{X}_{\alpha\widetilde{\beta}_0}-\delta_{\alpha\widetilde{\beta}_0}Y,\widetilde{X}_{\widetilde{\beta}_0\widetilde{\beta}_1}-\delta_{\widetilde{\beta}_0\widetilde{\beta}_1}Y,\ldots,\\&\qquad\qquad\widetilde{X}_{\widetilde{\beta}_{\ell-2}\widetilde{\beta}_{\ell-1}}-\delta_{\widetilde{\beta}_{\ell-2}\widetilde{\beta}_{\ell-1}}Y)S_{\widetilde{\beta}_{\ell-1}\gamma},\end{aligned}$$

where $\delta_{ij}$ is the Kronecker symbol and $f_0 \in \mathcal{W}^{s\times s}$. Applying (4.4) and (4.5), we obtain that

$$\begin{aligned}
L_{\alpha\gamma} = f_0 S_{\alpha\gamma} + f_1(S_{\alpha\gamma}Y - YS_{\alpha\gamma}) + \sum_{\beta_0=1}^{m} f_1(S_{\alpha\beta_0}X_{\beta_0\gamma} - S_{\alpha\beta_0}\delta_{\beta_0\gamma}Y) \\
+ \sum_{\ell=2}^{N} \sum_{\beta_0,\beta_1,\ldots,\beta_{\ell-1}=1}^{m} f_\ell(S_{\alpha\beta_0}X_{\beta_0\beta_1} - S_{\alpha\beta_0}\delta_{\beta_0\beta_1}Y, X_{\beta_1\beta_2} - \delta_{\beta_1\beta_2}Y, \ldots, \\
X_{\beta_{\ell-2}\beta_{\ell-1}} - \delta_{\beta_{\ell-2}\beta_{\ell-1}}Y, X_{\beta_{\ell-1}\gamma} - \delta_{\beta_{\ell-1}\gamma}Y) \\
+ \sum_{\ell=1}^{N} \sum_{\beta_0,\beta_1,\ldots,\beta_{\ell-1}=1}^{m} f_{\ell+1}(S_{\alpha\beta_0}Y - YS_{\alpha\beta_0}, X_{\beta_0\beta_1} - \delta_{\beta_0\beta_1}Y, \ldots, \\
X_{\beta_{\ell-2}\beta_{\ell-1}} - \delta_{\beta_{\ell-2}\beta_{\ell-1}}Y, X_{\beta_{\ell-1}\gamma} - \delta_{\beta_{\ell-1}\gamma}Y).
\end{aligned}$$

Using the assumption that $SX = \widetilde{X}S$ and the multilinearity of the mappings $f_\ell$, we obtain that

$$\begin{aligned}
L_{\alpha\gamma} = f_0 S_{\alpha\gamma} + f_1(S_{\alpha\gamma}Y) - f_1(YS_{\alpha\gamma}) + \sum_{\widetilde{\beta}_0=1}^{\widetilde{m}} f_1(\widetilde{X}_{\alpha\widetilde{\beta}_0}S_{\widetilde{\beta}_0\gamma}) - f_1(S_{\alpha\gamma}Y) \\
+ \sum_{\ell=2}^{N} \sum_{\widetilde{\beta}_0=1}^{\widetilde{m}} \sum_{\beta_1,\ldots,\beta_{\ell-1}=1}^{m} f_\ell(\widetilde{X}_{\alpha\widetilde{\beta}_0}S_{\widetilde{\beta}_0\beta_1}, X_{\beta_1\beta_2} - \delta_{\beta_1\beta_2}Y, \ldots, X_{\beta_{\ell-2}\beta_{\ell-1}} - \delta_{\beta_{\ell-2}\beta_{\ell-1}}Y, \\
X_{\beta_{\ell-1}\gamma} - \delta_{\beta_{\ell-1}\gamma}Y) \\
- \sum_{\ell=2}^{N} \sum_{\beta_1,\ldots,\beta_{\ell-1}=1}^{m} f_\ell(S_{\alpha\beta_1}Y, X_{\beta_1\beta_2} - \delta_{\beta_1\beta_2}Y, \ldots, X_{\beta_{\ell-2}\beta_{\ell-1}} - \delta_{\beta_{\ell-2}\beta_{\ell-1}}Y, \\
X_{\beta_{\ell-1}\gamma} - \delta_{\beta_{\ell-1}\gamma}Y) \\
+ \sum_{\ell=1}^{N} \sum_{\beta_0,\beta_1,\ldots,\beta_{\ell-1}=1}^{m} f_{\ell+1}(S_{\alpha\beta_0}Y, X_{\beta_0\beta_1} - \delta_{\beta_0\beta_1}Y, \ldots, X_{\beta_{\ell-2}\beta_{\ell-1}} - \delta_{\beta_{\ell-2}\beta_{\ell-1}}Y, \\
X_{\beta_{\ell-1}\gamma} - \delta_{\beta_{\ell-1}\gamma}Y) \\
- \sum_{\ell=1}^{N} \sum_{\beta_0,\beta_1,\ldots,\beta_{\ell-1}=1}^{m} f_{\ell+1}(YS_{\alpha\beta_0}, X_{\beta_0\beta_1} - \delta_{\beta_0\beta_1}Y, \ldots, X_{\beta_{\ell-2}\beta_{\ell-1}} - \delta_{\beta_{\ell-2}\beta_{\ell-1}}Y, \\
X_{\beta_{\ell-1}\gamma} - \delta_{\beta_{\ell-1}\gamma}Y).
\end{aligned}$$

After obvious cancellations and using the multilinearity of $f_\ell$ again, we obtain

$$\begin{aligned}
L_{\alpha\gamma} = f_0 S_{\alpha\gamma} + \sum_{\widetilde{\beta}_0=1}^{\widetilde{m}} f_1((\widetilde{X}_{\alpha\widetilde{\beta}_0} - \delta_{\alpha\widetilde{\beta}_0}Y)S_{\widetilde{\beta}_0\gamma}) \\
+ \sum_{\ell=2}^{N} \sum_{\widetilde{\beta}_0=1}^{\widetilde{m}} \sum_{\beta_1,\ldots,\beta_{\ell-1}=1}^{m} f_\ell((\widetilde{X}_{\alpha\widetilde{\beta}_0} - \delta_{\alpha\widetilde{\beta}_0}Y)S_{\widetilde{\beta}_0\beta_1}, X_{\beta_1\beta_2} - \delta_{\beta_1\beta_2}Y, \ldots, \\
X_{\beta_{\ell-2}\beta_{\ell-1}} - \delta_{\beta_{\ell-2}\beta_{\ell-1}}Y, X_{\beta_{\ell-1}\gamma} - \delta_{\beta_{\ell-1}\gamma}Y)
\end{aligned}$$

$$+ \sum_{\beta_0,\beta_1,\ldots,\beta_{N-1}=1}^{m} f_{N+1}(S_{\alpha\beta_0}Y - YS_{\alpha\beta_0}, X_{\beta_0\beta_1} - \delta_{\beta_0\beta_1}Y, \ldots, X_{\beta_{N-2}\beta_{N-1}} - \delta_{\beta_{N-2}\beta_{N-1}}Y, X_{\beta_{N-1}\gamma} - \delta_{\beta_{N-1}\gamma}Y).$$

Using (4.7) and (4.6), we obtain

$$\begin{aligned}
L_{\alpha\gamma} &= f_0 S_{\alpha\gamma} \\
&\quad + \sum_{\widetilde{\beta}_0=1}^{\widetilde{m}} f_1(\widetilde{X}_{\alpha\widetilde{\beta}_0} - \delta_{\alpha\widetilde{\beta}_0}Y)S_{\widetilde{\beta}_0\gamma} + \sum_{\widetilde{\beta}_0=1}^{\widetilde{m}} f_2(\widetilde{X}_{\alpha\widetilde{\beta}_0} - \delta_{\alpha\widetilde{\beta}_0}Y, S_{\widetilde{\beta}_0\gamma}Y - YS_{\widetilde{\beta}_0\gamma}) \\
&\quad + \sum_{\widetilde{\beta}_0=1}^{\widetilde{m}} \sum_{\beta_1=1}^{m} f_2(\widetilde{X}_{\alpha\widetilde{\beta}_0} - \delta_{\alpha\widetilde{\beta}_0}Y, S_{\widetilde{\beta}_0\beta_1}(X_{\beta_1\gamma} - \delta_{\beta_1\gamma}Y)) \\
&\quad + \sum_{\ell=3}^{N} \sum_{\widetilde{\beta}_0=1}^{\widetilde{m}} \sum_{\beta_1,\ldots,\beta_{\ell-1}=1}^{m} f_\ell(\widetilde{X}_{\alpha\widetilde{\beta}_0} - \delta_{\alpha\widetilde{\beta}_0}Y, S_{\widetilde{\beta}_0\beta_1}(X_{\beta_1\beta_2} - \delta_{\beta_1\beta_2}Y), \\
&\qquad X_{\beta_2\beta_3} - \delta_{\beta_2\beta_3}Y, \ldots, X_{\beta_{\ell-2}\beta_{\ell-1}} - \delta_{\beta_{\ell-2}\beta_{\ell-1}}Y, X_{\beta_{\ell-1}\gamma} - \delta_{\beta_{\ell-1}\gamma}Y) \\
&\quad + \sum_{\ell=2}^{N} \sum_{\widetilde{\beta}_0=1}^{\widetilde{m}} \sum_{\beta_1,\ldots,\beta_{\ell-1}=1}^{m} f_{\ell+1}(\widetilde{X}_{\alpha\widetilde{\beta}_0} - \delta_{\alpha\widetilde{\beta}_0}Y, S_{\widetilde{\beta}_0\beta_1}Y - YS_{\widetilde{\beta}_0\beta_1}, X_{\beta_1\beta_2} - \delta_{\beta_1\beta_2}Y, \ldots, \\
&\qquad X_{\beta_{\ell-2}\beta_{\ell-1}} - \delta_{\beta_{\ell-2}\beta_{\ell-1}}Y, X_{\beta_{\ell-1}\gamma} - \delta_{\beta_{\ell-1}\gamma}Y) \\
&\quad + \sum_{\beta_0,\beta_1,\ldots,\beta_{N-1}=1}^{m} f_{N+1}(S_{\alpha\beta_0}Y - YS_{\alpha\beta_0}, X_{\beta_0\beta_1} - \delta_{\beta_0\beta_1}Y, \ldots, \\
&\qquad X_{\beta_{N-2}\beta_{N-1}} - \delta_{\beta_{N-2}\beta_{N-1}}Y, X_{\beta_{N-1}\gamma} - \delta_{\beta_{N-1}\gamma}Y).
\end{aligned}$$

Using $SX = \widetilde{X}S$ and the multilinearity of the mappings $f_\ell$, we obtain that

$$\begin{aligned}
L_{\alpha\gamma} &= f_0 S_{\alpha\gamma} + \sum_{\widetilde{\beta}_0=1}^{\widetilde{m}} f_1(\widetilde{X}_{\alpha\widetilde{\beta}_0} - \delta_{\alpha\widetilde{\beta}_0}Y)S_{\widetilde{\beta}_0\gamma} \\
&\quad + \sum_{\widetilde{\beta}_0=1}^{\widetilde{m}} f_2(\widetilde{X}_{\alpha\widetilde{\beta}_0} - \delta_{\alpha\widetilde{\beta}_0}Y, S_{\widetilde{\beta}_0\gamma}Y) - \sum_{\widetilde{\beta}_0=1}^{\widetilde{m}} f_2(\widetilde{X}_{\alpha\widetilde{\beta}_0} - \delta_{\alpha\widetilde{\beta}_0}Y, YS_{\widetilde{\beta}_0\gamma}) \\
&\quad + \sum_{\widetilde{\beta}_0,\widetilde{\beta}_1=1}^{\widetilde{m}} f_2(\widetilde{X}_{\alpha\widetilde{\beta}_0} - \delta_{\alpha\widetilde{\beta}_0}Y, \widetilde{X}_{\widetilde{\beta}_0\widetilde{\beta}_1}S_{\widetilde{\beta}_1\gamma}) - \sum_{\widetilde{\beta}_0=1}^{\widetilde{m}} f_2(\widetilde{X}_{\alpha\widetilde{\beta}_0} - \delta_{\alpha\widetilde{\beta}_0}Y, S_{\widetilde{\beta}_0\gamma}Y) \\
&\quad + \sum_{\ell=3}^{N} \sum_{\widetilde{\beta}_0,\widetilde{\beta}_1=1}^{\widetilde{m}} \sum_{\beta_2,\ldots,\beta_{\ell-1}=1}^{m} f_\ell(\widetilde{X}_{\alpha\widetilde{\beta}_0} - \delta_{\alpha\widetilde{\beta}_0}Y, \widetilde{X}_{\widetilde{\beta}_0\widetilde{\beta}_1}S_{\widetilde{\beta}_1\beta_2}, X_{\beta_2\beta_3} - \delta_{\beta_2\beta_3}Y, \ldots, \\
&\qquad X_{\beta_{\ell-2}\beta_{\ell-1}} - \delta_{\beta_{\ell-2}\beta_{\ell-1}}Y, X_{\beta_{\ell-1}\gamma} - \delta_{\beta_{\ell-1}\gamma}Y) \\
&\quad - \sum_{\ell=3}^{N} \sum_{\widetilde{\beta}_0=1}^{\widetilde{m}} \sum_{\beta_2,\ldots,\beta_{\ell-1}=1}^{m} f_\ell(\widetilde{X}_{\alpha\widetilde{\beta}_0} - \delta_{\alpha\widetilde{\beta}_0}Y, S_{\widetilde{\beta}_0\beta_2}Y, X_{\beta_2\beta_3} - \delta_{\beta_2\beta_3}Y, \ldots, \\
&\qquad X_{\beta_{\ell-2}\beta_{\ell-1}} - \delta_{\beta_{\ell-2}\beta_{\ell-1}}Y, X_{\beta_{\ell-1}\gamma} - \delta_{\beta_{\ell-1}\gamma}Y)
\end{aligned}$$

$$+\sum_{\ell=2}^{N}\sum_{\widetilde{\beta}_0=1}^{\widetilde{m}}\sum_{\beta_1,\ldots,\beta_{\ell-1}=1}^{m} f_{\ell+1}(\widetilde{X}_{\alpha\widetilde{\beta}_0}-\delta_{\alpha\widetilde{\beta}_0}Y, S_{\widetilde{\beta}_0\beta_1}Y, X_{\beta_1\beta_2}-\delta_{\beta_1\beta_2}Y,\ldots,$$
$$X_{\beta_{\ell-2}\beta_{\ell-1}}-\delta_{\beta_{\ell-2}\beta_{\ell-1}}Y, X_{\beta_{\ell-1}\gamma}-\delta_{\beta_{\ell-1}\gamma}Y)$$
$$-\sum_{\ell=2}^{N}\sum_{\widetilde{\beta}_0=1}^{\widetilde{m}}\sum_{\beta_1,\ldots,\beta_{\ell-1}=1}^{m} f_{\ell+1}(\widetilde{X}_{\alpha\widetilde{\beta}_0}-\delta_{\alpha\widetilde{\beta}_0}Y, YS_{\widetilde{\beta}_0\beta_1}, X_{\beta_1\beta_2}-\delta_{\beta_1\beta_2}Y,\ldots,$$
$$X_{\beta_{\ell-2}\beta_{\ell-1}}-\delta_{\beta_{\ell-2}\beta_{\ell-1}}Y, X_{\beta_{\ell-1}\gamma}-\delta_{\beta_{\ell-1}\gamma}Y)$$
$$+\sum_{\beta_0,\beta_1,\ldots,\beta_{N-1}=1}^{m} f_{N+1}(S_{\alpha\beta_0}Y-YS_{\alpha\beta_0}, X_{\beta_0\beta_1}-\delta_{\beta_0\beta_1}Y,\ldots,$$
$$X_{\beta_{N-2}\beta_{N-1}}-\delta_{\beta_{N-2}\beta_{N-1}}Y, X_{\beta_{N-1}\gamma}-\delta_{\beta_{N-1}\gamma}Y).$$

After obvious cancellations and using the multilinearity of $f_\ell$ again, we obtain

$$L_{\alpha\gamma}=f_0S_{\alpha\gamma}+\sum_{\widetilde{\beta}_0=1}^{\widetilde{m}} f_1(\widetilde{X}_{\alpha\widetilde{\beta}_0}-\delta_{\alpha\widetilde{\beta}_0}Y)S_{\widetilde{\beta}_0\gamma}$$
$$+\sum_{\widetilde{\beta}_0,\widetilde{\beta}_1=1}^{\widetilde{m}} f_2(\widetilde{X}_{\alpha\widetilde{\beta}_0}-\delta_{\alpha\widetilde{\beta}_0}Y, (\widetilde{X}_{\widetilde{\beta}_0\widetilde{\beta}_1}-\delta_{\widetilde{\beta}_0\widetilde{\beta}_1}Y)S_{\widetilde{\beta}_1\gamma})$$
$$+\sum_{\ell=3}^{N}\sum_{\widetilde{\beta}_0,\widetilde{\beta}_1=1}^{\widetilde{m}}\sum_{\beta_2,\ldots,\beta_{\ell-1}=1}^{m} f_\ell(\widetilde{X}_{\alpha\widetilde{\beta}_0}-\delta_{\alpha\widetilde{\beta}_0}Y, (\widetilde{X}_{\widetilde{\beta}_0\widetilde{\beta}_1}-\delta_{\widetilde{\beta}_0\widetilde{\beta}_1}Y)S_{\widetilde{\beta}_1\beta_2}, X_{\beta_2\beta_3}-\delta_{\beta_2\beta_3}Y,$$
$$\ldots, X_{\beta_{\ell-2}\beta_{\ell-1}}-\delta_{\beta_{\ell-2}\beta_{\ell-1}}Y, X_{\beta_{\ell-1}\gamma}-\delta_{\beta_{\ell-1}\gamma}Y)$$
$$+\sum_{\beta_0,\beta_1,\ldots,\beta_{N-1}=1}^{m} f_{N+1}(S_{\alpha\beta_0}Y-YS_{\alpha\beta_0}, X_{\beta_0\beta_1}-\delta_{\beta_0\beta_1}Y,\ldots, X_{\beta_{N-2}\beta_{N-1}}-\delta_{\beta_{N-2}\beta_{N-1}}Y,$$
$$X_{\beta_{N-1}\gamma}-\delta_{\beta_{N-1}\gamma}Y)$$
$$+\sum_{\widetilde{\beta}_0=1}^{\widetilde{m}}\sum_{\beta_1,\ldots,\beta_{N-1}=1}^{m} f_{N+1}(\widetilde{X}_{\alpha\widetilde{\beta}_0}-\delta_{\alpha\widetilde{\beta}_0}Y, S_{\widetilde{\beta}_0\beta_1}Y-YS_{\widetilde{\beta}_0\beta_1}, X_{\beta_1\beta_2}-\delta_{\beta_1\beta_2}Y,\ldots,$$
$$X_{\beta_{N-2}\beta_{N-1}}-\delta_{\beta_{N-2}\beta_{N-1}}Y, X_{\beta_{N-1}\gamma}-\delta_{\beta_{N-1}\gamma}Y).$$

Continuing this argument, we obtain that

$$L_{\alpha\gamma}=f_0S_{\alpha\gamma}+\sum_{\ell=1}^{N}\sum_{\widetilde{\beta}_0,\widetilde{\beta}_1,\ldots,\widetilde{\beta}_{\ell-1}=1}^{\widetilde{m}} f_\ell(\widetilde{X}_{\alpha\widetilde{\beta}_0}-\delta_{\alpha\widetilde{\beta}_0}Y, \widetilde{X}_{\widetilde{\beta}_0\widetilde{\beta}_1}-\delta_{\widetilde{\beta}_0\widetilde{\beta}_1}Y,\ldots,$$
$$\widetilde{X}_{\widetilde{\beta}_{\ell-2}\widetilde{\beta}_{\ell-1}}-\delta_{\widetilde{\beta}_{\ell-2}\widetilde{\beta}_{\ell-1}}Y)S_{\widetilde{\beta}_{\ell-1}\gamma}$$
$$+\sum_{k=0}^{N}\sum_{\widetilde{\beta}_0,\ldots,\widetilde{\beta}_{k-1}=1}^{\widetilde{m}}\sum_{\beta_k,\ldots,\beta_{N-1}=1}^{m} f_{N+1}(\widetilde{X}_{\alpha\widetilde{\beta}_0}-\delta_{\alpha\widetilde{\beta}_0}Y, \widetilde{X}_{\widetilde{\beta}_0\widetilde{\beta}_1}-\delta_{\widetilde{\beta}_0\widetilde{\beta}_1}Y,\ldots,$$
$$\widetilde{X}_{\widetilde{\beta}_{k-2}\widetilde{\beta}_{k-1}}-\delta_{\widetilde{\beta}_{k-2}\widetilde{\beta}_{k-1}}Y, S_{\widetilde{\beta}_{k-1}\beta_k}Y-YS_{\widetilde{\beta}_{k-1}\beta_k}, X_{\beta_k\beta_{k+1}}-\delta_{\beta_k\beta_{k+1}}Y,\ldots,$$
$$X_{\beta_{N-2}\beta_{N-1}}-\delta_{\beta_{N-2}\beta_{N-1}}Y, X_{\beta_{N-1}\gamma}-\delta_{\beta_{N-1}\gamma}Y)$$

$$= \Big[\widetilde{L} + \sum_{k=0}^{N} \Big(\widetilde{X} - \bigoplus_{\nu=1}^{\widetilde{m}} Y\Big)^{\odot_s k} \odot_s \Big(S \bigoplus_{\nu=1}^{m} Y - \bigoplus_{\nu=1}^{\widetilde{m}} YS\Big) \odot_s \Big(X - \bigoplus_{\nu=1}^{m} Y\Big)^{\odot_s N-k} f_{N+1}\Big]_{\alpha\gamma}.$$

□

THEOREM 5.13. *The sum $f$ of the series* (5.11) *is a nc function on* $\mathrm{Nilp}(\mathcal{M}, Y)$ *with values in* $\mathcal{N}_{\mathrm{nc}}$.

REMARK 5.14. By Theorem 5.9, the coefficients $f_\ell$ (i.e., the linear mapping $f\colon \mathbf{T}(\mathcal{M}^{s\times s}) \to \mathcal{N}^{s\times s}$) are uniquely determined by the sum of the corresponding series, which is why we use the same notation for both the nc function and the linear mapping.

PROOF OF THEOREM 5.13. Let

$$X \in \mathrm{Nilp}(\mathcal{M}, Y; sm, \kappa), \quad \widetilde{X} \in \mathrm{Nilp}(\mathcal{M}, Y; s\widetilde{m}, \widetilde{\kappa}),$$

and $S \in \mathcal{R}^{s\widetilde{m}\times sm}$. Let $N \geq \kappa + \widetilde{\kappa}$. Then

$$f(X) = \sum_{\ell=0}^{N} \Big(X - \bigoplus_{\alpha=1}^{m} Y\Big)^{\odot_s \ell} f_\ell, \quad f(\widetilde{X}) = \sum_{\ell=0}^{N} \Big(\widetilde{X} - \bigoplus_{\alpha=1}^{\widetilde{m}} Y\Big)^{\odot_s \ell} f_\ell.$$

By Lemma 5.12, $Sf(X) - f(X)S = 0$, since for every $k\colon 0 \leq k \leq N$ either $k \geq \widetilde{\kappa}$ or $N - k \geq \kappa$, hence the right-hand side of (5.12) vanishes. Thus, $f$ respects intertwinings, i.e., is a nc function. □

In the special case $\mathcal{M} = \mathcal{R}^d$, we can consider instead of (5.11) a nc power series along $\boldsymbol{\mathcal{G}}_d$, i.e., a series of the form

$$\sum_{w\in\boldsymbol{\mathcal{G}}_d} \Big(X - \bigoplus_{\alpha=1}^{m} Y\Big)^{\odot_s w} f_w \tag{5.13}$$

on jointly nilpotent $d$-tuples of matrices $X$ about $Y$ (so that the sum has finitely many nonzero terms). Here $Y \in (\mathcal{R}^{s\times s})^d$ is a given center; $X \in \mathrm{Nilp}_d(\mathcal{R}, Y; sm)$, $m = 1, 2, \ldots$; $f_w\colon (\mathcal{R}^{s\times s})^{|w|} \to \mathcal{N}^{s\times s}$, $w \in \boldsymbol{\mathcal{G}}_d$, is a given sequence of $|w|$-linear mappings; and conditions (4.14)–(4.17) are satisfied. The series (5.13) can be alternatively written as

$$\sum_{w\in\boldsymbol{\mathcal{G}}_d} \bigg(\Big(\bigoplus_{\alpha=1}^{m} A_{w,(0)}\Big) \otimes \Big(\bigoplus_{\alpha=1}^{m} A_{w,(1)}\Big) \otimes \cdots \otimes \Big(\bigoplus_{\alpha=1}^{m} A_{w,(|w|)}\Big)\bigg) \star \Big(X - \bigoplus_{\alpha=1}^{m} Y\Big)^{[w]}.$$

Here

$$f_w = A_{w,(0)} \otimes A_{w,(1)} \otimes \cdots \otimes A_{w,(|w|)} \in \mathcal{N}^{s\times s} \otimes \underbrace{\mathcal{R}^{s\times s} \otimes \cdots \otimes \mathcal{R}^{s\times s}}_{|w| \text{ times}},$$

with the tensor product interpretation for the multilinear mappings $f_w$ (Remark 3.4), the sumless Sweedler notation (4.19), and the pseudo-power notation (4.21). Notice that conditions (4.14)–(4.17) are trivially satisfied for any sequence $f_w$ in the case of $s = 1$. In this case, (5.13) is a genuine nc power series

$$\sum_{w\in\boldsymbol{\mathcal{G}}} (X - I_m\mu)^w f_w,$$

where $Y = \mu \in \mathcal{R}^d$ and $f_w \in \mathcal{N}$.

We define a sequence of $\ell$-linear mappings $f_\ell\colon \left((\mathcal{R}^{s\times s})^d\right)^\ell \to \mathcal{N}^{s\times s}$, $\ell = 0, 1, \ldots$, by (5.5), so that (5.6) and (5.7) hold. Hence, the sum of a series of the form (5.13) coincides with the sum of the corresponding series of the form (5.11). Similarly to Remark 4.5, conditions (4.14)–(4.17) on the multilinear mappings $f_w$ are equivalent to conditions (4.4)–(4.7) on the multilinear mappings $f_\ell$. Therefore, Theorem 5.13 implies that:

THEOREM 5.15. *The sum $f$ of the series* (5.13) *is a nc function on* $\mathrm{Nilp}_d(\mathcal{R}, Y)$ *with values in* $\mathcal{N}_{\mathrm{nc}}$.

CHAPTER 6

# NC polynomials vs. polynomials in matrix entries

Let

$$p = \sum_{w \in \mathcal{G}_d \colon |w| \le L} p_w x^w \in \mathcal{N}\langle x_1, \dots, x_d \rangle \tag{6.1}$$

be a nc polynomial of degree $L$ in $x_1, \dots, x_d$ with coefficients in a module $\mathcal{N}$ over a ring $\mathcal{R}$. Evaluating $p$ on $d$-tuples $X = (X_1, \dots, X_d) \in (\mathcal{R}^{n \times n})^d$,

$$p(X) = \sum_{w \in \mathcal{G}_d \colon |w| \le L} X^w p_w \in \mathcal{N}^{n \times n}, \quad n = 1, 2, \dots, \tag{6.2}$$

defines a nc function on $(\mathcal{R}^d)_{\rm nc}$ with values in $\mathcal{N}_{\rm nc}$. Notice that the nc polynomial $p$ is determined by this nc function uniquely: at least when the module $\mathcal{N}$ is free, this follows from the absence of identities for $n \times n$ matrices over $\mathcal{R}$ (i.e., nc polynomials vanishing on $n \times n$ matrices over $\mathcal{R}$) for all $n = 1, 2, \dots$ — see [**91**, Example 1.4.4 and Theorem 1.4.5, page 22], and also follows from Theorem 5.10 by restricting a nc polynomial to $\mathrm{Nilp}_d(\mathcal{R})$. We will often identify the two nc objects, (6.1) and (6.2), and use the same notation for both. We observe that for every $n$, the nc function $p$ is a polynomial in $dn^2$ commuting variables $(X_i)_{jk}$, $i = 1, \dots, d$; $j, k = 1, \dots, n$, of degree at most $L$ with values in $\mathcal{N}^{n \times n}$.

THEOREM 6.1. *Let $f$ be a nc function on $(\mathbb{K}^d)_{\rm nc}$, where $\mathbb{K}$ is an infinite field, with values in $\mathcal{N}_{\rm nc}$ (so that $\mathcal{N}$ is a vector space over $\mathbb{K}$). Assume that for each $n$, $f(X_1, \dots, X_d)$ is a polynomial function in $dn^2$ commuting variables $(X_i)_{jk}$, $i = 1, \dots, d$; $j, k = 1, \dots, n$, with values in $\mathcal{N}^{n \times n}$. Assume also that the degrees of these polynomial functions of the commuting variables $(X_i)_{jk}$ are bounded. Then $f$ is a nc polynomial with coefficients in $\mathcal{N}$.*

PROOF. First, we observe that for every $n \in \mathbb{N}$ and $\ell \in \mathbb{N}$, the function

$$\Delta_R^\ell f(X^0, \dots, X^\ell)(Z^1, \dots, Z^\ell)$$

is a polynomial in the commuting variables

$$(X_i^\alpha)_{jk}, (Z_i^\beta)_{jk} \colon \alpha = 0, \dots, \ell, \ \beta = 1, \dots, \ell; \ i = 1, \dots, d; \ j, k = 1, \dots, n,$$

with the coefficients in $\mathcal{N}^{n \times n}$. This follows immediately from (3.29).

Second, the TT formula (4.1) implies that

$$\begin{aligned} f(Y + tZ) = \sum_{\ell=0}^{N} t^\ell \Delta_R^\ell f(Y, \dots, Y)(Z, \dots, Z) \\ + t^{N+1} \Delta_R^{N+1} f(Y, \dots, Y, Y + tZ)(Z, \dots, Z) \end{aligned} \tag{6.3}$$

for $n \in \mathbb{N}$, $Y, Z \in (\mathbb{K}^{n\times n})^d$, $t \in \mathbb{K}$, and for $N = 0, 1, \ldots$. For fixed $n$, $Y$, and $Z$, both $f_{Y,Z}(t) := f(Y + tZ)$ and $\Delta_R^{N+1}(Y, \ldots, Y, Y + tZ)(Z, \ldots, Z)$ are polynomials in $t$ with coefficients in $\mathcal{N}^{n\times n}$.

Let $L$ be the maximal degree of $f(X_1, \ldots, X_d)$ as a polynomial function in $dn^2$ commuting variables $(X_i)_{jk}$, $i = 1, \ldots, d$; $j, k = 1, \ldots, n$, as $n$ varies from 1 to $\infty$. Then for any fixed $n$, $Y$, and $Z$, the degree of the polynomial $f_{Y,Z}(t)$ is at most $L$. Therefore, for $N = L$ the remainder term in (6.3) vanishes identically as a function of $t$, and

$$f_{Y,Z}(t) = \sum_{\ell=0}^{L} t^\ell \Delta_R^\ell f(Y, \ldots, Y)(Z, \ldots, Z).$$

Setting $t = 1$ and $X = Y + Z$ (so that $f(X) = f_{Y,Z}(1)$), we obtain

$$f(X) = \sum_{\ell=0}^{L} \Delta_R^\ell f(\underbrace{Y, \ldots, Y}_{\ell+1 \text{ times}})(\underbrace{X - Y, \ldots, X - Y}_{\ell \text{ times}}). \tag{6.4}$$

Notice that this is a (uniformly in $n$) finite version of the TT formula (4.1). In particular, setting $Y = 0$ and using (4.10), we obtain

$$f(X) = \sum_{|w|\le L} X^w \, \Delta_R^{w^\top} f(\underbrace{0, \ldots, 0}_{|w|+1 \text{ times}}) \tag{6.5}$$

(cf. (5.2)). Therefore, $f$ is a nc polynomial with coefficients in $\mathcal{N}$. □

REMARK 6.2. Notice that if $\mathbb{K}$ is a field of characteristic zero, then by (6.3) and the classical single-variable Taylor formula, for all $\ell$: $0 \le \ell \le ML$ one has

$$\Delta_R^\ell f(Y, \ldots, Y)(Z, \ldots, Z) = \frac{1}{\ell!} \left.\frac{d^\ell f_{Y,Z}}{dt^\ell}\right|_{t=0}. \tag{6.6}$$

The following example shows that the assumption of boundedness of degrees in Theorem 6.1 cannot be omitted.

EXAMPLE 6.3. Suppose that we have a sequence of homogeneous polynomials $p_n \in \mathbb{K}\langle x_1, x_2\rangle$ of strictly increasing degrees $\alpha_n$ such that $p_n$ vanishes on $(\mathbb{K}^{n\times n})^2$. For example, we may take

$$p_n = \sum_{\pi \in S_{n+1}} \operatorname{sign}(\pi)\, x_1^{\pi(1)-1} x_2 \cdots x_1^{\pi(n+1)-1} x_2$$

as in [**91**] or [**41**], where $S_{n+1}$ is the symmetric group on $n+1$ elements, and $\operatorname{sign}(\pi)$ is the sign of the permutation $\pi$; clearly, $\alpha_n = \frac{(n+1)(n+2)}{2}$ in this case. Define a nc function $f\colon (\mathbb{K}^2)_{\rm nc} \to \mathbb{K}_{\rm nc}$ by

$$f(X_1, X_2) = \sum_{k=1}^{\infty} p_k(X_1, X_2),$$

where the sum is well defined: for $n \in \mathbb{N}$ and $(X_1, X_2) \in (\mathbb{K}^{n\times n})^2$ the terms $p_k(X_1, X_2)$, $k \ge n$, all vanish. Clearly, for all $n$, $f(X_1, X_2)$ is a polynomial function in $2n^2$ commuting variables $(X_1)_{jk}$, $(X_2)_{jk}$, $j, k = 1, \ldots, n$, with coefficients in

$\mathbb{K}^{n\times n}$. However, $f$ is not a nc polynomial. Indeed, suppose that $f = q$, where $q \in \mathbb{K}\langle x_1, x_2\rangle$. Let $\deg q = L$, and let

$$q = \sum_{j=0}^{L} q_j,$$

with $q_j \in \mathbb{K}\langle x_1, x_2\rangle$ homogeneous of degree $j$. Define a sequence of homogeneous polynomials

$$h_j = \begin{cases} p_k, & j = \alpha_k \text{ for some } k \in \mathbb{N}, \\ 0, & \text{otherwise}, \end{cases}$$

of degree $j$. Then we can write

$$f(X_1, X_2) = \sum_{j=0}^{\infty} h_j(X_1, X_2) = \sum_{j=0}^{\alpha_{n-1}} h_j(X_1, X_2)$$

for $(X_1, X_2) \in (\mathbb{K}^{n\times n})^2$. Comparing

$$f(tX_1, tX_2) = \sum_{j=0}^{\infty} h_j(X_1, X_2)\, t^j$$

with

$$q(tX_1, tX_2) = \sum_{j=0}^{L} q_j(X_1, X_2)\, t^j$$

as polynomials in $t \in \mathbb{K}$ for fixed $(X_1, X_2)$, we conclude that $h_j(X_1, X_2) = 0$ for all $j > L$. By the absence of polynomial identities for matrices of all sizes, $h_j = 0$ for all $j > L$. Since the sequence $\{\alpha_n\}$ is strictly increasing, this is a contradiction.

We will show now that Example 6.3 reflects the general situation: any nc function on $(\mathbb{K}^d)_{\rm nc}$ which, for each matrix size, is a polynomial in matrix entries, can be written as a sum of a nc polynomial and an infinite series of nc polynomials vanishing on matrices of increasing sizes.

THEOREM 6.4. *Let $f$ be a nc function on $(\mathbb{K}^d)_{\rm nc}$, where $\mathbb{K}$ is an infinite field, with values in $\mathcal{N}_{\rm nc}$ (so that $\mathcal{N}$ is a vector space over $\mathbb{K}$). Assume that for each $n$, $f(X_1, \ldots, X_d)$ is a polynomial function of degree $L_n$ in $dn^2$ commuting variables $(X_i)_{jk}$, $i = 1, \ldots, d$; $j, k = 1, \ldots, n$, with values in $\mathcal{N}^{n\times n}$. Then there exists a unique sequence of homogeneous nc polynomials $f_j \in \mathcal{N}\langle x_1, \ldots, x_d\rangle$ of degree $j$, $j = 0, 1, \ldots$, such that for all $n \in \mathbb{N}$,*

- *$f_j$ vanishes on $(\mathbb{K}^{n\times n})^d$ for all $j > L_n$,*
- *if $L_n > -\infty$ (where we use the usual convention* $\deg 0 = -\infty$*) then $f_{L_n}$ does not vanish identically on $(\mathbb{K}^{n\times n})^d$,*

*and for $X \in (\mathbb{K}^{n\times n})^d$,*

$$f(X) = \sum_{j=0}^{\infty} f_j(X) = \sum_{j=0}^{L_n} f_j(X). \tag{6.7}$$

REMARK 6.5. Since $n \times n$ matrices can be embedded into $(n+1) \times (n+1)$ matrices by adding zero row and column, the sequence of degrees $\{L_n\}$ is nondecreasing. It eventually stabilizes at some $L$ if and only if it is bounded, in which case we see that $f_j$, $j > L$, vanish on $(\mathbb{K}^{n\times n})^d$ for all $n$; this means that $f$ is a nc polynomial, i.e., Theorem 6.1 follows from Theorem 6.4.

REMARK 6.6. The expansion (6.7) can be written in the form

$$f(X) = \sum_{j=0}^{L_1} f_j(X) + \sum_{k=1}^{\infty} \sum_{j=L_k+1}^{L_{k+1}} f_j(X). \tag{6.8}$$

Here the $k$-th term in the infinite series is a nc polynomial of degree $L_{k+1}$ which vanishes on $\left(\mathbb{K}^{k\times k}\right)^d$ and does not vanish identically on $\left(\mathbb{K}^{(k+1)\times(k+1)}\right)^d$.

REMARK 6.7. Applying Theorem 6.4 to the nc function $f$ in Example 6.3, we see that $f_j = h_j$, $j = 0, 1, \ldots$. If the polynomials $p_n$ are chosen so that for each $n$, $p_n$ does not vanish identically on $\left(\mathbb{K}^{(n+1)\times(n+1)}\right)^d$ then $L_n = \alpha_{n-1}$, $n = 2, 3, \ldots$ (and $L_1 = -\infty$).

PROOF OF THEOREM 6.4. The proof is similar to that of Theorem 6.1. We first obtain, exactly as in the proof of Theorem 6.1, that for any fixed $n$, $Y$, and $Z$, the degree of the polynomial $f_{Y,Z}(t)$ is at most $L_n$. Therefore, we have

$$f_{Y,Z}(t) = \sum_{\ell=0}^{L_n} t^\ell \Delta_R^\ell f(Y, \ldots, Y)(Z, \ldots, Z).$$

Setting $t = 1$ and $X = Y + Z$ (so that $f(X) = f_{Y,Z}(1)$), we obtain

$$f(X) = \sum_{\ell=0}^{L_n} \Delta_R^\ell f(\underbrace{Y, \ldots, Y}_{\ell+1 \text{ times}})(\underbrace{X - Y, \ldots, X - Y}_{\ell \text{ times}}). \tag{6.9}$$

Notice that for each $n$, this is a finite version of the TT formula (4.1). In particular, setting $Y = 0$ and using (4.10), we obtain that

$$f(X) = \sum_{\ell=0}^{L_n} \Delta_R^\ell f(\underbrace{0, \ldots, 0}_{\ell+1 \text{ times}})(\underbrace{X, \ldots, X}_{\ell \text{ times}}) = \sum_{\ell=0}^{L_n} \sum_{|w|=\ell} X^w \, \Delta_R^{w^\top} f(\underbrace{0, \ldots, 0}_{l+1 \text{ times}}). \tag{6.10}$$

This yields the expansion (6.7) with

$$f_j = \sum_{|w|=j} \Delta_R^{w^\top} f(\underbrace{0, \ldots, 0}_{j+1 \text{ times}})\, x^w. \tag{6.11}$$

The fact that $f_j$ vanishes on $(\mathbb{K}^{n\times n})^d$ for all $j > L_n$ and the fact that $f_{LM_n}$ does not vanish identically on $(\mathbb{K}^{n\times n})^d$ follow from the assumption on the degree of $f(X_1, \ldots, X_d)$ as a polynomial function of matrix entries $(X_i)_{ab}$, $i = 1, \ldots, d$; $a, b = 1, \ldots, n$. Finally, nc polynomials $f_j$ in the expansion (6.7) are determined uniquely, because their evaluations $f_j(X)$ are the homogeneous components of degree $j$ of $f(X)$ viewed as a polynomial in the entries of the matrices $X_i$. □

A nc function $f \colon (\mathbb{K}^d)_{\text{nc}} \to \mathcal{N}_{\text{nc}}$, for $\mathcal{N}$ a vector space over a field $\mathbb{K}$, will be called *locally polynomial* if for each $n$, $f(X_1, \ldots, X_d)$ is a polynomial function in $dn^2$ commuting variables $(X_i)_{jk}$, $i = 1, \ldots, d$; $j, k = 1, \ldots, n$, with values in $\mathcal{N}^{n\times n}$. We denote the $\mathbb{K}$-vector space of locally polynomial nc functions from $\mathbb{K}^d_{\text{nc}}$ to $\mathcal{N}_{\text{nc}}$ by $\mathcal{N}_{\text{loc}}\langle x_1, \ldots, x_d\rangle$; it is obviously a module over the $\mathbb{K}$-algebra $\mathbb{K}_{\text{loc}}\langle x_1, \ldots, x_d\rangle$. For each $n$, we define a submodule $\mathcal{J}^n_{\mathcal{N}} \subseteq \mathcal{N}_{\text{loc}}\langle x_1, \ldots, x_d\rangle$ consisting of locally polynomial nc functions vanishing on $(\mathbb{K}^{n\times n})^d$. The submodules $\mathcal{J}_{\mathcal{N}}^n$, $n=1,2,\ldots$, form a decreasing sequence with zero intersection. In particular, we obtain a decreasing sequence of ideals $\mathcal{J}_{\mathbb{K}}^n$, $n=1,2,\ldots$, with zero intersection in the algebra $\mathbb{K}_{\rm loc}\langle x_1,\ldots,x_d\rangle$. Taking the ideals $\mathcal{J}_{\mathbb{K}}^n$ as a fundamental system of neighbourhoods of zero makes $\mathbb{K}_{\rm loc}\langle x_1,\ldots,x_d\rangle$ into a topological ring. Analogously, taking the submodules $\mathcal{J}_{\mathcal{N}}^n$ as a fundamental system of neighbourhoods of zero makes $\mathcal{N}_{\rm loc}\langle x_1,\ldots,x_d\rangle$ into a topological module over the topological ring $\mathbb{K}_{\rm loc}\langle x_1,\ldots,x_d\rangle$. Formulae (6.7) and (6.11) mean that the TT series of $f\in\mathcal{N}_{\rm loc}\langle x_1,\ldots,x_d\rangle$ converges to $f$ in this topology as a series of homogeneous polynomials; therefore, the space $\mathcal{N}\langle x_1,\ldots,x_d\rangle$ of nc polynomials is dense in the space $\mathcal{N}_{\rm loc}\langle x_1,\ldots,x_d\rangle$ of locally polynomial nc functions. In fact, $\mathcal{N}_{\rm loc}\langle x_1,\ldots,x_d\rangle$ is complete as a topological module — it is a completion of $\mathcal{N}\langle x_1,\ldots,x_d\rangle$; we leave the proof to the reader as an easy exercise.

We can extend Theorem 6.1 to the setting of arbitrary (possibly, infinite-dimensional) vector spaces over an infinite field $\mathbb{K}$. Let $\mathcal{M}$ and $\mathcal{N}$ be vector spaces over $\mathbb{K}$. A nc function $f\colon\mathcal{M}_{\rm nc}\to\mathcal{N}_{\rm nc}$ is said to be *polynomial on slices* if $f|_{\mathcal{M}^{n\times n}}$ is polynomial on slices of some degree $L_n$ for every $n=1,2,\ldots$; i.e., for every $Y,Z\in\mathcal{M}^{n\times n}$, $f_{Y,Z}(t)=f(Y+tZ)$ is a single-variable polynomial with coefficients in $\mathcal{N}^{n\times n}$ of degree at most $L_n$, and $\deg f_{Y,Z}=L_n$ for some $Y,Z\in\mathcal{M}^{n\times n}$ (for the case $\mathbb{K}=\mathbb{C}$, see [**61**, Section 26.2] or [**72**, Section 1]).

A nc polynomial over $\mathcal{M}$ with coefficients in $\mathcal{N}$ is a finite formal sum

$$p=\sum_{\ell=0}^{L}p_\ell,\tag{6.12}$$

where $p_\ell\colon\mathcal{M}\times\cdots\times\mathcal{M}\to\mathcal{N}$ are $\ell$-linear mappings; we say that $p$ is of degree $L$ if $p_L\neq 0$. We can evaluate $p$ on $X\in\mathcal{M}^{n\times n}$ by

$$p(X)=\sum_{\ell=0}^{L}X^{\odot\ell}p_\ell\in\mathcal{N}^{n\times n},\quad n=1,2,\ldots,\tag{6.13}$$

yielding a nc function on $\mathcal{M}_{\rm nc}$ with values in $\mathcal{N}_{\rm nc}$. Restricting this nc function to $\operatorname{Nilp}(\mathcal{M})$, we deduce from Theorem 5.9 that the coefficients $p_\ell$ are uniquely determined by $p$, so that we can identify the nc polynomial (6.12) with the nc function (6.13) and use the same notation for both. Notice that a nc polynomial over $\mathbb{K}^d$ can be identified with a nc polynomial (6.1) in $d$ noncommuting indeterminates, as in (4.9), (5.5)–(5.7) (with $s=1$).

If $p$ is a nc polynomial over $\mathcal{M}$ of degree $L$, then the nc function $p$ on $\mathcal{M}_{\rm nc}$ is polynomial on slices, with $\max_{n\in\mathbb{N}}L_n=L$. As in the special case $\mathcal{M}=\mathbb{K}^d$, the converse is also true; the proof is identical, up to notations, to that of Theorem 6.1.

THEOREM 6.8. *Let $f\colon\mathcal{M}_{\rm nc}\to\mathcal{N}_{\rm nc}$ be a nc function which is polynomial on slices. Assume that the degrees $L_n$ of $f|_{\mathcal{M}^{n\times n}}$, $n=1,2,\ldots$, are bounded. Then $f$ is a nc polynomial over $\mathcal{M}$ with coefficients in $\mathcal{N}$.*

There is an analogous generalization of Theorem 6.4 — we leave the details to the reader.

CHAPTER 7

# NC analyticity and convergence of TT series

In this section we discuss analyticity of nc functions and convergence of TT series. It turns out that very mild boundedness assumptions imply analyticity. There are two essentially different settings: one where the boundedness assumptions hold in every matrix dimension separately, and one where the boundedness assumptions hold uniformly for all matrix dimensions. We assume in this section that the ground ring $\mathcal{R}$ is the field $\mathbb{C}$; many results have analogues for the field $\mathbb{R}$, see Remarks 7.11, 7.35, and 7.56.

## 7.1. Analytic nc functions

In this section we will use the theory of analytic functions from vectors to vectors; our basic reference is [**61**, Chapter III, 3.16–3.19 and Chapter XXVI], see also [**72**].

Let $\mathcal{V}$ be a vector space over $\mathbb{C}$. A set $\Omega \subseteq \mathcal{V}_{\rm nc}$ is called *finitely open* if for every $n \in \mathbb{N}$ the set $\Omega_n \subseteq \mathcal{V}^{n\times n}$ is finitely open, i.e., the intersection of $\Omega_n$ with any finite-dimensional subspace $\mathcal{U}$ of $\mathcal{V}^{n\times n}$ is open in the Euclidean topology of $\mathcal{U}$. We notice that a finitely open nc set $\Omega$ is right and left admissible. Indeed, if $X \in \Omega_n$, $Y \in \Omega_m$, $Z \in \mathcal{V}^{n\times m}$, and $W \in \mathcal{V}^{m\times n}$, then the sets

$$\Omega_{n+m} \cap \operatorname{span}\left\{\begin{bmatrix} X & 0 \\ 0 & Y \end{bmatrix}, \begin{bmatrix} 0 & Z \\ 0 & 0 \end{bmatrix}\right\}$$

and

$$\Omega_{n+m} \cap \operatorname{span}\left\{\begin{bmatrix} X & 0 \\ 0 & Y \end{bmatrix}, \begin{bmatrix} 0 & 0 \\ W & 0 \end{bmatrix}\right\}$$

are open in the corresponding two-dimensional vector spaces and contain $\begin{bmatrix} X & 0 \\ 0 & Y \end{bmatrix}$, thus $\begin{bmatrix} X & tZ \\ 0 & Y \end{bmatrix}, \begin{bmatrix} X & 0 \\ tW & Y \end{bmatrix} \in \Omega_{n+m}$ for all sufficiently small $t \in \mathbb{C}$.

Let $\mathcal{W}$ be a Banach space over $\mathbb{C}$. An *admissible system of matrix norms over* $\mathcal{W}$ is a sequence of norms $\|\cdot\|_n$ on $\mathcal{W}^{n\times n}$, $n = 1, 2, \ldots$, satisfying the following two conditions:

- For every $n, m \in \mathbb{N}$ there exist $C_1(n,m)$, $C_1'(n,m) > 0$ such that for all $X \in \mathcal{W}^{n\times n}$ and $Y \in \mathcal{W}^{m\times m}$,

$$C_1(n,m)^{-1}\max\{\|X\|_n, \|Y\|_m\} \le \|X \oplus Y\|_{n+m} \le C_1'(n,m)\max\{\|X\|_n, \|Y\|_m\}. \tag{7.1}$$

- For every $n \in \mathbb{N}$ there exists $C_2(n) > 0$ such that for all $X \in \mathcal{W}^{n\times n}$ and $S, T \in \mathbb{C}^{n\times n}$,

$$\|SXT\|_n \le C_2(n)\|S\|\,\|X\|_n\|T\|, \tag{7.2}$$

where $\|\cdot\|$ denotes the operator norm of $\mathbb{C}^{n\times n}$ with respect to the standard Euclidean norm of $\mathbb{C}^n$.

PROPOSITION 7.1. *The conditions* (7.1)–(7.2) *are equivalent to the boundedness of the* injections

$$\iota_{ij}\colon \mathcal{W} \longrightarrow \mathcal{W}^{n\times n}, \quad w \longmapsto E_{ij}w, \tag{7.3}$$

*and of the* projections

$$\pi_{ij}\colon \mathcal{W}^{n\times n} \longrightarrow \mathcal{W}, \quad W = [w_{i'j'}]_{i',j'=1,\dots,n} \longmapsto w_{ij}, \tag{7.4}$$

*for all $n \in \mathbb{N}$ and all $i,j = 1,\dots,n$; here $E_{ij} \in \mathbb{C}^{n\times n}$ has the $(i,j)$-th entry* 1 *and all the other entries* 0*. Moreover, the conditions* (7.1)–(7.2) *imply that for any $s \in \mathbb{N}$, the* block injections

$$\iota^s_{ij}\colon \mathcal{W}^{s\times s} \longrightarrow \mathcal{W}^{ns\times ns} \cong \left(\mathcal{W}^{s\times s}\right)^{n\times n}, \quad W \longmapsto E_{ij}\otimes W, \tag{7.5}$$

*and the block projections*

$$\begin{aligned}\pi^s_{ij}\colon \mathcal{W}^{ns\times ns} &\cong \left(\mathcal{W}^{s\times s}\right)^{n\times n} \longrightarrow \mathcal{W}^{s\times s},\\ W = [([w_{ab}]_{a,b=1,\dots,s})_{i'j'}]_{i',j'=1,\dots,n} &\longmapsto ([w_{ab}]_{a,b=1,\dots,s})_{ij} = W_{ij},\end{aligned} \tag{7.6}$$

*are bounded.*

PROOF. Let $s,n \in \mathbb{N}$ and $W \in \mathcal{W}^{s\times s}$. Using (7.1) and (7.2), we obtain

$$\begin{aligned}\|\iota^s_{ij}(W)\|_{ns} = \|E_{ij}\otimes W\|_{ns} &= \|(P_{i1}\otimes I_s)(E_{11}\otimes W)(P_{1j}\otimes I_s\|_{ns}\\ &\le C_2(ns)\|E_{11}\otimes W\|_{ns} = C_2(ns)\|W\oplus 0_{(ns-s)\times(ns-s)}\|_{ns}\\ &\qquad\qquad \le C_2(ns)C_1'(s,ns-s)\|W\|_s.\end{aligned}$$

Here $P_{ij} = I_n - E_{ii} - E_{jj} + E_{ij} + E_{ji} \in \mathbb{C}^{n\times n}$ is a permutation matrix ($i \ne j$). Therefore the block injections $\iota^s_{ij}$ (in particular, the injections $\iota_{ij}$) are bounded.

Next, let $s,n \in \mathbb{N}$ and $W \in \mathcal{W}^{ns\times ns} \cong \left(\mathcal{W}^{s\times s}\right)^{n\times n}$. Using (7.1) and (7.2), we obtain

$$\begin{aligned}\|(\pi^s_{ij}(W)\|_s = \|W_{ij}\|_s &\le C_1(s,ns-s)\|W_{ij}\oplus 0_{(ns-s)\times(ns-s)}\|_{ns}\\ &= C_1(s,ns-s)\|(E_{1i}\otimes I_{\mathcal{W}^{s\times s}})W(E_{j1}\otimes I_{\mathcal{W}^{s\times s}})\|_{ns}\\ &\qquad\qquad \le C_1(s,ns-s)C_2(ns)\|W\|_{ns}.\end{aligned}$$

Therefore the block projections $\pi^s_{ij}$ (in particular, the projections $\pi_{ij}$) are bounded.

Assume now that the injections (7.3) and the projections (7.4) are all bounded. Let $n, m \in \mathbb{N}$ and $X \in \mathcal{W}^{n\times n}$, $Y \in \mathcal{W}^{m\times m}$. Then we obtain

$$\begin{aligned}
\|X \oplus Y\|_{n+m} &= \Big\| \sum_{i,j=1}^{n} \iota_{ij}(x_{ij}) + \sum_{i,j=1}^{m} \iota_{n+i,n+j}(y_{ij}) \Big\|_{n+m} \\
&\le \sum_{i,j=1}^{n} \|\iota_{ij}(x_{ij})\|_{n+m} + \sum_{i,j=1}^{m} \|\iota_{n+i,n+j}(y_{ij})\|_{n+m} \\
&\le \sum_{i,j=1}^{n} \|\iota_{ij}\| \cdot \|x_{ij}\|_1 + \sum_{i,j=1}^{m} \|\iota_{n+i,n+j}\| \cdot \|y_{ij}\|_1 \\
&= \sum_{i,j=1}^{n} \|\iota_{ij}\| \cdot \|\pi_{ij}(X)\|_1 + \sum_{i,j=1}^{m} \|\iota_{n+i,n+j}\| \cdot \|\pi_{ij}(Y)\|_1 \\
&\le \Big( \sum_{i,j=1}^{n} \|\iota_{ij}\| \cdot \|\pi_{ij}\| + \sum_{i,j=1}^{m} \|\iota_{n+i,n+j}\| \cdot \|\pi_{ij}\| \Big) \max\{\|X\|_n, \|Y\|_m\}
\end{aligned}$$

and

$$\begin{aligned}
\max\{\|X\|_n, \|Y\|_m\} &\le \max\Big\{ \Big\| \sum_{i,j=1}^{n} \iota_{ij}(x_{ij}) \Big\|_n, \Big\| \sum_{i,j=1}^{m} \iota_{ij}(y_{ij}) \Big\|_m \Big\} \\
&\le \max\Big\{ \sum_{i,j=1}^{n} \|\iota_{ij}\| \cdot \|x_{ij}\|_1, \sum_{i,j=1}^{m} \|\iota_{ij}\| \cdot \|y_{ij}\|_1 \Big\} \\
&= \max\Big\{ \sum_{i,j=1}^{n} \|\iota_{ij}\| \cdot \|\pi_{ij}(X \oplus Y)\|_1, \sum_{i,j=1}^{m} \|\iota_{ij}\| \cdot \|\pi_{n+i,n+j}(X \ominus Y)\|_1 \Big\} \\
&\le \max\Big\{ \sum_{i,j=1}^{n} \|\iota_{ij}\| \cdot \|\pi_{ij}\|, \sum_{i,j=1}^{m} \|\iota_{ij}\| \cdot \|\pi_{n+i,n+j}\| \Big\} \|X \oplus Y\|_{n+m}.
\end{aligned}$$

Therefore, (7.1) follows.

Assume again that the injections (7.3) and the projections (7.4) are all bounded. Let $n \in \mathbb{N}$, $X \in \mathcal{W}^{n\times n}$, and $S, T \in \mathbb{C}^{n\times n}$. Then

$$\begin{aligned}
\|SXT\|_n &= \Big\| \sum_{i,j=1}^{n} \iota_{ij}((SXT)_{ij}) \Big\|_n = \Big\| \sum_{i,j=1}^{n} \iota_{ij}\Big( \sum_{k,\ell=1}^{n} s_{ik} x_{k\ell} t_{\ell j} \Big) \Big\|_n \\
&\le \sum_{i,j,k,\ell=1}^{n} \|\iota_{ij}\| \cdot |s_{ik}| \cdot \|x_{k\ell}\|_1 \cdot |t_{\ell j}| = \sum_{i,j,k,\ell=1}^{n} \|\iota_{ij}\| \cdot |s_{ik}| \cdot \|\pi_{k\ell}(X)\|_1 \cdot |t_{\ell j}| \\
&\le \Big( \sum_{i,j=1}^{n} \|\iota_{ij}\| \Big) \Big( \sum_{k,\ell=1}^{n} \|\pi_{k\ell}\| \Big) \Big( \max_{i,k=1,\dots,n} |s_{ik}| \Big) \|X\|_n \Big( \max_{\ell,j=1,\dots,n} |t_{\ell j}| \Big) \\
&\le \Big( \sum_{i,j=1}^{n} \|\iota_{ij}\| \Big) \Big( \sum_{k,\ell=1}^{n} \|\pi_{k\ell}\| \Big) \|S\| \|X\|_n \|T\|,
\end{aligned}$$

i.e., (7.2) holds. □

Proposition 7.1 means that a sequence of norms $\|\cdot\|_n$ on $\mathcal{W}^{n\times n}$ is an admissible system of matrix norms over $\mathcal{W}$ if and only if $\mathcal{W}^{n\times n}$ is homeomorphic to the direct product of $n^2$ copies of $\mathcal{W}$ for each $n$. In particular, any two admissible systems of matrix norms over $\mathcal{W}$ yield equivalent norms on $\mathcal{W}^{n\times n}$ for each $n$, and for any admissible system of matrix norms over $\mathcal{W}$ the spaces $\mathcal{W}^{n\times n}$ are all complete.

Let $\mathcal{V}$ be a vector space, let $\Omega \subseteq \mathcal{V}_{\rm nc}$ be a finitely open nc set, and let $\mathcal{W}$ be a Banach space with an admissible system of matrix norms over $\mathcal{W}$. A nc function $f\colon \Omega \to \mathcal{W}_{\rm nc}$ is called *locally bounded on slices* if for every $n \in \mathbb{N}$ the function $f|_{\Omega_n}$ is locally bounded on slices, i.e., for every $X \in \Omega_n$ and $Z \in \mathcal{V}^{n\times n}$ there exists $\epsilon > 0$ such that $f(X + tZ)$ is bounded for $|t| < \epsilon$. A nc function $f\colon \Omega \to \mathcal{W}_{\rm nc}$ is called *Gâteaux (G-) differentiable* if for every $n \in \mathbb{N}$ the function $f|_{\Omega_n}$ is G-differentiable, i.e., for every $X \in \Omega_n$ and $Z \in \mathcal{V}^{n\times n}$ the G-derivative of $f$ at $X$ in direction $Z$,

$$\delta f(X)(Z) = \lim_{t\to 0} \frac{f(X + tZ) - f(X)}{t} = \frac{d}{dt} f(X + tZ)\Big|_{t=0},$$

exists. It follows that $f$ is *analytic on slices*, i.e., for every $X \in \Omega_n$ and $Z \in \mathcal{V}^{n\times n}$, $f(X+tZ)$ is an analytic function of $t$ in a neighbourhood of 0. By Hartogs' theorem [**97**, Page 28], $f$ is analytic on $\mathcal{U} \cap \Omega_n$ as a function of several complex variables for every $n$ and every finite-dimensional subspace $\mathcal{U}$ of $\mathcal{V}^{n\times n}$. We also note that $\delta f(X)\colon \mathcal{V}^{n\times n} \to \mathcal{W}^{n\times n}$ is a linear operator [**61**, Theorem 26.3.2].

Let $\mathcal{X}$ be a vector space, and let $Y \in \mathcal{X}$. A set $\Upsilon \subseteq \mathcal{X}$ is called *complete circular* (or a *c-star about* $Y$ in the terminology of [**61**, Definition 3.16.1]) if for every $X \in \Upsilon$ we have that $Y + t(X - Y) \in \Upsilon$ for all $t \in \mathbb{C}$ with $|t| \le 1$.

THEOREM 7.2. *Let a nc function $f\colon \Omega \to \mathcal{W}_{\rm nc}$ be locally bounded on slices. Then*

(1) *$f$ is G-differentiable.*

(2) *For every $n \in \mathbb{N}$, $Y \in \Omega_n$, $Z \in \mathcal{V}^{n\times n}$, and each $N \in \mathbb{N}$,*

$$\frac{1}{N!}\frac{d^N}{dt^N} f(Y + tZ)\Big|_{t=0} = \Delta_R^N f(\underbrace{Y, \ldots, Y}_{N+1 \text{ times}})(\underbrace{Z, \ldots, Z}_{N \text{ times}}).$$

(3) *For every $n \in \mathbb{N}$ and $Y \in \Omega_n$, let*

$$\Upsilon(Y) = \{X \in \Omega_n \colon Y + t(X - Y) \in \Omega_n \text{ for all } t \in \mathbb{C} \text{ with } |t| \le 1\}.$$

*Then $\Upsilon(Y)$ is the maximal complete circular set about $Y$ contained in $\Omega_n$ and is finitely open. For every $X \in \Upsilon(Y)$,*

$$f(X) = \sum_{\ell=0}^{\infty} \Delta_R^{\ell} f(\underbrace{Y, \ldots, Y}_{\ell+1 \text{ times}})(\underbrace{X - Y, \ldots, X - Y}_{\ell \text{ times}}), \tag{7.7}$$

*where the series converges absolutely.*

(4) *The series in (7.7) converges uniformly on compact subsets of $\mathcal{U} \cap \Upsilon(Y)$ for every finite-dimensional subspace $\mathcal{U}$ of $\mathcal{V}^{n\times n}$. Moreover, for every such compact set $K$ there exists a finitely open complete circular set $\Upsilon_K$ about $Y$, $K \subseteq \Upsilon_K \subseteq \Upsilon(Y)$ such that*

$$\sum_{\ell=0}^{\infty} \sup_{X\in\Upsilon_K} \|\Delta_R^{\ell} f(\underbrace{Y, \ldots, Y}_{\ell+1 \text{ times}})(\underbrace{X - Y, \ldots, X - Y}_{\ell \text{ times}})\|_n < \infty. \tag{7.8}$$

PROOF. (1),(2). Let $n, N \in \mathbb{N}$, $Y \in \Omega_n$, and $Z \in \mathcal{V}^{n\times n}$ be fixed. Since the nc set $\Omega$ is finitely open, $\underbrace{Y \oplus \ldots \oplus Y}_{N+2 \text{ times}} \in \Omega_{n(N+2)}$ and there exists $r > 0$ such that the

$(N+2) \times (N+2)$ matrix over $\mathcal{V}^{n\times n}$,

$$\begin{bmatrix} Y & rZ & 0 & \dots & 0 \\ 0 & \ddots & \ddots & \ddots & \vdots \\ \vdots & \ddots & \ddots & \ddots & 0 \\ \vdots & & \ddots & \ddots & rZ \\ 0 & \dots & \dots & 0 & Y \end{bmatrix},$$

belongs to $\Omega_{n(N+2)}$. Since $\Omega_{n(N+2)}$ is finitely open and $f$ is locally bounded on slices, there exists $\epsilon_1 > 0$ such that

$$\begin{bmatrix} Y & rZ & 0 & \dots & 0 \\ 0 & \ddots & \ddots & \ddots & \vdots \\ \vdots & \ddots & \ddots & \ddots & 0 \\ \vdots & & \ddots & Y & rZ \\ 0 & \dots & \dots & 0 & Y+trZ \end{bmatrix} \in \Omega_{n(N+2)}$$

and

$$f\left(\begin{bmatrix} Y & rZ & 0 & \dots & 0 \\ 0 & \ddots & \ddots & \ddots & \vdots \\ \vdots & \ddots & \ddots & \ddots & 0 \\ \vdots & & \ddots & Y & rZ \\ 0 & \dots & \dots & 0 & Y+trZ \end{bmatrix}\right)$$

is bounded for $t \in \mathbb{C}\colon |t| < \epsilon_1$. Since the Banach space $\mathcal{W}$ is equipped with an admissible system of matrix norms over $\mathcal{W}$, by Theorem 3.11, $\Delta_R^{N+1} f(Y, \dots, Y, Y + trZ)(rZ, \dots, rZ)$ is bounded for $t \in \mathbb{C}\colon |t| < \epsilon_1$. Next, since $\Omega_n$ is finitely open, there exists $\epsilon_2 > 0$ such that $Y + trZ \in \Omega_n$ for all $t \in \mathbb{C}\colon |t| < \epsilon_2$. According to the TT formula (4.1), we have for $\epsilon = \min\{\epsilon_1, \epsilon_2\}$ and $|t| < \epsilon$ that

$$f(Y+trZ) = \sum_{\ell=0}^{N} \Delta_R^{\ell} f(Y, \dots, Y)(trZ, \dots, trZ) \\ + \Delta_R^{N+1} f(Y, \dots, Y, Y + trZ)(trZ, \dots, trZ),$$

or, for $|t| < \epsilon r$, that

$$f(Y+tZ) = \sum_{\ell=0}^{N} t^{\ell} \Delta_R^{\ell} f(Y, \dots, Y)(Z, \dots, Z) \\ + t^{N+1} \Delta_R^{N+1} f(Y, \dots, Y, Y + tZ)(Z, \dots, Z). \tag{7.9}$$

Since $\Delta_R^{N+1} f(Y, \dots, Y, Y + tZ)(Z, \dots, Z)$ is bounded for $|t| < \epsilon r$, $f(Y + tZ)$ is $N$ times differentiable as a function of $t$ at $t = 0$, and the equality in part (2) holds. Clearly, part (1) follows as the special case of (2) where $N = 1$.

(3). The fact that $\Upsilon(Y)$ is the maximal complete circular set about $Y$ contained in $\Omega_n$ is immediate from the definition of $\Upsilon(Y)$.

The fact that $\Upsilon(Y)$ is finitely open is obtained as follows. Let $\mathcal{U}$ be a finite-dimensional subspace of $\mathcal{V}^{n\times n}$. Then

$$\mathcal{U}\cap\Upsilon(Y)=\{X\in\mathcal{U}\cap\Omega_n\colon Y+t(X-Y)\in\Omega_n \text{ for all } t\in\mathbb{C} \text{ with } |t|\le 1\}.$$

Suppose that we are given a norm $\|\cdot\|$ on the space $\operatorname{span}\{Y\}+\mathcal{U}$. (Since this space is finite-dimensional, all norms on it are equivalent.) Let $X\in\mathcal{U}\cap\Upsilon(Y)$. Then for every $t\in\overline{\mathbb{D}}$ there is an open ball $B_t$ centered at $Y+t(X-Y)$ in the space $\operatorname{span}\{Y\}+\mathcal{U}$ which is contained in the open set $(\operatorname{span}\{Y\}+\mathcal{U})\cap\Omega_n$. Since the set $S:=\{Y+t(X-Y)\colon |t|\le 1\}$ is compact in $\operatorname{span}\{Y\}+\mathcal{U}$, there exists a finite number of balls $B_{t_k}$, $k=1,\ldots,m$, which covers $S$. Let $R_k$ be the radius of $B_{t_k}$. Since for every $t\in\overline{\mathbb{D}}$ there is a $k\colon 1\le k\le m$ such that $Y+t(X-Y)\in B_{t_k}$, we have that

$$\phi(t):=\max_{1\le k\le m}\{R_k-|t-t_k|\|X-Y\|\}>0.$$

Clearly, the function $\phi$ is continuous on $\overline{\mathbb{D}}$, hence

$$R:=\min_{t\in\overline{\mathbb{D}}}\phi(t)>0.$$

Then for every $t\in\overline{\mathbb{D}}$ there is $k_0\colon 1\le k_0\le m$ such that

$$R_{k_0}-|t-t_{k_0}|\|X-Y\|=\phi(t).$$

Therefore, we have

$$Y+t(X+Z-Y)=Y+t_{k_0}(X-Y)+(t-t_{k_0})(X-Y)+tZ\in(\operatorname{span}\{Y\}+\mathcal{U})\cap\Omega_n$$

for every $Z\in\operatorname{span}\{Y\}+\mathcal{U}$ with $\|Z\|<R$, since

$$\|(t-t_{k_0})(X-Y)+tZ\|\le|t-t_{k_0}|\|X-Y\|+|t|\|Z\|<R_{k_0}$$

and $Y+t(X+Z-Y)\in B_{t_{k_0}}$. In particular, $Y+t(X+Z-Y)\in(\operatorname{span}\{Y\}+\mathcal{U})\cap\Omega_n$ is true for every $t\in\overline{\mathbb{D}}$ and $Z\in\mathcal{U}$ with $\|Z\|<R$. (Clearly, the induced norm on $\mathcal{U}$ is equivalent to any other norm on $\mathcal{U}$.) Therefore, $X+Z\in\mathcal{U}\cap\Upsilon(Y)$. We conclude that the set $\mathcal{U}\cap\Upsilon(Y)$ is open in $\mathcal{U}$ and thus the set $\Upsilon(Y)$ is finitely open.

It follows from parts (1) and (2) that for every $X\in\Upsilon(Y)$ the function $f(Y+t(X-Y))$ is analytic in $t$ on the closed disk $\{t\in\mathbb{C}\colon |t|\le 1\}$ (see the remark on the analyticity on slices preceding this theorem), and

$$f(Y+t(X-Y))=\sum_{\ell=0}^{\infty}t^\ell\Delta_R^\ell f(Y,\ldots,Y)(X-Y,\ldots,X-Y),\quad |t|\le 1,$$

where the series converges absolutely. In particular, this is true for $t=1$, which yields the last statement in part (3).

(4). Let $\mathcal{U}$ be a finite-dimensional subspace of $\mathcal{V}^{n\times n}$. Then, clearly, $(\operatorname{span}\{Y\}+\mathcal{U})\cap\Upsilon(Y)$ is a complete circular set about $Y$, which is open in $\operatorname{span}\{Y\}+\mathcal{U}$. By part (1), $f$ is G-differentiable in $\Omega$, and in particular in $(\operatorname{span}\{Y\}+\mathcal{U})\cap\Upsilon(Y)$. By the remark preceding this theorem, $f$ is analytic in $(\operatorname{span}\{Y\}+\mathcal{U})\cap\Upsilon(Y)$ as a function of several complex variables. More precisely, if we pick up some basis $\{E_j\}_{j=1,\ldots,d}$ in $\operatorname{span}\{Y\}+\mathcal{U}$ and write $X-Y=\sum_{j=1}^d z_jE_j$, then $g(z):=f(X)$ is an analytic function in several complex variables, $z=(z_1,\ldots,z_d)$, in a complete circular domain with center $z=0$. Since $\Delta_R^\ell f(Y,\ldots,Y)$ is a $\ell$-linear form, the series in (7.7) in $z$-coordinates coincides with the series of homogeneous polynomials in the Taylor expansion of $g(z)$. By [**97**, Theorem I.3.3], this series converges uniformly on any compact subset $K_Y$ of $(\operatorname{span}\{Y\}+\mathcal{U})\cap\Upsilon(Y)$. In particular, it converges uniformly on the compact set $K := K_Y \cap \mathcal{U}$ (unless the latter set is empty). On the other hand, given any compact subset $K$ of $\mathcal{U} \cap \Upsilon(Y)$, there exists $\delta > 0$ such that the set

$$K_Y := \{X + sY : X \in K,\ s \in \mathbb{C}, |s| \leq \delta\}$$

is contained in $(\operatorname{span}\{Y\} + \mathcal{U}) \cap \Upsilon(Y)$. Clearly, $K_Y$ is compact and $K = K_Y \cap \mathcal{U}$. Therefore we can conclude that the series in (7.7) converges uniformly on compact subsets of $\mathcal{U} \cap \Upsilon(Y)$.

The second statement in part (4) follows from [**61**, Theorem 26.3.8]. $\square$

REMARK 7.3. The fact that the *G-differential* $\delta f(X) \colon \mathcal{V}^{n\times n} \to \mathcal{W}^{n\times n}$ is a linear operator follows also from Theorem 7.2(2) for $N = 1$, since $\delta f(X) = \Delta_R f(X, X)$. More generally, one defines the *N-th G-differential* $\delta^N f(X) \colon \mathcal{V}^{n\times n} \to \mathcal{W}^{n\times n}$ by

$$\delta^N f(X)(Z) = \frac{d^N}{dt^N} f(X + tZ)\Big|_{t=0}.$$

It is known [**61**, Theorem 26.3.5] that $\delta^N f(X)(Z)$ is a homogeneous polynomial of degree $N$ in $Z$, that is, $\delta^N f(X)(\alpha Z + \beta W)$ is a homogeneous polynomial of degree $N$ in two complex variables $\alpha$ and $\beta$ for any $Z$ and $W$. We observe that this also follows from Theorem 7.2(2), since

$$\frac{1}{N!}\delta^N f(X) = \Delta_R^N f(\underbrace{X, \ldots, X}_{N+1 \text{ times}}). \tag{7.10}$$

Let $\mathcal{V}$ be a Banach space equipped with an admissible system of matrix norms over $\mathcal{V}$. We will say that a set $\Omega \subseteq \mathcal{V}_{\rm nc}$ is *open* if for every $n \in \mathbb{N}$ and $Y \in \Omega_n$ there exists $\delta_n > 0$ such that the open ball

$$B(Y, \delta_n) := \{X \in \mathcal{V}^{n\times n} \colon \|X - Y\|_n < \delta_n\}$$

is contained in $\Omega_n$. Clearly, an open set is finitely open.

Let $\mathcal{V}$ and $\mathcal{W}$ be Banach spaces equipped with admissible systems of matrix norms over $\mathcal{V}$ and over $\mathcal{W}$, and let $\Omega \subseteq \mathcal{V}_{\rm nc}$ be an open nc set. A nc function $f \colon \Omega \to \mathcal{W}_{\rm nc}$ is called *locally bounded* if for every $n \in \mathbb{N}$ the function $f|_{\Omega_n}$ is locally bounded, i.e., for every $Y \in \Omega_n$ there exists $\delta_n > 0$ such that $f$ is bounded on $B(Y, \delta_n)$. Clearly, a locally bounded nc function is locally bounded on slices.

Let $\Omega \subseteq \mathcal{V}_{\rm nc}$ be an open nc set. A nc function $f \colon \Omega \to \mathcal{W}_{\rm nc}$ is called *Fréchet (F-) differentiable* if for every $n \in \mathbb{N}$ the function $f|_{\Omega_n}$ is F-differentiable, i.e., $f|_{\Omega_n}$ is G-differentiable and for any $X \in \Omega_n$ the linear operator $\delta f(X) \colon \mathcal{V}^{n\times n} \to \mathcal{W}^{n\times n}$ is bounded. It was shown by Zorn in [**110**] that in this case

$$\lim_{\|Z\|_n \to 0} \frac{\|f(X + Z) - f(X) - \delta f(X)(Z)\|_n}{\|Z\|_n} = 0$$

for all $X \in \Omega_n$. A nc function $f \colon \Omega \to \mathcal{W}_{\rm nc}$ is called *analytic* if for every $n \in \mathbb{N}$ the function $f|_{\Omega_n}$ is analytic, i.e., $f|_{\Omega_n}$ is locally bounded and G-differentiable. By [**61**, Theorem 3.17.1], in this case $f|_{\Omega_n}$ is also continuous and F-differentiable. It follows that an analytic nc function $f \colon \Omega \to \mathcal{W}_{\rm nc}$ is continuous (i.e., all its restrictions $f|_{\Omega_n}$ are continuous) and F-differentiable.

THEOREM 7.4. *Let a nc function $f \colon \Omega \to \mathcal{W}_{\rm nc}$ be locally bounded. Then, in addition to the conclusions of Theorem 7.2, $f$ is analytic. Let $n \in \mathbb{N}$, $Y \in \Omega_n$, and* *let $\Upsilon$ be an open complete circular set about $Y$ such that $\Upsilon \subseteq \Omega_n$ and $f$ is bounded on $\Upsilon$. For every $\epsilon > 0$, the TT series in (7.7) converges uniformly on the set*

$$\Upsilon_\epsilon := \{X \in \Upsilon \colon Y + (1+\epsilon)(X - Y) \in \Upsilon\}.$$

*Moreover,*

$$\sum_{\ell=0}^{\infty} \sup_{X \in \Upsilon_\epsilon} \|\Delta_R^\ell f(\underbrace{Y, \ldots, Y}_{\ell+1 \text{ times}})(\underbrace{X - Y, \ldots, X - Y}_{\ell \text{ times}})\|_n < \infty. \tag{7.11}$$

PROOF. Since $f$ is locally bounded, $f$ is also locally bounded on slices. By Theorem 7.2, $f$ is G-differentiable. Therefore $f$ is analytic.

Assume that $\|f(X)\|_n \leq M$ for some $M > 0$ and all $X \in \Upsilon$. Then for any $Z \in \mathcal{V}^{n \times n}$ with $Y + Z \in \Upsilon$ the function $f(Y + tZ)$ is analytic in $t$ on the closed disk $\{t \in \mathbb{C} \colon |t| \leq 1\}$, and the Cauchy inequalities

$$\frac{1}{N!} \left\| \frac{d^N}{dt^N} f(Y + tZ)\Big|_{t=0} \right\|_n \leq M$$

hold for every $N \in \mathbb{N}$. By Theorem 7.2(2) and the homogeneity of degree $N$ in $Z$ of $\Delta_R^N f(Y, \ldots, Y)(Z, \ldots, Z)$, we obtain the following estimate for all $X \in \Upsilon_\epsilon$:

$$\|\Delta_R^N f(Y, \ldots, Y)(X - Y, \ldots, X - Y)\|_n \leq \frac{M}{(1+\epsilon)^N}.$$

This implies the convergence of the series in (7.11), and hence the uniform convergence of the TT series in (7.7). $\square$

We notice an important special case of Theorem 7.4.

COROLLARY 7.5. *Let a nc function $f \colon \Omega \to \mathcal{W}_{\rm nc}$ be locally bounded. For every $n \in \mathbb{N}$ and $Y \in \Omega_n$, let $\delta_n = \sup\{r > 0 \colon f \text{ is bounded on } B(Y, r)\}$. Then the TT series in (7.7) converges absolutely and uniformly on every open ball $B(Y, r) \subsetneq B(Y, \delta_n)$. Moreover,*

$$\sum_{\ell=0}^{\infty} \sup_{X \in B(Y,r)} \|\Delta_R^\ell f(\underbrace{Y, \ldots, Y}_{\ell+1 \text{ times}})(\underbrace{X - Y, \ldots, X - Y}_{\ell \text{ times}})\|_n < \infty. \tag{7.12}$$

COROLLARY 7.6. *Let $\Omega \subseteq \mathcal{V}_{\rm nc}$ be an open nc set. Then a nc function $f \colon \Omega \to \mathcal{W}_{\rm nc}$ is locally bounded if and only if $f$ is continuous if and only if $f$ is F-differentiable if and only if $f$ is analytic.*

REMARK 7.7. The fact that $\delta f(X) \colon \mathcal{V}^{n \times n} \to \mathcal{W}^{n \times n}$ is a bounded linear operator for any $X \in \Omega_n$ (and thus $f$ is F-differentiable) follows also from the equality $\delta f(X) = \Delta_R f(X, X)$. Indeed, $\Delta_R f(X, X)(Z)$ is the $(1, 2)$ block entry of the matrix $f\left(\begin{bmatrix} X & Z \\ 0 & X \end{bmatrix}\right)$. Since $f$ is locally bounded, there exists $\delta > 0$ such that $f\left(\begin{bmatrix} X & Z \\ 0 & X \end{bmatrix}\right)$ is bounded when $\|Z\|_n < \delta$. But then so is $\Delta_R f(X, X)(Z)$, since the system of matrix norms over $\mathcal{W}$ is admissible. More generally, it is known [**61**, Theorems 26.3.5 and 26.3.6] that the $N$-th G-differential $\delta^N f(X) \colon \mathcal{V}^{n \times n} \to \mathcal{W}^{n \times n}$ is a bounded homogeneous polynomial of degree $N$, i.e.,

$$\|\delta^N f(X)(Z)\|_n \leq C \|Z\|_n^N$$

for some $C > 0$. This also follows from (7.10), since $\Delta_R^N f(X,\dots,X)(Z,\dots,Z)$ is the $(1, N+1)$ block entry of the $(N+1)\times(N+1)$ matrix

$$f\left(\begin{bmatrix} X & Z & 0 & \dots & 0 \\ 0 & \ddots & \ddots & \ddots & \vdots \\ \vdots & \ddots & \ddots & \ddots & 0 \\ \vdots & & \ddots & \ddots & Z \\ 0 & \dots & \dots & 0 & X \end{bmatrix}\right),$$

see Theorem 3.11. It follows from the proof of Theorem 7.4 that, moreover, there exist positive constants $K$ and $\rho$ such that

$$\|\Delta_R^N f(X,\dots,X)(Z,\dots,Z)\|_n \le K\rho^N \|Z\|_n^N. \tag{7.13}$$

The convergence results, Theorems 7.2 and 7.4, allow us to write infinite nc power expansions centered at a matrix $Y \in \Omega_s$ valid in matrix dimensions which are multiples of $s$, as in Theorem 4.2.

THEOREM 7.8. *Suppose that $\mathcal{V}$ is a vector space over $\mathbb{C}$, $\Omega \subseteq \mathcal{V}_{\mathrm{nc}}$ is a finitely open nc set, and $\mathcal{W}$ is a Banach space with an admissible system of matrix norms over $\mathcal{W}$. Let $f\colon \Omega \to \mathcal{W}_{\mathrm{nc}}$ be a nc function which is locally bounded on slices, and let $Y \in \Omega_s$. Then for an arbitrary $m \in \mathbb{N}$,*

$$f(X) = \sum_{\ell=0}^{\infty} \left(X - \bigoplus_{\alpha=1}^{m} Y\right)^{\odot_s \ell} \Delta_R^\ell f(\underbrace{Y,\dots,Y}_{\ell+1 \text{ times}}), \tag{7.14}$$

*where the series converges absolutely and uniformly on compact subsets of $\mathcal{U} \cap \Upsilon(\bigoplus_{\alpha=1}^m Y)$ for any finite-dimensional subspace $\mathcal{U}$ of $\mathcal{V}^{sm\times sm}$. Here $\Upsilon(\bigoplus_{\alpha=1}^m Y)$ is the maximal complete circular set about $\bigoplus_{\alpha=1}^m Y$ contained in $\Omega_{sm}$, as in Theorem 4.2(3). (Notice that $\coprod_{m=1}^\infty \Upsilon\left(\bigoplus_{\alpha=1}^m Y\right)$ is a nc set.)*

*If, furthermore, $\mathcal{V}$ is a Banach space with an admissible system of matrix norms over $\mathcal{V}$, $\Omega$ is an open nc set, and $f$ is locally bounded, then the TT series in (7.14) converges uniformly on $\Upsilon_\epsilon$ for every $\epsilon > 0$ and every $\Upsilon$ an open complete circular set about $\bigoplus_{\alpha=1}^m Y$ contained in $\Omega_{sm}$ and such that $f$ is bounded on $\Upsilon$, where $\Upsilon_\epsilon$ is defined as in Theorem 7.4. As a special case, the TT series converges uniformly on every open ball $B(\bigoplus_{\alpha=1}^m Y, r) \subsetneq B(\bigoplus_{\alpha=1}^m Y, \delta_{sm})$, where $B(\bigoplus_{\alpha=1}^m Y, \delta_{sm})$ is defined as in Corollary 7.5.*

*In particular, if $\mu \in \Omega_1$ then (7.14) becomes*

$$f(X) = \sum_{\ell=0}^{\infty} (X - I_m\mu)^{\odot \ell}\, \Delta_R^\ell f(\underbrace{\mu,\dots,\mu}_{\ell+1 \text{ times}}). \tag{7.15}$$

We have the following uniqueness theorem for convergent nc power expansion of a nc function.

THEOREM 7.9. *Suppose that $\mathcal{V}$ is a vector space over $\mathbb{C}$, $\Omega \subseteq \mathcal{V}_{\mathrm{nc}}$ is a finitely open nc set, and $\mathcal{W}$ is a Banach space with an admissible system of matrix norms over $\mathcal{W}$. Let $f\colon \Omega \to \mathcal{W}_{\mathrm{nc}}$ be a G-differentiable nc function and $Y \in \Omega_s$. Suppose that*

$$f(X) = \sum_{\ell=0}^{\infty} \left(X - \bigoplus_{\alpha=1}^{m} Y\right)^{\odot_s \ell} f_\ell,$$

*for* $X \in \Gamma_{sm}$, $m = 1, 2, \dots$, *where* $\Gamma$ *is a finitely open subset of* $\Omega$ *which contains* $\bigoplus_{\alpha=1}^{m} Y$ *for every* $m = 1, 2, \dots$, *and* $f_\ell \colon (\mathcal{V}^{s\times s})^\ell \to \mathcal{W}^{s\times s}$ *is a* $\ell$*-linear mapping,* $\ell = 0, 1, \dots$. *Then*

$$f_\ell = \Delta_R^\ell f(\underbrace{Y, \dots, Y}_{\ell+1 \text{ times}}).$$

PROOF. Since $\Gamma$ is finitely open, for every $Z^1, \dots, Z^\ell \in \mathcal{V}^{s\times s}$ there exists a real $r > 0$ such that

$$X = \begin{bmatrix} Y & rZ^1 & 0 & \dots & 0 \\ 0 & \ddots & \ddots & \ddots & \vdots \\ \vdots & \ddots & \ddots & \ddots & 0 \\ \vdots & & \ddots & \ddots & rZ^\ell \\ 0 & \dots & \dots & 0 & Y \end{bmatrix} \in \Gamma_{\ell+1}.$$

As in the proof of Theorem 5.9, we see that $f_\ell(rZ^1, \dots, rZ^\ell)$, and thus

$$f_\ell(Z^1, \dots, Z^\ell) = r^{-\ell} f_\ell(rZ^1, \dots, rZ^\ell),$$

is uniquely determined by $f(X)$. On the other hand, if $r > 0$ is sufficiently small, we have $X \in \Upsilon(\bigoplus_{\alpha=1}^{\ell+1} Y)$ and the series (7.14) converges—see Theorem 7.2 and Theorem 7.8. Then we have by (3.29) that

$$f_\ell(Z^1, \dots, Z^\ell) = \Delta_R^\ell f(Y, \dots, Y)(Z^1, \dots, Z^\ell).$$

□

In the case where the space $\mathcal{V}$ is finite-dimensional, so that we may assume without loss of generality that $\mathcal{V} = \mathbb{C}^d$ (equipped with any admissible system of matrix norms), we can expand each term of the convergent TT series using higher order partial nc difference-differential operators, as in Corollary 4.4 and Theorem 4.6. We notice that in this case a finitely open nc set is the same as an open nc set, and a G-differentiable nc function is the same as a F-differentiable nc function.

THEOREM 7.10. *Suppose that* $\Omega \subseteq (\mathbb{C}^d)_{\rm nc}$ *is an open nc set, and* $\mathcal{W}$ *is a Banach space with an admissible system of matrix norms over* $\mathcal{W}$. *Let* $f \colon \Omega \to \mathcal{W}_{\rm nc}$ *be a nc function which is locally bounded on slices. Then* $f$ *is analytic, and for any* $s \in \mathbb{N}$, $Y \in \Omega_s$, *and arbitrary* $m \in \mathbb{N}$,

$$\begin{aligned} (7.16)\quad f(X) &= \sum_{\ell=0}^{\infty} \Big(X - \bigoplus_{\alpha=1}^{m} Y\Big)^{\odot_s \ell} \Delta_R^\ell f(\underbrace{Y, \dots, Y}_{\ell+1 \text{ times}}) \\ &= \sum_{\ell=0}^{\infty} \left( \sum_{|w|=\ell} \left( \bigoplus_{\alpha=1}^{m} A_{w,(0)} \otimes \dots \otimes \bigoplus_{\alpha=1}^{m} A_{w,(\ell)} \right) \star \left( X - \bigoplus_{\alpha=1}^{m} Y \right)^{[w]} \right), \end{aligned}$$

*where the series converges absolutely and uniformly on compact subsets of the set* $\Upsilon(\bigoplus_{\alpha=1}^{m} Y)$. *Here* $\Upsilon(\bigoplus_{\alpha=1}^{m} Y)$ *is the maximal complete circular set about* $\bigoplus_{\alpha=1}^{m} Y$ *contained in* $\Omega_{sm}$, *as in Theorem 4.2(3),*

$$A_{w,(0)} \otimes \dots \otimes A_{w,(\ell)} = \Delta_R^{w^\top} f(\underbrace{Y, \dots, Y}_{\ell+1 \text{ times}}),$$

*with the tensor product interpretation for the values of* $\Delta_R^{w^\top} f$ *(Remark* 3.5*), the sumless Sweedler notation* (4.19)*, and the pseudo-power notation* (4.21)*.*

*In particular, if* $\mu \in \Omega_1$ *and* $Y = \mu$*, then* (7.16) *becomes*

$$f(X) = \sum_{\ell=0}^{\infty} \Big( \sum_{|w|=\ell} (X - I_m\mu)^w \, \Delta_R^{w^\top} f(\underbrace{\mu, \ldots, \mu}_{\ell+1 \text{ times}}) \Big). \tag{7.17}$$

REMARK 7.11. The results of this section are of a mixed nature: noncommutative and complex analytic. Therefore, they admit only partial analogues in the real case. Let $\mathcal{V}$ be a vector space over $\mathbb{R}$, let $\Omega \subseteq \mathcal{V}_{\rm nc}$ be a finitely open nc set, and let $\mathcal{W}$ be a Banach space over $\mathbb{R}$ with an admissible system of matrix norms over $\mathcal{W}$. (All the notions above are defined exactly as in the complex case.) If a nc function $f \colon \Omega \to \mathcal{W}_{\rm nc}$ is locally bounded on slices then, similar to Theorem 7.2(1), $f$ is G-differentiable. Moreover, similar to Theorem 7.2(2), $f(Y + tZ)$ is infinitely many times differentiable at 0 as a function of $t \in \mathbb{R}$ for every $n \in \mathbb{N}$, $Y \in \Omega_n$, and $Z \in \mathcal{V}^{n \times n}$, and

$$\frac{1}{N!} \frac{d^N}{dt^N} f(Y + tZ)\Big|_{t=0} = \Delta_R^N f(Y, \ldots, Y)(Z, \ldots, Z), \qquad N = 1, 2, \ldots,$$

holds.

It does not follow that $f(Y + tZ)$ is analytic at 0 as a function of $t \in \mathbb{R}$, i.e., one cannot guarantee the convergence of the TT series of $f$ as in (7.7). It also does not follow that $f$ is infinitely many times differentiable on $\mathcal{U} \cap \Omega_n$ as a function of several real variables, for a finite-dimensional subspace $\mathcal{U}$ of $\mathcal{V}^{n \times n}$.

If, in addition, $f$ is *locally bounded in* $\Omega$ *on affine finite-dimensional subspaces*, i.e., if for every $n, M \in \mathbb{N}$, $Y \in \Omega_n$, and $Z_1, \ldots, Z_M \in \mathcal{V}^{n \times n}$ there exists $\epsilon > 0$ such that $f(Y + t_1 Z_1 + \cdots + t_M Z_M)$ is bounded on the set $\{t \in \mathbb{R} \colon \max_{1 \le j \le M} |t_j| < \epsilon\}$, then $f(Y + t_1 Z_1 + \cdots + t_M Z_M)$ is infinitely many times differentiable at 0 as a function of $t_1, \ldots, t_M \in \mathbb{R}$, and the identity

$$\begin{aligned} \frac{1}{k!} \frac{\partial^{|k|}}{\partial t_1^{k_1} \cdots \partial t_M^{k_M}} f(Y + t_1 Z_1 + \cdots + t_M Z_M)\Big|_{t_1 = \cdots = t_M = 0} \\ = \sum_{\pi} \Delta_R^{|k|} f(Y, \ldots, Y)(Z_{\pi(1)}, \ldots, Z_{\pi(N)}) \end{aligned} \tag{7.18}$$

holds. Here $k = (k_1, \ldots, k_M)$ is an arbitrary $M$-tuple of nonnegative integers,

$$|k| := k_1 + \cdots + k_M, \quad k! := k_1! \cdots k_M!,$$

and $\pi$ runs over the set of all permutations with repetitions of the set $\{1, \ldots, M\}$, where the element $j$ appears exactly $k_j$ times. The identity (7.18) is of course also true in the complex case, and in both real and complex cases its proof is an obvious modification of the proof of part (2) of Theorem 7.2.

If $f$ is *real analytic on slices*, i.e., if $f(Y + tZ)$ is analytic at 0 as a function of $t \in \mathbb{R}$ for every $n \in \mathbb{N}$, $Y \in \Omega_n$, and $Z \in \mathcal{V}^{n \times n}$ then the TT series of $f$, as in (7.7) or as in (7.14) (in particular, (7.15)), converges absolutely in some neighborhood of 0 in every slice. In the case where $\mathcal{V} = \mathbb{R}^d$, if $f$ is *real analytic*, i.e., $f|_{\Omega_n}$ is real analytic for each $n \in \mathbb{N}$, one can write the TT series expansions of $f$ about $Y \in \Omega_n$, as in (7.16) (in particular, (7.17)), with the series converging absolutely and uniformly on compact subsets of $\Upsilon(Y) = \{X \in \Omega_n \colon Y + t(X - Y) \in \Omega_n \text{ for all } t \in [-1, 1]\}$.

## 7.2. Uniformly-open topology over an operator space

Recall that a Banach space $\mathcal{W}$ over $\mathbb{C}$ equipped with an admissible system of matrix norms such that (7.1) and (7.2) hold with $C_1 = C_1' = C_2 = 1$ independent of $n$ and $m$, is an *operator space*; see [**37, 79, 78**]. Let $\mathcal{W}$ be an operator space. For $Y \in \mathcal{W}^{s\times s}$ and $r > 0$, define a *nc ball centered at* $Y$ *of radius* $r$ as

$$B_{\rm nc}(Y,r) = \coprod_{m=1}^{\infty} B\Big(\bigoplus_{\alpha=1}^{m} Y, r\Big) = \coprod_{m=1}^{\infty} \Big\{X \in \mathcal{W}^{sm\times sm} \colon \Big\|X - \bigoplus_{\alpha=1}^{m} Y\Big\|_{sm} < r\Big\}.$$

PROPOSITION 7.12. *Let $Y \in \mathcal{W}^{s\times s}$ and $r > 0$. For any $X \in B_{\rm nc}(Y,r)$ there is a $\rho > 0$ such that $B_{\rm nc}(X,\rho) \subseteq B_{\rm nc}(Y,r)$. Hence, nc balls form a basis for a topology on $\mathcal{W}_{\rm nc}$. This topology will be called the* uniformly-open topology.

PROOF. Let $X \in B_{\rm nc}(Y,r)_{sm} = B(\bigoplus_{\alpha=1}^{m} Y, r)$, i.e., $\delta := \|X - \bigoplus_{\alpha=1}^{m} Y\|_{sm} < r$. Set $\rho := r - \delta$. Let $Z \in B_{\rm nc}(X,\rho)_{smn} = B(\bigoplus_{\beta=1}^{n} X, \rho)$. Then, by (7.1) with $C_1' = 1$, we have

$$\begin{aligned}\Big\|Z - \bigoplus_{\alpha=1}^{mn} Y\Big\|_{smn} &\le \Big\|Z - \bigoplus_{\beta=1}^{n} X\Big\|_{smn} + \Big\|\bigoplus_{\beta=1}^{n} X - \bigoplus_{\alpha=1}^{mn} Y\Big\|_{smn} \\ &= \Big\|Z - \bigoplus_{\beta=1}^{n} X\Big\|_{smn} + \Big\|\bigoplus_{\beta=1}^{n}\Big(X - \bigoplus_{\alpha=1}^{m} Y\Big)\Big\|_{smn} \\ &\le \Big\|Z - \bigoplus_{\beta=1}^{n} X\Big\|_{smn} + \Big\|X - \bigoplus_{\alpha=1}^{m} Y\Big\|_{sm} < \rho + \delta = r,\end{aligned}$$

i.e., $Z \in B_{\rm nc}(Y,r)_{smn}$, which proves the first statement.

Let $Y^1 \in \mathcal{W}^{s_1\times s_1}$, $Y^2 \in \mathcal{W}^{s_2\times s_2}$, and $r_1, r_2 > 0$ be such that $B_{\rm nc}(Y^1,r_1) \cap B_{\rm nc}(Y^2,r_2) \neq \emptyset$. Let $m$ be the smallest common multiple of $s_1$ and $s_2$ for which $B_{\rm nc}(Y^1,r_1)_m \cap B_{\rm nc}(Y^2,r_2)_m \neq \emptyset$. If $X \in B_{\rm nc}(Y^1,r_1)_m \cap B_{\rm nc}(Y^2,r_2)_m$. Setting

$$\rho := \min\Big\{r_1 - \Big\|X - \bigoplus_{\alpha=1}^{m/s_1} Y^1\Big\|_m,\ r_2 - \Big\|X - \bigoplus_{\alpha=1}^{m/s_2} Y^2\Big\|_m\Big\}$$

and using the same argument as in the preceding paragraph, we obtain that

$$B_{\rm nc}(X,\rho) \subseteq B_{\rm nc}(Y^1,r_1) \cap B_{\rm nc}(Y^2,r_2).$$

Thus, nc balls form a basis for a topology on $\mathcal{W}_{\rm nc}$. □

REMARK 7.13. Notice that multiplication by a scalar as a mapping $\mathbb{C}\times\mathcal{W}_{\rm nc} \to \mathcal{W}_{\rm nc}$ is continuous in the uniformly-open topology .

Open sets in the uniformly-open topology on $\mathcal{W}_{\rm nc}$ will be called *uniformly open.* The uniformly-open topology is never Hausdorff: the points $\bigoplus_{\alpha=1}^{m} Y$ and $\bigoplus_{\beta=1}^{m'} Y$, $m \neq m'$, cannot be separated by nc balls. However, the following Hausdorff-like property holds: if $X \in \mathcal{W}^{n\times n}$ and $X' \in \mathcal{W}^{n'\times n'}$ are such that there does not exist $Y \in \mathcal{W}_{\rm nc}$ for which $X = \bigoplus_{\alpha=1}^{m} Y$ and $X' = \bigoplus_{\beta=1}^{m'} Y$ for some $m$ and $m'$, then there exist $\delta$ and $\delta'$ such that $B_{\rm nc}(X,\delta) \cap B_{\rm nc}(X',\delta') = \emptyset$. One can choose $\delta$ and $\delta'$ such that $\delta + \delta' < \Big\|\bigoplus_{\alpha=1}^{M} X - \bigoplus_{\beta=1}^{M'} X'\Big\|_{Mn}$, where $Mn = M'n'$ is the least common multiple of $n$ and $n'$. We only have to show that the norm above is nonzero in this case.

PROPOSITION 7.14. *For $X \in \mathcal{W}^{n\times n}$ and $X' \in \mathcal{W}^{n'\times n'}$, one has*

$$\Big\| \bigoplus_{\alpha=1}^{M} X - \bigoplus_{\beta=1}^{M'} X' \Big\|_{Mn} = 0$$

*if and only if there exists $Y \in \mathcal{W}_{\rm nc}$ for which $X = \bigoplus_{\alpha=1}^{m} Y$ and $X' = \bigoplus_{\beta=1}^{m'} Y$.*

PROOF. The "if" part is obvious. Suppose

$$(\overline{X} :=) \bigoplus_{\alpha=1}^{M} X = \bigoplus_{\beta=1}^{M'} X'.$$

For the purpose of this proof, let us use the convention that the rows and columns of a $N \times N$ matrix $Z$ over $\mathcal{W}$ are enumerated from 0 to $N-1$. Define the diagonal shift $S$ which acts on such matrices as follows:

$$(SZ)_{ij} = Z_{(i-1) \bmod N, (j-1) \bmod N}, \quad i,j = 0,\dots,N-1.$$

Clearly, the inverse shift is given by

$$(S^{-1}Z)_{ij} = Z_{(i+1) \bmod N, (j+1) \bmod N}, \quad i,j = 0,\dots,N-1.$$

Clearly, we have $S^{\pm n}\overline{X} = \overline{X}$ and $S^{\pm n'}\overline{X} = \overline{X}$, where $N = Mn$. Let $d = \gcd(n,n')$. Since there exist $k,k' \in \mathbb{Z}$ such that $d = kn + k'n'$ (see, e.g., [**101**, Problem 1.1]), we obtain that $S^d\overline{X} = \overline{X}$. Therefore, all $d \times d$ blocks on the main block diagonal of the matrix $\overline{X}$ are equal, say $Y$. Since every other $d \times d$-block NW-to-SE diagonal of $\overline{X}$ has at least one zero block, it must have all zero blocks. We conclude that $\overline{X} = \bigoplus_{\gamma=1}^{Mn/d} Y$, and therefore $X = \bigoplus_{\alpha=1}^{m} Y$ and $X' = \bigoplus_{\beta=1}^{m'} Y$, with $m = n/d$ and $m' = n'/d$. The proof is complete. □

Alternatively, the closure of $X \in \mathcal{W}^{n\times n}$, $\operatorname{clos}\{X\}$, consists of all $X' = \bigoplus_{\beta=1}^{m'} Y$, $m' = 1,2,\dots$, where $Y \in \mathcal{W}^{s\times s}$ is such that $n = sm$ and $X = \bigoplus_{\alpha=1}^{m} Y$, with $Y$ not representable as a direct sum of matrices over $\mathcal{W}$ — such $Y$ is unique by Proposition 7.14; the quotient topology on $\widehat{\mathcal{W}_{\rm nc}} := \mathcal{W}_{\rm nc}/\sim$ where $X \sim X'$ whenever $X = \bigoplus_{\alpha=1}^{m} Y$ and $X' = \bigoplus_{\beta=1}^{m'} Y$ for some $Y \in \mathcal{W}_{\rm nc}$ is Hausdorff.

Let $\mathcal{M}$ be a module. For a set $\Omega \subseteq \mathcal{M}_{\rm nc}$, the *radical of* $\Omega$ is the set

$$\operatorname{rad}\Omega := \Big\{ Y \in \mathcal{M}_{\rm nc} \colon \bigoplus_{\alpha=1}^{m} Y \in \Omega \text{ for some } m \in \mathbb{N} \Big\}.$$

We will say that $\Omega \subseteq \mathcal{M}_{\rm nc}$ is a *radical set* if $\operatorname{rad}\Omega = \Omega$.

PROPOSITION 7.15. *For a set $\Omega \subseteq \mathcal{M}_{\rm nc}$, one has*

$$\widetilde{\operatorname{rad}\Omega} \subseteq \operatorname{rad}\widetilde{\Omega}, \tag{7.19}$$

*where $\widetilde{\Omega}$ and $\widetilde{\operatorname{rad}\Omega}$ are defined by* (A.1). *In particular, if $\Omega$ is similarity invariant, then so is* $\operatorname{rad}\Omega$. *If $\Omega$ is furthermore a nc set, then so is* $\operatorname{rad}\Omega$. *If $\Omega$ is furthermore right admissible, then so is* $\operatorname{rad}\Omega$.

PROOF. If $Y \in \widetilde{\operatorname{rad}\Omega}$, then $SYS^{-1} \in \operatorname{rad}\Omega$ for every invertible matrix over $\mathcal{R}$ of the same size as $Y$, i.e.,

$$\bigoplus_{\alpha=1}^{m} SYS^{-1} = \Big(\bigoplus_{\alpha=1}^{m} S\Big)\Big(\bigoplus_{\alpha=1}^{m} Y\Big)\Big(\bigoplus_{\alpha=1}^{m} S\Big)^{-1} \in \Omega$$

for some $m \in \mathbb{N}$. Therefore, $\bigoplus_{\alpha=1}^{m} Y \in \widetilde{\Omega}$ and thus $Y \in \operatorname{rad}\widetilde{\Omega}$. The inclusion (7.19) follows.

In particular, if $\widetilde{\Omega} = \Omega$, we obtain $\widetilde{\operatorname{rad}\Omega} \subseteq \operatorname{rad}\Omega$. Together with the obvious inclusion $\widetilde{\operatorname{rad}\Omega} \supseteq \operatorname{rad}\Omega$, this implies the equality $\widetilde{\operatorname{rad}\Omega} = \operatorname{rad}\Omega$, i.e., $\operatorname{rad}\Omega$ is similarity invariant.

Let $Y, Y' \in \operatorname{rad}\Omega$. Then $\bigoplus_{\alpha=1}^{m} Y \in \Omega$ and $\bigoplus_{\alpha=1}^{m'} Y' \in \Omega$ for some $m$ and $m'$. If $\Omega$ is a nc set, then $\bigoplus_{\alpha=1}^{\mathrm{lcm}(m,m')} Y \oplus \bigoplus_{\beta=1}^{\mathrm{lcm}(mm')} Y' \in \Omega$. Since $\Omega$ is similarity invariant, the permutation of blocks gives that $\bigoplus_{\alpha=1}^{\mathrm{lcm}(m,m')}(Y \oplus Y') \in \Omega$, and hence $Y \oplus Y' \in \operatorname{rad}\Omega$. We conclude that $\operatorname{rad}\Omega$ is a nc set in this case.

Suppose that $\Omega$ is a right admissible similarity invariant nc set. Let $Y \in (\operatorname{rad}\Omega)_n$, $Y' \in (\operatorname{rad}\Omega)_{n'}$, $Z \in \mathcal{R}^{n \times n'}$. Then $\bigoplus_{\alpha=1}^{m} Y \in \Omega_{nm}$ and $\bigoplus_{\alpha=1}^{m'} Y' \in \Omega_{n'm'}$ for some $m$ and $m'$. Moreover, $\bigoplus_{\alpha=1}^{\mathrm{lcm}(m,m')} Y \in \Omega_{\mathrm{lcm}(m,m')n}$ and $\bigoplus_{\alpha=1}^{\mathrm{lcm}(m,'m')'} Y' \in \Omega_{\mathrm{lcm}(m,m')n'}$. By Proposition A.2,

$$\begin{bmatrix} \bigoplus_{\alpha=1}^{\mathrm{lcm}(m,m')} Y & \bigoplus_{\alpha=1}^{\mathrm{lcm}(m,m')} Z \\ 0 & \bigoplus_{\alpha=1}^{\mathrm{lcm}(m,m')} Y' \end{bmatrix} \in \Omega_{(n+n')\mathrm{lcm}(m,m')}.$$

Since $\Omega$ is similarity invariant, the permutation of blocks gives that

$$\bigoplus_{\alpha=1}^{\mathrm{lcm}(m,m')} \begin{bmatrix} Y & Z \\ 0 & Y' \end{bmatrix} \in \Omega_{(n+n')\mathrm{lcm}(m,m')}.$$

Therefore, $\begin{bmatrix} Y & Z \\ 0 & Y' \end{bmatrix} \in \operatorname{rad}\Omega$. This implies that the nc set $\operatorname{rad}\Omega$ is right admissible.

□

We note that inclusion (7.19) can be proper, even in the case of nc sets. This example also shows that $\widetilde{\Omega}$ is not necessarily radical, even if $\Omega$ is.

EXAMPLE 7.16. Let $\Omega \subseteq \mathbb{C}_{\mathrm{nc}}$ be a nc set consisting of matrices $\bigoplus_{\alpha=1}^{m} Y$, $m = 1, 2, \ldots$, with $Y = \operatorname{diag}[0,1,1,0] \in \mathbb{C}^{4\times 4}$. Clearly, the set $\Omega$ is radical. Its similarity invariant envelope, $\widetilde{\Omega}$, consists of diagonalizable square matrices over $\mathbb{C}$ with eigenvalues 0 and 1 of the same even multiplicity. In particular, $\widetilde{\Omega}$ contains the matrix $\operatorname{diag}[0,1,0,1] \in \mathbb{C}^{4\times 4}$, so that the matrix $\operatorname{diag}[0,1] \in \mathbb{C}^{2\times 2}$ belongs to $\operatorname{rad}\widetilde{\Omega}$. Thus, $\widetilde{\operatorname{rad}\Omega} = \widetilde{\Omega}$ is a proper subset of $\operatorname{rad}\widetilde{\Omega}$.

We also note that $\operatorname{rad}\Omega$ is not necessarily a nc set, even if $\Omega$ is.

EXAMPLE 7.17. Let $\Omega \subseteq \mathbb{C}_{\mathrm{nc}}$ be a nc set consisting of matrices $0_{2\times 2}, I_2 \in \Omega_2$ and all their direct sums, in any possible order. Then $1 \times 1$ matrices 0 and 1 belong to $\operatorname{rad}\Omega$, however their direct sum, $0 \oplus 1 = \operatorname{diag}[0,1]$ does not.

Let $\mathcal{W}$ be an operator space. We now define another topology on $\mathcal{W}_{\mathrm{nc}}$ by means of the pseudometric $\tau\colon \mathcal{W}_{\mathrm{nc}} \times \mathcal{W}_{\mathrm{nc}} \to \mathbb{R}_{+}$:

$$\tau(X, X') := \Big\| \bigoplus_{\alpha=1}^{\mathrm{lcm}(n,n')/n} X - \bigoplus_{\beta=1}^{\mathrm{lcm}(n,n')/n'} X' \Big\|_{\mathrm{lcm}(n,n')} \tag{7.20}$$

for every $n, n' \in \mathbb{N}$, $X \in \mathcal{W}^{n\times n}$, and $X' \in \mathcal{W}^{n'\times n'}$. We will call this topology a *$\tau$-topology*, and the corresponding open sets *$\tau$-open*. *$\tau$-balls* centered at points $Y \in \mathcal{W}_{\rm nc}$ of radii $\delta > 0$,

$$B_\tau(Y, \delta) := \{X \in \mathcal{W}_{\rm nc} \colon \tau(X, Y) < \delta\},$$

form a base for the $\tau$-topology. Notice that $\tau$-balls, and thus $\tau$-open sets, are radical.

PROPOSITION 7.18. *Let $\mathcal{W}$ be an operator space.*

(1) *For every $s \in \mathbb{N}$, $Y \in \mathcal{W}^{s\times s}$, and $\delta > 0$, one has*

$$B_{\rm nc}(Y, \delta) = B_\tau(Y, \delta) \cap \coprod_{m=1}^{\infty} \mathcal{W}^{sm\times sm}. \tag{7.21}$$

(2) *If $\Omega \subseteq \mathcal{W}_{\rm nc}$ is a uniformly-open set, then* rad $\Omega$ *is the smallest $\tau$-open set that contains $\Omega$.*

(3) *$\Omega \subseteq \mathcal{W}_{\rm nc}$ is a $\tau$-open set if and only if $\Omega$ is uniformly-open and radical.*

PROOF. (1) follows from the definition of $\tau$-balls.

(2) It follows from (7.21) that rad $B_{\rm nc}(Y, \delta) = B_\tau(Y, \delta)$. Then

$$\operatorname{rad}\Omega = \operatorname{rad}\bigcup_{Y\in\Omega} B_{\rm nc}(Y, \delta_Y) = \bigcup_{Y\in\Omega} \operatorname{rad} B_{\rm nc}(Y, \delta_Y) = \bigcup_{Y\in\Omega} B_\tau(Y, \delta_Y),$$

where $\delta_Y$ is any positive real number such that $B_{\rm nc}(Y, \delta_Y) \subseteq \Omega$. Therefore, rad $\Omega$ is $\tau$-open. Since any $\tau$-open set is radical, if such a set contains $\Omega$, then it must contain rad $\Omega$ as well. We conclude that rad $\Omega$ is the smallest $\tau$-open set that contains $\Omega$.

(3) Part (2) implies that if $\Omega$ is uniformly-open and radical, i.e., rad $\Omega = \Omega$, then $\Omega$ is $\tau$-open. The converse is obvious. □

By part (1) of Proposition 7.18 that the $\tau$-topology on $\mathcal{W}_{\rm nc}$ is weaker than the uniformly-open topology. However, two points of $\mathcal{W}_{\rm nc}$ cannot be separated by two nc balls if and only if they cannot be separated by two $\tau$-balls. Indeed, Proposition 7.14 says that $X \sim X'$ if and only if $\tau(X, X') = 0$. Therefore, the quotient spaces of $\mathcal{W}_{\rm nc}$ with respect to the equivalence $\sim$ and the $\tau$-equivalence coincide as sets. Since the pre-image of any set in the quotient set $\widehat{\mathcal{W}_{\rm nc}}$ under the quotient map is radical, the corresponding quotient topologies on $\widehat{\mathcal{W}_{\rm nc}}$ coincide. In fact, $\widehat{\mathcal{W}_{\rm nc}}$ is a metric space with respect to the quotient metric $\widehat{\tau}$ which is defined by $\widehat{\tau}(\widehat{X}, \widehat{X'}) := \tau(X, X')$ for any representatives $X$ and $X'$ of the cosets $\widehat{X}$ and $\widehat{X'}$, respectively (the definition is correct due to the properties of the system of norms $\|\cdot\|_n$ on matrices over an operator space). The quotient topology on $\widehat{\mathcal{W}_{\rm nc}}$ is the metric topology with respect to the metric $\widehat{\tau}$.

## 7.3. Uniformly analytic nc functions

Let $\mathcal{V}$ and $\mathcal{W}$ be Banach spaces equipped with admissible systems of matrix norms. A linear mapping $\phi \colon \mathcal{V} \to \mathcal{W}$ is called *completely bounded* if

$$\|\phi\|_{\rm cb} := \sup_{n\in\mathbb{N}} \|\operatorname{id}_{\mathbb{C}^{n\times n}} \otimes \phi\|_{\mathcal{V}^{n\times n}\to\mathcal{W}^{n\times n}} < \infty,$$

*completely contractive* if $\|\phi\|_{\rm cb} \le 1$, and *completely isometric* if $\operatorname{id}_{\mathbb{C}^{n\times n}} \otimes \phi$ is an isometry for every $n \in \mathbb{N}$.

PROPOSITION 7.19. *If $\mathcal{W}$ is a Banach space equipped with an admissible system of matrix norms $\|\cdot\|_n$ such that (7.1) and (7.2) hold with $C_1$, $C_1'$, and $C_2$ independent of $n$ and $m$, then the injections (7.3) and the projections (7.4) are uniformly completely bounded, i.e., for every $n \in \mathbb{N}$ and $i, j = 1, \ldots, n$, the mappings $\iota_{ij}$ and $\pi_{ij}$ are completely bounded. Moreover,*

$$\|\iota_{ij}\|_{\rm cb} \leq C_1' C_2^2, \quad \|\pi_{ij}\|_{\rm cb} \leq C_1 C_2^2.$$

*If $\mathcal{W}$ is an operator space, then all $\iota_{ij}$ are complete isometries and all $\pi_{ij}$ are complete coisometries.*

PROOF. It follows from the proof of Proposition 7.1 that $\|\iota_{ij}^s\| \leq C_1' C_2$ and $\|\pi_{ij}^s\| \leq C_1 C_2$ for all $n, i, j, s$. We have, for any $n, i, j, s$ and $W \in \mathcal{W}^{s\times s}$, that

$$\begin{aligned}
\|(\mathrm{id}_{\mathbb{C}^{s\times s}} \otimes \iota_{ij})(W)\|_{sn} &= \Big\|\Big(\mathrm{id}_{\mathbb{C}^{s\times s}} \otimes \iota_{ij}\Big)\Big(\sum_{a,b=1}^{s} E_{ab} w_{ab}\Big)\Big\|_{sn} \\
&= \Big\|\sum_{a,b=1}^{s} E_{ab} \otimes \iota_{ij}(w_{ab})\Big\|_{sn} = \Big\|\sum_{a,b=1}^{s} E_{ab} \otimes (E_{ij} w_{ab})\Big\|_{sn} \\
&= \Big\|\sum_{a,b=1}^{s} (E_{ab} \otimes E_{ij}) w_{ab}\Big\|_{sn} = \Big\|\sum_{a,b=1}^{s} P(s,n)(E_{ij} \otimes E_{ab}) P(s,n)^\top w_{ab}\Big\|_{sn} \\
&= \Big\|\sum_{a,b=1}^{s} P(s,n)(E_{ij} \otimes E_{ab} w_{ab}) P(s,n)^\top\Big\|_{sn} = \|P(s,n)(E_{ij} \otimes W) P(s,n)^\top\|_{sn} \\
&\leq C_2 \|E_{ij} \otimes W\|_{sn} = C_2 \|\iota_{ij}^s(W)\|_{sn} \leq C_2 \|\iota_{ij}^s\| \|W\|_s.
\end{aligned}$$

Here we use the $sn \times sn$ permutation matrices $P(s,n) = [E_{ij}^T]_{i=1,\ldots,s;\, j=1,\ldots,n}$ which allow us to change the order of factors in tensor products:

$$A \otimes B = P(r,p)(B \otimes A) P(m,q)^\top \tag{7.22}$$

for any $A \in \mathbb{C}^{r\times m}$ and $B \in \mathbb{C}^{p\times q}$. See [**62**, Pages 259–261]. We conclude from the calculation above that

$$\|\iota_{ij}\|_{\rm cb} \leq C_2 \sup_{s\in\mathbb{N}} \|\iota_{ij}^s\| \leq C_1' C_2^2.$$

Suppose now that

$$W = [([w_{i'j'}]_{i',j'=1,\ldots,n})_{ab}]_{a,b=1,\ldots,s} \in \big(\mathcal{W}^{n\times n}\big)^{s\times s} \cong \mathbb{C}^{s\times s} \otimes \mathcal{W}^{n\times n}.$$

Then

$$\begin{aligned}
\widetilde{W} := P(n,s) W P(n,s)^\top &= [([\widetilde{w}_{ab}]_{a,b=1,\ldots,s})_{i'j'}]_{i',j'=1,\ldots,n} \\
&\in \big(\mathcal{W}^{s\times s}\big)^{n\times n} \cong \mathbb{C}^{n\times n} \otimes \mathcal{W}^{s\times s},
\end{aligned}$$

where

$$(\widetilde{w}_{ab})_{i'j'} = (w_{i'j'})_{ab}, \quad a, b = 1 \ldots, s;\ i', j' = 1, \ldots, n.$$

We have

$$\|(\mathrm{id}_{\mathbb{C}^{s\times s}}\otimes\pi_{ij})(W)\|_s = \Big\|\Big(\mathrm{id}_{\mathbb{C}^{s\times s}}\otimes\pi_{ij}\Big)\Big(\sum_{a,b=1}^{s}\sum_{i',j'=1}^{n} E_{ab}\otimes E_{i'j'}(w_{i'j'})_{ab}\Big)\Big\|_s$$
$$= \Big\|\sum_{a,b=1}^{s} E_{ab}(w_{ij})_{ab}\Big\|_s = \Big\|\sum_{a,b=1}^{s} E_{ab}(\widetilde{w}_{ab})_{ij}\Big\|_s = \|\widetilde{W}_{ij}\|_s = \|\pi^s_{ij}(\widetilde{W})\|_s$$
$$\le \|\pi^s_{ij}\|\,\|\widetilde{W}\|_{sn} \le C_2\|\pi^s_{ij}\|\,\|W\|_{sn}.$$

Therefore,

$$\|\pi_{ij}\|_{\mathrm{cb}} \le C_2\sup_{s\in\mathbb{N}}\|\pi^s_{ij}\| \le C_1C_2^2.$$

In the case of operator spaces, it follows that all the operators $\iota_{ij}$ and $\pi_{ij}$ are complete contractions. Moreover, since $\pi_{ij}\iota_{ij} = I_{\mathcal{W}}$, we conclude that all $\iota_{ij}$ are complete isometries, and all $\pi_{ij}$ are complete coisometries. □

Let $\mathcal{V},\mathcal{W}$ be operator spaces, and let $\Omega\subseteq\mathcal{V}_{\mathrm{nc}}$ be a uniformly open nc set. A nc function $f\colon\Omega\to\mathcal{W}_{\mathrm{nc}}$ is called *uniformly locally bounded* if for any $s\in\mathbb{N}$ and $Y\in\Omega_s$ there exists a $r>0$ such that $B_{\mathrm{nc}}(Y,r)\subseteq\Omega$ and $f$ is bounded on $B_{\mathrm{nc}}(Y,r)$, i.e., there is a $M>0$ such that $\|f(X)\|_{sm}\le M$ for all $m\in\mathbb{N}$ and $X\in B_{\mathrm{nc}}(Y,r)_{sm}$. A function $f\colon\Omega\to\mathcal{W}_{\mathrm{nc}}$ is called *uniformly analytic* if $f$ is uniformly locally bounded and analytic, i.e., $f$ is uniformly locally bounded and G-differentiable. Later on (see Example 8.14), we will present an example of analytic (in fact, entire) nc function which is not uniformly analytic in any neighborhood of $0_{1\times1}$.

PROPOSITION 7.20. *Let $\Omega\subseteq\mathcal{V}_{\mathrm{nc}}$ be a uniformly open nc set. If a nc function $f\colon\Omega\to\mathcal{W}_{\mathrm{nc}}$ is continuous with respect to the uniformly-open topologies on $\mathcal{V}_{\mathrm{nc}}$ and $\mathcal{W}_{\mathrm{nc}}$, then $f$ is uniformly locally bounded.*

PROOF. Given $s\in\mathbb{N}$ and $Y\in\Omega_s$, there exists a nc ball $B_{\mathrm{nc}}(Y,r)\subseteq\Omega$ such that $\|f(X)-f(\bigoplus_{\alpha=1}^m Y)\|_{sm}<1$ for all $m\in\mathbb{N}$ and $X\in B_{\mathrm{nc}}(Y,r)_{sm}$. Then

$$\|f(X)\|_{sm} < \Big\|f\Big(\bigoplus_{\alpha=1}^{m} Y\Big)\Big\|_{sm}+1 = \Big\|\bigoplus_{\alpha=1}^{m} f(Y)\Big\|_{sm}+1 \le \|f(Y)\|_s+1.$$

□

Let $l\colon\mathcal{V}\to\mathcal{W}$ be a linear mapping of operator spaces. We extend $l$ to a nc function $\widetilde{l}\colon\mathcal{V}_{\mathrm{nc}}\to\mathcal{W}_{\mathrm{nc}}$ such that $l_n:=\widetilde{l}|_{\mathcal{V}^{n\times n}}$ is the linear mapping from $\mathcal{V}^{n\times n}$ to $\mathcal{W}^{n\times n}$ defined by $l_n([v_{ij}])=[l(v_{ij})]$, $n\in\mathbb{N}$, i.e., $l_n=\mathrm{id}_{\mathbb{C}^{n\times n}}\otimes l$. Then it is clear that $\widetilde{l}$ is uniformly locally bounded if and only if $\widetilde{l}$ is continuous with respect to the uniformly-open topologies if and only if $l$ is completely bounded. We will see shortly (Corollary 7.28) that the first equivalence is true for arbitrary nc functions.

We will present two versions of the main result on the uniform convergence of the TT series in the uniformly-open topology on two different types of domains. The proof of the first theorem (like that of Theorem 7.4) is of a mixed nature: noncommutative and complex analytic. The second theorem admits a purely noncommutative proof; in particular, it admits an analogue in the real case, see Remark 7.35 below.

Let $\mathcal{X}$ be a vector space over $\mathbb{C}$ and let $Y \in \mathcal{X}^{s\times s}$. A nc set $\Upsilon_{\rm nc} \subseteq \mathcal{X}_{\rm nc}$ is called *complete circular about* $Y$ if for every $m \in \mathbb{N}$ the set $(\Upsilon_{\rm nc})_{sm} = \Upsilon_{\rm nc} \cap \mathcal{X}^{sm\times sm}$ is complete circular about $\bigoplus_{\alpha=1}^{m} Y$ (see the paragraph preceding Theorem 7.2). For any $\epsilon > 0$, we define

$$\Upsilon_{{\rm nc},\epsilon} := \coprod_{m=1}^{\infty} \Big\{X \in (\Upsilon_{\rm nc})_{sm} \colon \bigoplus_{\alpha=1}^{m} Y + (1+\epsilon)\Big(X - \bigoplus_{\alpha=1}^{m} Y\Big) \in (\Upsilon_{\rm nc})_{sm}\Big\}. \tag{7.23}$$

It is easily seen that if $\Upsilon_{\rm nc}$ is a uniformly open complete circular nc set about $Y$, then so is $\Upsilon_{{\rm nc},\epsilon}$, and $\bigcup_{\epsilon>0} \Upsilon_{{\rm nc},\epsilon} = \Upsilon_{\rm nc}$.

THEOREM 7.21. *Let a nc function $f\colon \Omega \to \mathcal{W}_{\rm nc}$ be uniformly locally bounded. Let $s \in \mathbb{N}$, $Y \in \Omega_s$, and let $\Upsilon_{\rm nc}$ be a uniformly open complete circular nc set about $Y$ such that $\Upsilon_{\rm nc} \subseteq \Omega$ and $f$ is bounded $\Upsilon_{\rm nc}$. Then, for every $\epsilon > 0$, $m \in \mathbb{N}$, and $X \in (\Upsilon_{{\rm nc},\epsilon})_{sm}$,*

$$f(X) = \sum_{\ell=0}^{\infty} \Big(X - \bigoplus_{\alpha=1}^{m} Y\Big)^{\odot_s \ell} \Delta_R^{\ell} f\big(\underbrace{Y, \ldots, Y}_{\ell+1 \text{ times}}\big), \tag{7.24}$$

*where the series converges absolutely and uniformly on $\Upsilon_{{\rm nc},\epsilon}$. Moreover,*

$$\sum_{\ell=0}^{\infty} \sup_{m\in\mathbb{N},\, X\in(\Upsilon_{{\rm nc},\epsilon})_{sm}} \Big\|\Big(X - \bigoplus_{\alpha=1}^{m} Y\Big)^{\odot_s \ell} \Delta_R^{\ell} f\big(\underbrace{Y, \ldots, Y}_{\ell+1 \text{ times}}\big)\Big\|_{sm} < \infty. \tag{7.25}$$

PROOF. Let $s \in \mathbb{N}$ and $Y \in \Omega_s$ be fixed. Let $\Upsilon_{\rm nc}$ be a uniformly open complete circular set about $Y$ on which $f$ is bounded, say by $M$. Given $\epsilon > 0$, $m \in \mathbb{N}$, and $X \in (\Upsilon_{{\rm nc},\epsilon})_{sm}$, as in the proof of Theorem 7.4 we have that

$$\begin{aligned} &\Big\|\Big(X - \bigoplus_{\alpha=1}^{m} Y\Big)^{\odot_s \ell} \Delta_R^{\ell} f\big(\underbrace{Y, \ldots, Y}_{\ell+1 \text{ times}}\big)\Big\|_{sm} \\ &= \Big\|\Delta_R^{\ell} f\Big(\underbrace{\bigoplus_{\alpha=1}^{m} Y, \ldots, \bigoplus_{\alpha=1}^{m} Y}_{\ell+1 \text{ times}}\Big)\Big(\underbrace{X - \bigoplus_{\alpha=1}^{m} Y, \ldots, X - \bigoplus_{\alpha=1}^{m} Y}_{\ell \text{ times}}\Big)\Big\|_{sm} \le \frac{M}{(1+\epsilon)^{\ell}}, \end{aligned} \tag{7.26}$$

but in this case the constant $M$ is independent on $m$. Therefore (7.25) holds and the TT series in (7.24) converges absolutely and uniformly on $\Upsilon_{{\rm nc},\epsilon}$. The fact that the sum of the series equals $f(X)$ follows from Theorem 7.2. □

Let $\mathcal{X}$ be a vector space over $\mathbb{C}$ and let $Y \in \mathcal{X}^{s\times s}$. A nc set $\Upsilon_{\rm nc} \subseteq \mathcal{X}_{\rm nc}$ is called *matrix circular about* $Y$ if for every $m \in \mathbb{N}$, $X \in (\Upsilon_{\rm nc})_{sm}$, and unitary matrices $U, V \in \mathbb{C}^{m\times m}$ we have

$$\bigoplus_{\alpha=1}^{m} Y + (U \otimes I_s)\Big(X - \bigoplus_{\alpha=1}^{m} Y\Big)(V \otimes I_s) \in (\Upsilon_{\rm nc})_{sm}.$$

For any $\epsilon > 0$, we define $\Upsilon_{\mathrm{nc},\epsilon}$ as in (7.23). It is easily seen that if $\Upsilon_{\mathrm{nc}}$ is a uniformly open matrix circular nc set about $Y$, then so is $\Upsilon_{\mathrm{nc},\epsilon}$.

LEMMA 7.22. *Let $\Upsilon_{\mathrm{nc}} \subseteq \mathcal{X}_{\mathrm{nc}}$ be a matrix circular nc set about $Y \in \mathcal{X}^{s\times s}$. Then, for every $m, N \in \mathbb{N}$, $X \in (\Upsilon_{\mathrm{nc}})_{sm}$, and $\zeta, \eta \in \mathbb{C}$ with $|\zeta|^2 + |\eta|^2 = 1$,*

$$A := \begin{bmatrix} \bigoplus_{\alpha=1}^{m} Y & X - \bigoplus_{\alpha=1}^{m} Y & 0 & \dots & 0 \\ 0 & \ddots & \ddots & \ddots & \vdots \\ \vdots & & \ddots & \ddots & 0 \\ \vdots & & & \ddots & X - \bigoplus_{\alpha=1}^{m} Y \\ 0 & \dots & \dots & 0 & \bigoplus_{\alpha=1}^{m} Y \end{bmatrix} \in (\Upsilon_{\mathrm{nc}})_{sm(N+1)} \tag{7.27}$$

*and*

$$B := \begin{bmatrix} \bigoplus_{\alpha=1}^{m} Y & X - \bigoplus_{\alpha=1}^{m} Y & 0 & \dots & 0 \\ 0 & \ddots & \ddots & \ddots & \vdots \\ \vdots & & \ddots & X - \bigoplus_{\alpha=1}^{m} Y & 0 \\ \vdots & & \ddots & \bigoplus_{\alpha=1}^{m} Y & \zeta\left(X - \bigoplus_{\alpha=1}^{m} Y\right) \\ 0 & \dots & \dots & 0 & \bigoplus_{\alpha=1}^{m} Y + \eta\left(X - \bigoplus_{\alpha=1}^{m} Y\right) \end{bmatrix} \in (\Upsilon_{\mathrm{nc}})_{sm(N+2)}. \tag{7.28}$$

PROOF. We observe that

$$\left(\begin{bmatrix} 0 & \dots & \dots & 0 & I_m \\ I_m & \ddots & & \vdots & 0 \\ 0 & \ddots & \ddots & \vdots & \vdots \\ \vdots & \ddots & \ddots & 0 & \vdots \\ 0 & \dots & 0 & I_m & 0 \end{bmatrix} \otimes I_s\right)\left(A - \bigoplus_{\alpha=1}^{m(N+1)} Y\right)$$
$$= \begin{bmatrix} 0 & 0 & \dots & 0 \\ 0 & X - \bigoplus_{\alpha=1}^{m} Y & \ddots & \vdots \\ \vdots & \ddots & \ddots & 0 \\ 0 & \dots & 0 & X - \bigoplus_{\alpha=1}^{m} Y \end{bmatrix}$$

and

$$\left(\begin{bmatrix} 0 & \dots & \dots & 0 & -\eta I_m & \zeta I_m \\ I_m & \ddots & & \vdots & 0 & 0 \\ 0 & \ddots & \ddots & \vdots & \vdots & \vdots \\ \vdots & \ddots & \ddots & 0 & \vdots & \vdots \\ \vdots & & \ddots & I_m & 0 & 0 \\ 0 & \dots & \dots & 0 & \overline{\zeta} I_m & \overline{\eta} I_m \end{bmatrix} \otimes I_s\right)\left(B - \bigoplus_{\alpha=1}^{m(N+2)} Y\right)$$

$$= \begin{bmatrix} 0 & 0 & \dots & 0 \\ 0 & X - \bigoplus_{\alpha=1}^{m} Y & \ddots & \vdots \\ \vdots & \ddots & \ddots & 0 \\ 0 & \dots & 0 & X - \bigoplus_{\alpha=1}^{m} Y \end{bmatrix},$$

with square matrices of size $sm(N+1)$ in the first equality and $sm(N+2)$ in the second equality. Hence the result follows. $\square$

THEOREM 7.23. *Let a nc function $f\colon \Omega \to \mathcal{W}_{\rm nc}$ be uniformly locally bounded. Let $s \in \mathbb{N}$, $Y \in \Omega_s$, and let $\Upsilon_{\rm nc}$ be a uniformly open matrix circular nc set about $Y$ that is contained in $\Omega$ and on which $f$ is bounded. Then, for every $\epsilon > 0$, (7.24) is true, with the TT series converging absolutely and uniformly, on $\Upsilon_{{\rm nc},\epsilon}$ and (7.25) holds.*

PROOF. Let $\Upsilon_{\rm nc}$ be a uniformly open matrix circular nc set about $Y$, with the norm of $f$ bounded on $\Upsilon_{\rm nc}$, say by $M > 0$. For $\epsilon > 0$, $m \in \mathbb{N}$, and $X \in (\Upsilon_{{\rm nc},\epsilon})_{sm}$, we have

$$(7.29)\quad \Big\|\Delta_R^{\ell} f\big(\underbrace{\bigoplus_{\alpha=1}^{m} Y, \dots, \bigoplus_{\alpha=1}^{m} Y}_{\ell+1 \text{ times}}\big)\big(\underbrace{X - \bigoplus_{\alpha=1}^{m} Y, \dots, X - \bigoplus_{\alpha=1}^{m} Y}_{\ell \text{ times}}\big)\Big\|_{sm}$$

$$= (1+\epsilon)^{-\ell}\Big\|\Delta_R^{\ell} f\big(\underbrace{\bigoplus_{\alpha=1}^{m} Y, \dots, \bigoplus_{\alpha=1}^{m} Y}_{\ell+1 \text{ times}}\big)\big(\underbrace{(1+\epsilon)\big(X - \bigoplus_{\alpha=1}^{m} Y\big), \dots, (1+\epsilon)\,\big(X - \bigoplus_{\alpha=1}^{m} Y\big)}_{\ell \text{ times}}\big)\Big\|_{sm}$$

$$= (1+\epsilon)^{-\ell}\left\| \pi_{1,\ell+1} f\left(\begin{bmatrix} \bigoplus_{\alpha=1}^{m} Y & (1+\epsilon)(X - \bigoplus_{\alpha=1}^{m} Y) & 0 & \dots & 0 \\ 0 & \ddots & \ddots & \ddots & \vdots \\ \vdots & \ddots & \ddots & \ddots & 0 \\ \vdots & & \ddots & \ddots & (1+\epsilon)(X - \bigoplus_{\alpha=1}^{m} Y) \\ 0 & \dots & \dots & 0 & \bigoplus_{\alpha=1}^{m} Y \end{bmatrix}\right)\right\|_{sm}$$

$$\le \frac{M}{(1+\epsilon)^{\ell}},$$

which implies that (7.25) holds and the TT series in (7.24) converges absolutely and uniformly on $\Upsilon_{\mathrm{nc},\epsilon}$. Here we used the fact that

$$\begin{bmatrix} \bigoplus_{\alpha=1}^m Y & (1+\epsilon)(X-\bigoplus_{\alpha=1}^m Y) & 0 & \ldots & 0 \\ 0 & \ddots & \ddots & \ddots & \vdots \\ \vdots & \ddots & \ddots & \ddots & 0 \\ \vdots & & \ddots & \ddots & (1+\epsilon)(X-\bigoplus_{\alpha=1}^m Y) \\ 0 & \ldots & \ldots & 0 & \bigoplus_{\alpha=1}^m Y \end{bmatrix} \in (\Upsilon_{\mathrm{nc}})_{sm(\ell+1)} \tag{7.30}$$

by (7.27), with $X$ replaced by $\bigoplus_{\alpha=1}^m Y + (1+\epsilon)(X-\bigoplus_{\alpha=1}^m Y)$ and $N$ replaced by $\ell$.

It remains to prove that the sum of the series equals $f(X)$, i.e., that the remainder term in the TT formula (see (4.1) and (4.2)),

$$\Delta_R^{N+1} f\Big(\underbrace{\bigoplus_{\alpha=1}^m Y, \ldots, \bigoplus_{\alpha=1}^m Y}_{N+1 \text{ times}}, X\Big)\Big(\underbrace{\Big(X-\bigoplus_{\alpha=1}^m Y\Big), \ldots, \Big(X-\bigoplus_{\alpha=1}^m Y\Big)}_{N+1 \text{ times}}\Big)$$
$$= \Big(X-\bigoplus_{\alpha=1}^m Y\Big)^{\odot_s (N+1)} \Delta_R^{N+1} f\big(\underbrace{Y, \ldots, Y}_{N+1 \text{ times}}, X\big),$$

tends to 0 as $N \to \infty$. We have

$$\Big\|\Delta_R^{N+1} f\Big(\underbrace{\bigoplus_{\alpha=1}^m Y, \ldots, \bigoplus_{\alpha=1}^m Y}_{N+1 \text{ times}}, X\Big)\Big(\underbrace{(X-\bigoplus_{\alpha=1}^m Y), \ldots, (X-\bigoplus_{\alpha=1}^m Y)}_{N+1 \text{ times}}\Big)\Big\|_{sm}$$
$$= \frac{1}{\delta(1+\epsilon)^N}\Big\|\Delta_R^{N+1} f\Big(\underbrace{\bigoplus_{\alpha=1}^m Y, \ldots, \bigoplus_{\alpha=1}^m Y}_{N+1 \text{ times}}, X\Big)$$
$$\Big(\underbrace{(1+\epsilon)(X-\bigoplus_{\alpha=1}^m Y), \ldots, (1+\epsilon)(X-\bigoplus_{\alpha=1}^m Y)}_{N \text{ times}}, \delta(X-\bigoplus_{\alpha=1}^m Y)\Big)\Big\|_{sm}$$
$$= \frac{1}{\delta(1+\epsilon)^N}$$
$$\cdot \left\| \pi_{1,N+2} f\left( \begin{bmatrix} \bigoplus_{\alpha=1}^m Y & (1+\epsilon)(X-\bigoplus_{\alpha=1}^m Y) & 0 & \ldots & 0 \\ 0 & \ddots & \ddots & \ddots & \vdots \\ \vdots & \ddots & \ddots & (1+\epsilon)(X-\bigoplus_{\alpha=1}^m Y) & 0 \\ \vdots & & \ddots & \bigoplus_{\alpha=1}^m Y & \delta(X-\bigoplus_{\alpha=1}^m Y) \\ 0 & \ldots & \ldots & 0 & X \end{bmatrix} \right) \right\|_{sm}$$
$$\leq \frac{M}{\delta(1+\epsilon)^N} \longrightarrow 0$$

as $N \to \infty$. The last inequality holds with $\delta = \sqrt{\epsilon^2 + 2\epsilon}$; indeed, for this choice of $\delta$ we have

(7.31)

$$\begin{bmatrix} \bigoplus_{\alpha=1}^{m} Y & (1+\epsilon)(X - \bigoplus_{\alpha=1}^{m} Y) & 0 & \dots & 0 \\ 0 & \ddots & \ddots & \ddots & \vdots \\ \vdots & \ddots & \ddots & (1+\epsilon)(X - \bigoplus_{\alpha=1}^{m} Y) & 0 \\ \vdots & & \ddots & \bigoplus_{\alpha=1}^{m} Y & \delta(X - \bigoplus_{\alpha=1}^{m} Y) \\ 0 & \dots & \dots & 0 & X \end{bmatrix} \in (\Upsilon_{\rm nc})_{sm(N+2)}$$

by (7.28) with $X$ replaced by $\bigoplus_{\alpha=1}^{m} Y + (1+\epsilon)(X - \bigoplus_{\alpha=1}^{m} Y)$, $\zeta = \frac{\delta}{1+\epsilon}$, $\eta = \frac{1}{1+\epsilon}$. □

REMARK 7.24. Notice that Theorem 7.23 holds under a weaker assumption that $\Upsilon_{\rm nc}$ is a uniformly open nc set satisfying (7.27) and (7.28).

REMARK 7.25. Notice that in the proof of Theorem 7.23 we actually established the following result. Let $\Upsilon_{\rm nc}$ be a uniformly open nc set, which contains $Y$, is contained in $\Omega$, and on which $f$ is bounded. Then, for every $\epsilon > 0$, (7.24) holds, with the series converging absolutely and uniformly, on the set of all $X \in \Omega$ such that

- (7.30) holds for all $\ell \in \mathbb{N}$;
- there exists $\delta > 0$ such that (7.31) holds for all $N \in \mathbb{N}$.

This gives even weaker assumptions on a domain where the TT series in (7.24) converges absolutely and uniformly.

We notice an important special case of the two previous theorems for the uniform convergence on open nc balls — clearly, nc balls are both complete circular and matrix circular.

COROLLARY 7.26. *Let a nc function $f : \Omega \to \mathcal{W}_{\rm nc}$ be uniformly locally bounded. For every $s \in \mathbb{N}$, $Y \in \Omega_s$, let $\delta := \sup\{r > 0 : f$ is bounded on $B_{\rm nc}(Y, r)\}$. Then* (7.24) *holds, with the TT series converging absolutely and uniformly, on every open nc ball $B_{\rm nc}(Y, r)$ with $r < \delta$. Moreover,*

$$\sum_{\ell=0}^{\infty} \sup_{m\in\mathbb{N},\, X\in B_{\rm nc}(Y,r)_{sm}} \left\| \left(X - \bigoplus_{\alpha=1}^{m} Y\right)^{\odot_s \ell} \Delta_R^{\ell} f(\underbrace{Y, \dots, Y}_{\ell+1 \text{ times}}) \right\|_{sm} < \infty. \tag{7.32}$$

REMARK 7.27. In the case where $Y = \mu \in \Omega_1$, Theorems 7.21 and 7.23, and Corollary 7.26, give results on the absolute and uniform convergence of the TT series (7.15) centered at a scalar point.

COROLLARY 7.28. *Let $\Omega \subseteq \mathcal{V}_{\rm nc}$ be a uniformly open nc set. Then a nc function $f : \Omega \to \mathcal{W}_{\rm nc}$ is uniformly locally bounded if and only if $f$ is continuous with respect to the uniformly-open topologies on $\mathcal{V}_{\rm nc}$ and $\mathcal{W}_{\rm nc}$ if and only if $f$ is uniformly analytic.*

PROOF. If $f$ is uniformly locally bounded, then $f$ is also locally bounded on slices, thus by Theorem 7.2 $f$ is G-differentiable. Therefore $f$ is uniformly analytic. Taking into account Proposition 7.20, it remains to show that if $f$ is uniformly locally bounded then $f$ is continuous with respect to the uniformly-open topologies.

Let the norm of $f$ be bounded by $M > 0$ on a nc ball $B_{\rm nc}(Y, r_1)$ centered at $Y \in \mathcal{V}^{s\times s}$. Then, using the estimate (7.26) or (7.29) with $\epsilon = \frac{r_1 - r}{r}$ for $r < r_1$, we obtain for $X \in B_{\rm nc}(Y, r)_{sm}$, $m \in \mathbb{N}$, that

$$\Big\| f(X) - f\Big(\bigoplus_{\alpha=1}^{m} Y\Big)\Big\|_{sm} \leq M \sum_{\ell=1}^{\infty} \Big(\frac{r}{r_1}\Big)^{\ell} = \frac{Mr}{r_1 - r}.$$

Hence, for any given $\eta > 0$ there exists $r < r_1$ such that for every $m \in \mathbb{N}$ and $X \in B_{\rm nc}(Y, r)_{sm}$ we have

$$\Big\| f(X) - f\Big(\bigoplus_{\alpha=1}^{m} Y\Big)\Big\|_{sm} < \eta.$$

In other words, $f$ is continuous at $Y$ with respect to the uniformly-open topologies. □

We consider now the case where the space $\mathcal{V}$ is finite dimensional, so that we can assume without loss of generality that $\mathcal{V} = \mathbb{C}^d$ with some operator space structure. Notice that any two operator space structures on $\mathbb{C}^d$ are completely isomorphic, i.e., the identity mapping of $\mathbb{C}^d$ is completely bounded both ways (see, e.g., [**79**, Proposition 1.10(iii)], thus the uniformly-open topology on $(\mathbb{C}^d)_{\rm nc}$ is uniquely determined. We can expand each term of the convergent TT series (of polylinear forms) using higher order nc difference-differential operators as in (4.12), (7.16), and (7.17), and consider the question of the convergence along the free monoid $\mathcal{G}_d$. This is a stronger notion of convergence, since we do not group together all the terms corresponding to the words of a given length. To define this convergence, we need to specify an order on $\mathcal{G}_d$; however, we will mostly deal with absolute convergence along $\mathcal{G}_d$, so that the choice of the order does not matter.

Let $\Upsilon_{\rm nc} \subseteq (\mathbb{C}^d)_{\rm nc}$ be a matrix circular nc set about $Y = (Y_1, \ldots, Y_d) \in \left(\mathbb{C}^{s\times s}\right)^d$, and let $\boldsymbol{\rho} = (\rho_1, \ldots, \rho_d)$ be a $d$-tuple of positive real numbers with $\sum_{j=1}^{d} \rho_j < 1$. We define

(7.33)
$$\Upsilon_{{\rm nc},\boldsymbol{\rho}} := \coprod_{m=1}^{\infty} \Big\{ X \in (\Upsilon_{\rm nc})_{sm} \colon \bigoplus_{\alpha=1}^{m} Y + \frac{1}{\rho_j}\Big(0, \ldots, 0, \underbrace{X_j - \bigoplus_{\alpha=1}^{m} Y_j}_{j-\text{th place}}, 0, \ldots, 0\Big) \in (\Upsilon_{\rm nc})_{sm} \quad \text{for all } j = 1, \ldots, d\Big\}.$$

If $\Upsilon_{\rm nc}$ is a uniformly open matrix circular nc set about $Y$, then so is $\Upsilon_{{\rm nc},\boldsymbol{\rho}}$; the fact that $\Upsilon_{{\rm nc},\boldsymbol{\rho}}$ is uniformly open follows from the fact that the linear mapping $Z \mapsto Z_j e_j$ on $\left(\mathbb{C}^{s\times s}\right)^d$ is completely bounded as a finite-rank mapping of operator spaces, see [**79**, Proposition 1.10(iii)].

THEOREM 7.29. *Suppose that $\Omega \subseteq (\mathbb{C}^d)_{\rm nc}$ is a uniformly open nc set, $\mathcal{W}$ is an operator space, $f \colon \Omega \to \mathcal{W}_{\rm nc}$ is a nc function which is uniformly locally bounded, $s \in \mathbb{N}$, $Y \in \Omega_s$. Let $\Upsilon_{\rm nc}$ be a uniformly open matrix circular nc set about $Y$ that is contained in $\Omega$ and on which $f$ is bounded. Then, for every $\boldsymbol{\rho} = (\rho_1, \ldots, \rho_d) \in$*

$(\mathbb{R}_+ \setminus \{0\})^d$ *with* $\sum_{j=1}^d \rho_j < 1$, $m \in \mathbb{N}$, *and* $X \in (\Upsilon_{\mathrm{nc},\boldsymbol{\rho}})_{sm}$,

$$f(X) = \sum_{w \in \mathcal{G}_d} \Big(X - \bigoplus_{\alpha=1}^m Y\Big) \Delta_R^{w^\top} f(\underbrace{Y, \ldots, Y}_{|w|+1 \text{ times}}) \tag{7.34}$$

$$= \sum_{w \in \mathcal{G}_d} \bigg(\Big(\bigoplus_{\alpha=1}^m A_{w,(0)}\Big) \otimes \Big(\bigoplus_{\alpha=1}^m A_{w,(1)}\Big) \otimes \cdots \otimes \Big(\bigoplus_{\alpha=1}^m A_{w,(|w|)}\Big)\bigg) \star \Big(X - \bigoplus_{\alpha=1}^m Y\Big)^{[w]},$$

*where the series converges absolutely and uniformly on* $\Upsilon_{\mathrm{nc},\boldsymbol{\rho}}$. *Moreover,*

$$\sum_{w \in \mathcal{G}_d} \sup_{m \in \mathbb{N},\, X \in (\Upsilon_{\mathrm{nc},\boldsymbol{\rho}})_{sm}} \bigg\| \bigg(\Big(\bigoplus_{\alpha=1}^m A_{w,(0)}\Big) \otimes \Big(\bigoplus_{\alpha=1}^m A_{w,(1)}\Big) \otimes \cdots \otimes \Big(\bigoplus_{\alpha=1}^m A_{w,(|w|)}\Big)\bigg) \star \Big(X - \bigoplus_{\alpha=1}^m Y\Big)^{[w]} \bigg\|_{sm} < \infty. \tag{7.35}$$

*Here* $A_{w,(0)} \otimes A_{w,(1)} \otimes \cdots \otimes A_{w,(|w|)} = \Delta_R^{w^\top} f(\underbrace{Y, \ldots, Y}_{|w|+1 \text{ times}})$, *with the tensor product interpretation for the values of* $\Delta_R^{w^\top} f$ *(Remark* 3.5*), the sumless Sweedler notation* (4.19)*, and the pseudo-power notation* (4.21)*.*

Proof. Let $\Upsilon_{\mathrm{nc}}$ be a uniformly open matrix circular nc set about $Y$ such that the norm of $f$ is bounded, say by $M > 0$, on $\Upsilon_{\mathrm{nc}}$. For $\boldsymbol{\rho} = (\rho_1, \ldots, \rho_d) \in (\mathbb{R}_+ \setminus \{0\})^d$ with $\sum_{j=1}^d \rho_j < 1$, $m \in \mathbb{N}$, and $X \in (\Upsilon_{\mathrm{nc},\boldsymbol{\rho}})_{sm}$, we obtain, using (3.54), that

$$\sum_{|w|=\ell} \Big\| \Delta_R^{w^\top} f(\underbrace{\bigoplus_{\alpha=1}^m Y, \ldots, \bigoplus_{\alpha=1}^m Y}_{\ell+1 \text{ times}}) \Big(X_{i_1} - \bigoplus_{\alpha=1}^m Y_{i_1}, \ldots, X_{i_\ell} - \bigoplus_{\alpha=1}^m Y_{i_\ell}\Big) \Big\|_{sm}$$

$$= \sum_{|w|=\ell} \boldsymbol{\rho}^w \Big\| \Delta_R^{w^\top} f(\underbrace{\bigoplus_{\alpha=1}^m Y, \ldots, \bigoplus_{\alpha=1}^m Y}_{\ell+1 \text{ times}}) \Big(\frac{1}{\rho_{i_1}}\Big(X_{i_1} - \bigoplus_{\alpha=1}^m Y_{i_1}\Big), \ldots, \frac{1}{\rho_{i_\ell}}\Big(X_{i_\ell} - \bigoplus_{\alpha=1}^m Y_{i_\ell}\Big)\Big) \Big\|_{sm}$$

$$= \sum_{|w|=\ell} \boldsymbol{\rho}^w$$

$$\cdot \left\| \pi_{1,\ell+1} f \left( \begin{bmatrix} \bigoplus_{\alpha=1}^m Y & \frac{1}{\rho_{i_1}}(X_{i_1} - \bigoplus_{\alpha=1}^m Y_{i_1}) e_{i_1} & 0 & \ldots & 0 \\ 0 & \ddots & \ddots & \ddots & \vdots \\ \vdots & \ddots & \ddots & \ddots & 0 \\ \vdots & & \ddots & \ddots & \frac{1}{\rho_{i_\ell}}(X_{i_\ell} - \bigoplus_{\alpha=1}^m Y_{i_\ell}) e_{i_\ell} \\ 0 & \ldots & \ldots & 0 & \bigoplus_{\alpha=1}^m Y \end{bmatrix} \right) \right\|_{sm}$$

$$\leq M \sum_{|w|=\ell} \boldsymbol{\rho}^w = M\Big(\sum_{j=1}^d \rho_j\Big)^\ell,$$

which implies that (7.35) holds and the TT series in (7.34) converges absolutely and uniformly on $\Upsilon_{\mathrm{nc},\boldsymbol{\rho}}$. Here we used the fact that

(7.36)

$$\begin{bmatrix} \bigoplus_{\alpha=1}^{m} Y & \frac{1}{\rho_{i_1}}(X_{i_1} - \bigoplus_{\alpha=1}^{m} Y_{i_1})e_{i_1} & 0 & \cdots & 0 \\ 0 & \ddots & \ddots & \ddots & \vdots \\ \vdots & & \ddots & \ddots & 0 \\ \vdots & & & \ddots & \frac{1}{\rho_{i_\ell}}(X_{i_\ell} - \bigoplus_{\alpha=1}^{m} Y_{i_\ell})e_{i_\ell} \\ 0 & \cdots & \cdots & 0 & \bigoplus_{\alpha=1}^{m} Y \end{bmatrix} \in (\Upsilon_{\mathrm{nc}})_{sm(\ell+1)},$$

which is obtained using the same argument as in the proof of (7.27) in Lemma 7.22.

In order to show that the sum of the (absolutely convergent) TT series in (7.34) equals $f(X)$, we observe that its sum must coincide with the sum of the TT series in (7.24) which is obtained by grouping together the terms corresponding to the words of the same length. But the remark made in the paragraph preceding Lemma 7.22, together with Theorem 7.23, means that the TT series (7.24) converges to $f(X)$ everywhere in $\Upsilon_{\mathrm{nc}}$. □

REMARK 7.30. Notice that in the proof of Theorem 7.29 we actually established the following result. Let $\Upsilon_{\mathrm{nc}}$ be a uniformly open nc set, which contains $Y$, is contained in $\Omega$, and on which $f$ is bounded. Then, for every $\boldsymbol{\rho} = (\rho_1, \ldots, \rho_d) \in (\mathbb{R}_+ \setminus \{0\})^d$ with $\sum_{j=1}^{d} \rho_j < 1$, (7.34) holds, with the series converging absolutely and uniformly, on the set of all $X \in \Omega$ such that (7.36) holds for all $\ell \in \mathbb{N}$ and such that the TT series in (7.24), which is obtained by grouping together the terms corresponding to the words of the same length, converges to $f(X)$. This gives a weaker assumption on a domain where the TT series in (7.34) converges absolutely and uniformly.

Unlike in the setting of Theorem 7.21 or of Theorem 7.23, it may happen that for a uniformly open matrix circular nc set $\Upsilon_{\mathrm{nc}}$ we have $\bigcup_{\boldsymbol{\rho}} \Upsilon_{\mathrm{nc},\boldsymbol{\rho}} \subsetneq \Upsilon_{\mathrm{nc}}$, i.e., we cannot guarantee the absolute convergence along $\mathcal{G}_d$ of the TT series in (7.34) on all of $\Upsilon_{\mathrm{nc}}$. We elucidate the situation in the special case of open nc balls.

COROLLARY 7.31. *Let a nc function $f\colon \Omega \to \mathcal{W}_{\mathrm{nc}}$ be uniformly locally bounded. For every $s \in \mathbb{N}$, $Y \in \Omega_s$, let $\delta :=$* $\sup\{r > 0\colon f$ is bounded on $B_{\mathrm{nc}}(Y, r)\}$. *Then (7.34) holds, with the TT series converging absolutely and uniformly, on every* open nc diamond about $Y$

$$\Diamond_{\mathrm{nc}}(Y, r) := \coprod_{m=1}^{\infty} \Big\{X \in \Omega_{sm}\colon \sum_{j=1}^{d} \|e_j\|_1 \Big\|X_j - \bigoplus_{\alpha=1}^{m} Y_j\Big\| < r\Big\} \tag{7.37}$$

*with* $r < \delta$. *Moreover,*

(7.38)

$$\sum_{w\in\mathcal{G}_d} \sup_{m\in\mathbb{N},\, X\in\Diamond_{\mathrm{nc}}(Y,r)_{sm}} \left\| \left( \left(\bigoplus_{\alpha=1}^{m} A_{w,(0)}\right) \otimes \left(\bigoplus_{\alpha=1}^{m} A_{w,(1)}\right) \otimes \cdots \otimes \left(\bigoplus_{\alpha=1}^{m} A_{w,(|w|)}\right) \right) \star \left( X - \bigoplus_{\alpha=1}^{m} Y \right)^{[w]} \right\|_{sm} < \infty.$$

PROOF. We first observe that for any $n \in \mathbb{N}$, $A \in \mathbb{C}^{n\times n}$, and $j = 1, \ldots, d$, we have $\|Ae_j\|_n = \|e_j\|_1 \|A\|$, since $\operatorname{span}\{e_j\}$ is a one-dimensional subspace in $\mathbb{C}^d$ and therefore there is only one operator space structure on $\operatorname{span} e_j$, up to a constant factor $\|e_j\|_1$; see [**79**, Proposition 1.10(ii)].

Let $r$ be given such that $0 < r < \delta$. Let $r'$ be a real number with $r < r' < \delta$, and let the norm of $f$ be bounded by $M > 0$ on $B_{\mathrm{nc}}(Y, r')$. Given $X \in \Diamond_{\mathrm{nc}}(Y,r)_{sm}$, one can find $\boldsymbol{\rho} = (\rho_1, \ldots, \rho_d)$ with

$$\rho_j > \frac{1}{r'}\|e_j\|_1 \left\| X_j - \bigoplus_{\alpha=1}^{m} Y \right\|, \quad j = 1, \ldots, d,$$

such that $\sum_{j=1}^{d} \rho_j < \frac{r}{r'}$. Since open nc balls are uniformly open and matrix circular nc sets, so is $\Upsilon_{\mathrm{nc}} := B_{\mathrm{nc}}(Y, r')$. For $X$ and $\boldsymbol{\rho}$ as above, we observe that $X \in \Upsilon_{\mathrm{nc},\boldsymbol{\rho}}$. Therefore, (7.34) holds with the series absolutely convergent. Using the estimate in the proof of Theorem 7.29 we obtain that

$$\sum_{|w|=\ell} \Big\| \Delta_R^{w^\top} f\big(\underbrace{\bigoplus_{\alpha=1}^{m} Y, \ldots, \bigoplus_{\alpha=1}^{m} Y}_{\ell+1 \text{ times}}\big)\Big(X_{i_1} - \bigoplus_{\alpha=1}^{m} Y_{i_1}, \ldots, X_{i_\ell} - \bigoplus_{\alpha=1}^{m} Y_{i_\ell}\Big)\Big\|_{sm} \le M\Big(\frac{r}{r'}\Big)^{\ell}.$$

Since the right-hand side is independent of $\boldsymbol{\rho}$, $m$, and $X$, (7.38) follows. $\square$

REMARK 7.32. Since $\bigcup_{0<r<\delta} \Diamond_{\mathrm{nc}}(Y,r) = \Diamond_{\mathrm{nc}}(Y,\delta)$, the TT series in (7.34) converges absolutely along $\mathcal{G}_d$ on the nc diamond $\Diamond_{\mathrm{nc}}(Y,\delta)$.

REMARK 7.33. It is clear that open nc diamonds are uniformly open and that, for all $Y$ and $r$, $\Diamond_{\mathrm{nc}}(Y,r) \subseteq B_{\mathrm{nc}}(Y,r)$. This inclusion is always proper. The equality would occur if and only if for all $m \in \mathbb{N}$ and $X \in (\mathbb{C}^{sm\times sm})^d$ one has $\|X\|_{sm} = \sum_{j=1}^{d} \|e_j\|_1 \|X_j\|$. Indeed, we have $X = \sum_{j=1}^{d} X_j e_j$ and $\|X\|_{sm} \le \sum_{j=1}^{d} \|X_j e_j\|_{sm} = \sum_{j=1}^{d} \|e_j\|_1 \|X_j\|$ for all $m$ and $X$, so that the inequality is always an equality if and only if $\Diamond_{\mathrm{nc}}(Y,r) = B_{\mathrm{nc}}(Y,r)$ for some, and hence for all, $Y \in (\mathbb{C}^{s\times s})^d$ and $r$. However $\sum_{j=1}^{d} \alpha_j \|X_j\|$ with some positive coefficients $\alpha_j$ is not an operator space norm on $((\mathbb{C}^d)^{s\times s})_{\mathrm{nc}}$.

REMARK 7.34. Instead of considering operator spaces, we can consider Banach spaces equipped with admissible systems of matrix norms such that (7.1) and (7.2) hold with constants $C_1$, $C_1'$, and $C_2$ independent of $n$ and $m$, as in Proposition 7.19. We have to require $C_1' = 1$ to ensure that Proposition 7.12 holds, i.e., the nc balls form a base for a topology on the corresponding nc space. In this case, we can still consider uniformly locally bounded nc functions and Proposition 7.20, Theorem 7.21, Corollary 7.26, and Corollary 7.28 remain true. The condition $C_1 = 1$ is natural to ensure the Hausdorff-like property of the uniformly-open topology, see Proposition 7.14 together with the paragraph that precedes it. The condition $C_2 = 1$ is essential for the proofs of Theorem 7.23, Theorem 7.29, and Corollary 7.31 (in particular, it guarantees that nc balls are matrix circular nc sets).

REMARK 7.35. The results of this section, unlike those of Section 7.1 (see Remark 7.11), are of a purely noncommutative nature (except for Theorem 7.21). Therefore, they admit full analogues in the real case. An operator space over $\mathbb{R}$ is a Banach space over $\mathbb{R}$ equipped with an admissible system of matrix norms such that the real analogues of (7.1) and (7.2) hold with $C_1 = C_1' = C_2 = 1$ independent of $n$ and $m$ (see [**92**]). Theorem 7.23 and Corollaries 7.26, 7.28 are true for the case where $\mathcal{V}$ and $\mathcal{W}$ are operator spaces over $\mathbb{R}$, with all the notions defined exactly as in the complex case. Also, Theorem 7.29 and Corollary 7.31 are true with $\mathbb{C}^d$ replaced by $\mathbb{R}^d$ and $\mathcal{W}$ an operator space over $\mathbb{R}$.

## 7.4. Analytic and uniformly analytic higher order nc functions

We will discuss now analyticity of higher order nc functions in various settings. Our main goal here is to prove analogues of Theorem 3.10 for the corresponding classes of higher order nc functions.

Let $\mathcal{W}$ be a Banach space over $\mathbb{C}$. An *admissible system of rectangular matrix norms over* $\mathcal{W}$ is a doubly-indexed sequence of norms $\|\cdot\|_{n,m}$ on $\mathcal{W}^{n\times m}$, $n, m = 1, 2, \ldots$, satisfying the following condition: for every $n, p, q, m \in \mathbb{N}$ there exists $C(n, p, q, m) > 0$ such that for all $X \in \mathcal{W}^{p\times q}$, $S \in \mathbb{C}^{n\times p}$, and $T \in \mathbb{C}^{q\times m}$,

$$\|SXT\|_{n,m} \leq C(n,p,q,m)\|S\|\,\|X\|_{p,q}\|T\|, \tag{7.39}$$

where $\|\cdot\|$ denotes the operator norm of rectangular matrices over $\mathbb{C}$ with respect to the standard Euclidean norms; cf. (7.2). Next we show that (7.39) implies the analogue of (7.1) for rectangular matrices,

$$\begin{aligned} C_1(n,m,r,s)^{-1}\max\{\|X\|_{n,m}, \|Y\|_{r,s}\} &\leq \|X \oplus Y\|_{n+r,m+s} \\ &\leq C_1'(n,m,r,s)\max\{\|X\|_{n,m}, \|Y\|_{r,s}\}, \end{aligned} \tag{7.40}$$

for every $n, m, r, s \in \mathbb{N}$, $X \in \mathcal{W}^{n\times m}$, $Y \in \mathcal{W}^{r\times s}$ and some positive constants $C_1(n, m, r, s)$ and $C_1'(n, m, r, s)$.

PROPOSITION 7.36. *Let $\mathcal{W}$ be a Banach space equipped with an admissible system of rectangular matrix norms over $\mathcal{W}$, i.e., (7.39) holds. Then (7.40) holds for every $n, m, r, s \in \mathbb{N}$, $X \in \mathcal{W}^{n\times m}$, and $Y \in \mathcal{W}^{r\times s}$ with*

$$C_1(n,m,r,s) = \max\{C(n, n+r, m+s, m), C(r, n+r, m+s, s)\},$$

$$C_1'(n,m,r,s) = C(n+r, n, m, m+s) + C(n+r, r, s, m+s).$$

PROOF. Let $X \in \mathcal{W}^{n\times m}$ and $Y \in \mathcal{W}^{r\times s}$. Then

$$\begin{aligned} \|X \oplus Y\|_{n+r,m+s} &= \|(X \oplus 0_{r\times s}) + (0_{n\times m} \oplus Y)\|_{n+r,m+s} \\ &\leq \left\| \begin{bmatrix} I_n \\ 0_{r\times n} \end{bmatrix} X \begin{bmatrix} I_m & 0_{m\times s} \end{bmatrix} + \begin{bmatrix} 0_{n\times r} \\ I_r \end{bmatrix} Y \begin{bmatrix} 0_{s\times m} & I_s \end{bmatrix} \right\| \\ &\leq \Big( C(n+r, n, m, m+s) + C(n+r, r, s, m+s) \Big) \max\{\|X\|_{n,m}, \|Y\|_{r,s}\} \end{aligned}$$

and

$$\begin{aligned}\max\{\|X\|_{n,m}, \|Y\|_{r,s}\} &= \max\Big\{\Big\| \begin{bmatrix} I_n & 0_{n\times r}\end{bmatrix} \begin{bmatrix} X & 0_{n\times s} \\ 0_{r\times m} & Y\end{bmatrix} \begin{bmatrix} I_m \\ 0_{s\times m}\end{bmatrix} \Big\|_{n,m}, \\ &\qquad \Big\| \begin{bmatrix} 0_{r\times n} & I_r\end{bmatrix} \begin{bmatrix} X & 0_{n\times s} \\ 0_{r\times m} & Y\end{bmatrix} \begin{bmatrix} 0_{m\times s} \\ I_s\end{bmatrix} \Big\|_{r,s}\Big\} \\ &\le \max\{C(n, n+r, m+s, m), C(r, n+r, m+s, s)\}\|X\oplus Y\|_{n+r,m+s}.\end{aligned}$$

□

Clearly, an admissible system of rectangular matrix norms over $\mathcal{W}$ induces an admissible system of (square) matrix norms over $\mathcal{W}$. Notice that if $\mathcal{W}$ is an operator space, then an admissible system of rectangular matrix norms over $\mathcal{W}$ is unique and one can choose

$$C(n,p,q,m) = C_1(n,p,q,m) = C_1'(n,p,q,m) = 1$$

for every $n, p, q, m \in \mathbb{N}$ [**78**, Exercises 13.1 and 13.2].

We can also introduce rectangular block injections and projections as follows. Let $\mathcal{R}$ be a unital commutative ring, let $\mathcal{M}$ be a module over $\mathcal{R}$, and let $s$, $s_1$, …, $s_m \in \mathbb{N}$ be such that $s = s_1 + \cdots + s_m$. Set

$$E_i := \operatorname{col}[0_{s_1\times s_i}, \ldots, 0_{s_{i-1}\times s_i}, I_{s_i}, 0_{s_{i+1}\times s_{i+1}}, \cdots, 0_{s_m\times s_i}] \in \mathcal{R}^{s\times s_i}, \quad i = 1, \ldots, m.$$

The *rectangular block injections* are defined by

$$\iota_{ij}^{(s_1,\ldots,s_m)}\colon \mathcal{M}^{s_i\times s_j} \to \mathcal{M}^{s\times s}, \quad \iota_{ij}^{(s_1,\ldots,s_m)}\colon Z \mapsto E_i Z E_j^\top, \quad i,j = 1,\ldots,m, \tag{7.41}$$

and the *rectangular block projections* are defined by

$$\pi_{ij}^{(s_1,\ldots,s_m)}\colon \mathcal{M}^{s\times s} \to \mathcal{M}^{s_i\times s_j}, \quad \pi_{ij}^{(s_1,\ldots,s_m)}\colon Z \mapsto E_i^\top Z E_j, \quad i,j = 1,\ldots,m, \tag{7.42}$$

The following analogue of Proposition 7.1, in conjunction with Proposition 7.36, holds.

PROPOSITION 7.37. *Condition* (7.39) *(valid for all $n$, $p$, $q$, $m \in \mathbb{N}$, $X \in \mathcal{W}^{p\times q}$, $S \in \mathbb{C}^{n\times p}$, and $T \in \mathbb{C}^{q\times m}$) is equivalent to the boundedness of the rectangular block injections $\iota_{ij}^{(s_1,\ldots,s_m)}$ and rectangular block projections $\pi_{ij}^{(s_1,\ldots,s_m)}$, $m \in \mathbb{N}$, $s_1$, …, $s_m \in \mathbb{N}$, $i, j = 1, \ldots, m$. In particular,*

$$\|\iota_{ij}^{(s_1,\ldots,s_m)}\| \le C(s, s_i, s_j, s), \quad \|\pi_{ij}^{(s_1,\ldots,s_m)}\| \le C(s_i, s, s, s_j). \tag{7.43}$$

PROOF. If (7.39) holds for all $n$, $p$, $q$, $m \in \mathbb{N}$, $X \in \mathcal{W}^{p\times q}$, $S \in \mathbb{C}^{n\times p}$, and $T \in \mathbb{C}^{q\times m}$, then (7.43) follows immediately from (7.41) and (7.42), thus $\iota_{ij}^{(s_1,\ldots,s_m)}$ and $\pi_{ij}^{(s_1,\ldots,s_m)}$ are bounded.

The converse direction can be shown analogously to that of Proposition 7.1. □

We also have the following analogue of Proposition 7.19.

PROPOSITION 7.38. *If $\mathcal{W}$ is a Banach space equipped with an admissible system of rectangular matrix norms over $\mathcal{W}$ such that* (7.39) *holds with $C$ independent of $n$, $p$, $q$, and $m$, then the rectangular block injections* (7.41) *and the rectangular block projections* (7.42) *are uniformly completely bounded, i.e., for every $m \in \mathbb{N}$*

*and* $s_1,\dots,s_m\in\mathbb{N}$*, the mappings* $\iota_{ij}^{(s_1,\dots,s_m)}$ *and* $\pi_{ij}^{(s_1,\dots,s_m)}$ *are completely bounded. Moreover,*

$$\|\iota_{ij}^{(s_1,\dots,s_m)}\|_{\rm cb}\le C,\quad \|\pi_{ij}^{(s_1,\dots,s_m)}\|_{\rm cb}\le C. \tag{7.44}$$

*If* $\mathcal{W}$ *is an operator space, then all* $\iota_{ij}^{(s_1,\dots,s_m)}$ *are complete isometries and all* $\pi_{ij}^{(s_1,\dots,s_m)}$ *are complete coisometries.*

PROOF. Let $n,m\in\mathbb{N}$, $1\le i,j\le m$, $s_1,\dots,s_m\in\mathbb{N}$, $W=(w_{ab})_{a,b=1,\dots,n}\in(\mathcal{W}^{s_i\times s_j})^{n\times n}\cong\mathcal{W}^{s_in\times s_jn}$. For $a,b=1,\dots,n$, let $E_{ab}\in\mathbb{C}^{n\times n}$ be defined by $(E_{ab})_{\alpha\beta}=\delta_{(a,b),(\alpha,\beta)}$, $\alpha,\beta=1,\dots,n$. Then

$$\begin{aligned}
&\|(\mathrm{id}_{\mathbb{C}^{n\times n}}\otimes\iota_{ij}^{(s_1,\dots,s_m)})(W)\|_{(s_1+\cdots+s_m)n}\\
&\qquad=\Big\|(\mathrm{id}_{\mathbb{C}^{n\times n}}\otimes\iota_{ij}^{(s_1,\dots,s_m)})\Big(\sum_{a,b=1}^n E_{ab}\otimes w_{ab}\Big)\Big\|_{(s_1+\cdots+s_m)n}\\
&=\Big\|\sum_{a,b=1}^n E_{ab}\otimes\iota_{ij}^{(s_1,\dots,s_m)}(w_{ab})\Big\|_{(s_1+\cdots+s_m)n}=\Big\|\sum_{a,b=1}^n E_{ab}\otimes E_iw_{ab}E_j^\top\Big\|_{(s_1+\cdots+s_m)n}\\
&\qquad=\|(I_n\otimes E_i)W(I_n\otimes E_j^\top)\|_{(s_1+\cdots+s_m)n}\le C\|W\|_{s_in,s_jn}.
\end{aligned}$$

This proves the first inequality in (7.44).

Suppose now that

$$W=(w_{ab})_{a,b=1,\dots,n}\in\Big(\mathcal{W}^{(s_1+\cdots+s_m)\times(s_1+\cdots+s_m)}\Big)^{n\times n}\cong\mathcal{W}^{(s_1+\cdots+s_m)n\times(s_1+\cdots+s_m)n}.$$

Then

$$\begin{aligned}
&\|(\mathrm{id}_{\mathbb{C}^{n\times n}}\otimes\pi_{ij}^{(s_1,\dots,s_m)})(W)\|_{s_in,s_jn}\\
&\qquad=\Big\|(\mathrm{id}_{\mathbb{C}^{n\times n}}\otimes\pi_{ij}^{(s_1,\dots,s_m)})\Big(\sum_{a,b=1}^n E_{ab}\otimes w_{ab}\Big)\Big\|_{s_in,s_jn}\\
&=\Big\|\sum_{a,b=1}^n E_{ab}\otimes\pi_{ij}^{(s_1,\dots,s_m)}(w_{ab})\Big\|_{s_in,s_jn}=\Big\|\sum_{a,b=1}^n E_{ab}\otimes E_i^\top w_{ab}E_j\Big\|_{s_in,s_jn}\\
&\qquad=\|(I_n\otimes E_i^\top)W(I_n\otimes E_j)\|_{s_in,s_jn}\le C\|W\|_{(s_1+\cdots+s_m)n}.
\end{aligned}$$

This proves the second inequality in (7.44).

If $\mathcal{W}$ is an operator space, we obtain that the operators $\iota_{ij}^{(s_1,\dots,s_m)}$ and $\pi_{ij}^{(s_1,\dots,s_m)}$ are complete contractions. Moreover, since $\pi_{ij}^{(s_1,\dots,s_m)}\iota_{ij}^{(s_1,\dots,s_m)}=\mathrm{id}_{\mathcal{W}^{s_i\times s_j}}$, all $\iota_{ij}^{(s_1,\dots,s_m)}$ are complete isometries and all $\pi_{ij}^{(s_1,\dots,s_m)}$ are complete coisometries. □

Let $\mathcal{V}_0$, …, $\mathcal{V}_k$, $\mathcal{W}_1$, …, $\mathcal{W}_k$ be vector spaces over $\mathbb{C}$, let $\Omega^{(j)}\subseteq\mathcal{V}_{j,\rm nc}$, $j=0,\dots,k$, be finitely open nc sets, and let $\mathcal{W}_0$ be a Banach space with an admissible system of rectangular matrix norms over $\mathcal{W}_0$. A higher order nc function $f\in\mathcal{T}^k(\Omega^{(0)},\dots,\Omega^{(k)};\mathcal{W}_{0,\rm nc},\dots,\mathcal{W}_{k,\rm nc})$ is called *W-locally bounded on slices* if for every $n_0,\dots,n_k\in\mathbb{N}$, $Y^0\in\Omega^{(0)}_{n_0}$, …, $Y^k\in\Omega^{(k)}_{n_k}$, $Z^0\in{\mathcal{V}_0}^{n_0\times n_0}$, …, $Z^k\in{\mathcal{V}_k}^{n_k\times n_k}$, $W^1\in{\mathcal{W}_1}^{n_0\times n_1}$, …, $W^k\in{\mathcal{W}_k}^{n_{k-1}\times n_k}$ there exists $\epsilon>0$ such that $f(Y^0+tZ^0,\dots,Y^k+tZ^k)(W^1,\dots,W^k)$ is bounded in ${\mathcal{W}_0}^{n_0\times n_k}$ for $|t|<\epsilon$, i.e., the function $f|_{\Omega^{(0)}_{n_0}\times\cdots\times\Omega^{(k)}_{n_k}}$ is locally bounded on slices for every fixed $W^1$, …, $W^k$.

A higher order nc function $f \in \mathcal{T}^k(\Omega^{(0)}, \ldots, \Omega^{(k)}; \mathcal{W}_{0,\mathrm{nc}}, \ldots, \mathcal{W}_{k,\mathrm{nc}})$ is called *$W$-Gâteaux ($G_W$-) differentiable* if for every $n_0, \ldots, n_k \in \mathbb{N}$, $Y^0 \in \Omega^{(0)}_{n_0}, \ldots, Y^k \in \Omega^{(k)}_{n_k}$, $Z^0 \in {\mathcal{V}_0}^{n_0 \times n_0}, \ldots, Z^k \in {\mathcal{V}_k}^{n_k \times n_k}$, $W^1 \in {\mathcal{W}_1}^{n_0 \times n_1}, \ldots, W^k \in {\mathcal{W}_k}^{n_{k-1} \times n_k}$ the G-derivative,

(7.45)
$$\lim_{t \to 0} \frac{f(Y^0 + tZ^0, \ldots, Y^k + tZ^k)(W^1, \ldots, W^k) - f(Y^0, \ldots, Y^k)(W^1, \ldots, W^k)}{t} = \frac{d}{dt} f(Y^0 + tZ^0, \ldots, Y^k + tZ^k)(W^1, \ldots, W^k)\Big|_{t=0},$$

exists, i.e., the function $f|_{\Omega^{(0)}_{n_0} \times \cdots \times \Omega^{(k)}_{n_k}}$ is G-differentiable for every fixed $W^1, \ldots, W^k$. It follows that $f$ is *$W$-analytic on slices*, i.e., for every $Y^0 \in \Omega^{(0)}_{n_0}, \ldots, Y^k \in \Omega^{(k)}_{n_k}$, $Z^0 \in {\mathcal{V}_0}^{n_0 \times n_0}, \ldots, Z^k \in {\mathcal{V}_k}^{n_k \times n_k}$, $W^1 \in {\mathcal{W}_1}^{n_0 \times n_1}, \ldots, W^k \in {\mathcal{W}_k}^{n_{k-1} \times n_k}$, $f(Y^0 + tZ^0, \ldots, Y^k + tZ^k)(W^1, \ldots, W^k)$ is an analytic function of $t$ in a neighbourhood of 0. By Hartogs' theorem [**97**, Page 28], $f$ is analytic on

$$(\mathcal{U}_0 \cap \Omega^{(0)}_{n_0}) \times \cdots \times (\mathcal{U}_k \cap \Omega^{(k)}_{n_k}) \times \mathcal{Y}_1 \times \cdots \times \mathcal{Y}_k$$

as a function of several complex variables for every $n_0, \ldots, n_k \in \mathbb{N}$ and for all finite-dimensional subspaces $\mathcal{U}_0 \subseteq {\mathcal{V}_0}^{n_0 \times n_0}, \ldots, \mathcal{U}_k \subseteq {\mathcal{V}_k}^{n_k \times n_k}$, $\mathcal{Y}_1 \subseteq {\mathcal{W}_1}^{n_0 \times n_1}, \ldots, \mathcal{Y}_k \subseteq {\mathcal{W}_k}^{n_{k-1} \times n_k}$.

THEOREM 7.39. *Suppose that $f \in \mathcal{T}^k(\Omega^{(0)}, \ldots, \Omega^{(k)}; \mathcal{W}_{0,\mathrm{nc}}, \ldots, \mathcal{W}_{k,\mathrm{nc}})$ is $W$-locally bounded on slices. Then so is*

$$\Delta_R f \in \mathcal{T}^{k+1}(\Omega^{(0)}, \ldots, \Omega^{(k)}, \Omega^{(k)}; \mathcal{W}_{0,\mathrm{nc}}, \ldots, \mathcal{W}_{k,\mathrm{nc}}, \mathcal{V}_{k,\mathrm{nc}}).$$

PROOF. Let $n_0, \ldots, n_{k+1} \in \mathbb{N}$, $Y^0 \in \Omega^{(0)}_{n_0}, \ldots, Y^k \in \Omega^{(k)}_{n_k}$, $Y^{k+1} \in \Omega^{(k)}_{n_{k+1}}$, $Z^0 \in {\mathcal{V}_0}^{n_0 \times n_0}, \ldots, Z^k \in {\mathcal{V}_k}^{n_k \times n_k}$, $Z^{k+1} \in {\mathcal{V}_k}^{n_{k+1} \times n_{k+1}}$, $W^1 \in {\mathcal{W}_1}^{n_0 \times n_1}, \ldots, W^{k+1} \in (\mathcal{W}_{k+1})^{n_k \times n_{k+1}}$. Since $\Omega^{(k)} \subseteq \mathcal{V}_{k,\mathrm{nc}}$ is a right admissible nc set, there exists $r > 0$ such that $\begin{bmatrix} Y^k & rW^{k+1} \\ 0 & Y^{k+1} \end{bmatrix} \in \Omega^{(k)}_{n_k + n_{k+1}}$. Since $\Omega^{(k)}$ is finitely open, there exists $\epsilon_1 > 0$ such that $\begin{bmatrix} Y^k + tZ^k & rW^{k+1} \\ 0 & Y^{k+1} + tZ^{k+1} \end{bmatrix} \in \Omega_{n_k + n_{k+1}}$ for all $t$ with $|t| < \epsilon_1$. Since $f$ is $W$-locally bounded on slices, there exists $\epsilon_2 > 0$ such that

$$f\left(Y^0 + tZ^0, \ldots, Y^{k-1} + tZ^{k-1}, \begin{bmatrix} Y^k + tZ^k & rW^{k+1} \\ 0 & Y^{k+1} + tZ^{k+1} \end{bmatrix}\right) (W^1, \ldots, W^{k-1}, \mathrm{row}[W^k, 0])$$

is bounded for $|t| < \epsilon_2$. Therefore, using (3.19), we obtain that

$$\begin{aligned} &\Delta_R f(Y^0 + tZ^0, \ldots, Y^{k+1} + tZ^{k+1})(W^1, \ldots, W^{k+1}) \\ &\quad = r^{-1} f\left(Y^0 + tZ^0, \ldots, Y^{k-1} + tZ^{k-1}, \begin{bmatrix} Y^k + tZ^k & rW^{k+1} \\ 0 & Y^{k+1} + tZ^{k+1} \end{bmatrix}\right) \\ &\qquad (W^1, \ldots, W^{k-1}, \mathrm{row}[W^k, 0]) \begin{bmatrix} 0 \\ I_{n_{k+1}} \end{bmatrix} \end{aligned}$$

is bounded for $|t| < \min\{\epsilon_1, \epsilon_2\}$. We conclude that $\Delta_R f$ is $W$-locally bounded on slices. Notice that we used here the fact that $\mathcal{W}_{0,\mathrm{nc}}$ is equipped with an admissible system of rectangular matrix norms (more precisely, the fact that the block projection implemented by the multiplication with $\begin{bmatrix} 0 \\ I_{n_{k+1}} \end{bmatrix}$ on the right is bounded). □

REMARK 7.40. An analogue of Theorem 7.39 also holds, with an analogous proof, for ${}_j\Delta_R f$, see Remark 3.18.

THEOREM 7.41. *Suppose that* $f \in \mathcal{T}^k(\Omega^{(0)}, \ldots, \Omega^{(k)}; \mathcal{W}_{0,\mathrm{nc}}, \ldots, \mathcal{W}_{k,\mathrm{nc}})$ *is* $W$*-locally bounded on slices. Then* $f$ *is* $G_W$*-differentiable.*

PROOF. We will first show that $f$ is $W$*-continuous on slices*, i.e.,

$$\lim_{t \to 0} f(Y^0 + tZ^0, \ldots, Y^k + tZ^k)(W^1, \ldots, W^k) = f(Y^0, \ldots, Y^k)(W^1, \ldots, W^k),$$

for all $n_0, \ldots, n_k \in \mathbb{N}$, $Y^0 \in \Omega^{(0)}_{n_0}, \ldots, Y^k \in \Omega^{(k)}_{n_k}$, $Z^0 \in \mathcal{V}_0{}^{n_0 \times n_0}, \ldots, Z^k \in \mathcal{V}_k{}^{n_k \times n_k}$, $W^1 \in \mathcal{W}_1{}^{n_0 \times n_1}, \ldots, W^k \in \mathcal{W}_k{}^{n_{k-1} \times n_k}$. We have, by (3.44),

$$\begin{aligned}(7.46) \quad & f(Y^0 + tZ^0, \ldots, Y^k + tZ^k)(W^1, \ldots, W^k) - f(Y^0, \ldots, Y^k)(W^1, \ldots, W^k) \\ & \quad = \sum_{j=0}^{k} \big(f(Y^0 + tZ^0, \ldots, Y^j + tZ^j, Y^{j+1}, \ldots, Y^k)(W^1, \ldots, W^k) \\ & \quad\quad - f(Y^0 + tZ^0, \ldots, Y^{j-1} + tZ^{j-1}, Y^j, \ldots, Y^k)(W^1, \ldots, W^k)\big) \\ & = t \sum_{j=0}^{k} {}_j\Delta_R f(Y^0 + tZ^0, \ldots, Y^{j-1} + tZ^{j-1}, Y^j + tZ^j, Y^j, Y^{j+1}, \ldots, Y^k) \\ & \qquad\qquad (W^1, \ldots, W^{j-1}, Z^j, W^j, W^{j+1}, \ldots, W^k) \to 0\end{aligned}$$

as $t \to 0$ because the latter sum is bounded for small $t$, see Remark 7.40.

Since ${}_j\Delta_R f$ is $W$-continuous on slices by Remark 7.40 again and the paragraph above, it follows from (7.46) that $f$ is $\mathrm{G}_W$-differentiable and

$$\begin{aligned}(7.47) \quad & \frac{d}{dt} f(Y^0 + tZ^0, \ldots, Y^k + tZ^k)(W^1, \ldots, W^k)\Big|_{t=0} \\ & \quad = \sum_{j=0}^{k} {}_j\Delta_R f(Y^0, \ldots, Y^{j-1}, Y^j, Y^j, Y^{j+1}, \ldots, Y^k) \\ & \qquad\qquad (W^1, \ldots, W^{j-1}, W^j, Z^j, W^{j+1}, \ldots, W^k).\end{aligned}$$

□

The converse of Theorem 7.41 is clear. We denote by

$$\mathcal{T}^k_{G_W} = \mathcal{T}^k_{G_W}(\Omega^{(0)}, \ldots, \Omega^{(k)}; \mathcal{W}_{0,\mathrm{nc}}, \ldots, \mathcal{W}_{k,\mathrm{nc}})$$

the subclass of $\mathcal{T}^k(\Omega^{(0)}, \ldots, \Omega^{(k)}; \mathcal{W}_{0,\mathrm{nc}}, \ldots, \mathcal{W}_{k,\mathrm{nc}})$ consisting of nc functions of order $k$ that are $\mathrm{G}_W$-differentiable.

The following statement is an obvious corollary of Theorems 7.39 and 7.41.

COROLLARY 7.42. *Let* $f \in \mathcal{T}^k_{G_W}(\Omega^{(0)}, \ldots, \Omega^{(k)}; \mathcal{W}_{0,\mathrm{nc}}, \ldots, \mathcal{W}_{k,\mathrm{nc}})$. *Then*

$$\Delta_R f \in \mathcal{T}^{k+1}_{G_W}(\Omega^{(0)}, \ldots, \Omega^{(k)}, \Omega^{(k)}; \mathcal{W}_{0,\mathrm{nc}}, \ldots, \mathcal{W}_{k,\mathrm{nc}}, \mathcal{V}_{k,\mathrm{nc}}).$$

REMARK 7.43. An analogue of Corollary 7.42 also holds for ${}_j\Delta_R f$, see Remarks 3.18 and 7.40.

We recall that a $k$-linear mapping $\omega\colon \mathcal{W}_1 \times \cdots \times \mathcal{W}_k \to \mathcal{W}_0$, where $\mathcal{W}_0, \ldots, \mathcal{W}_k$ are Banach spaces, is called *bounded* if

$$\|\omega\|_{\mathcal{L}^k} := \sup_{\|w^1\|=\cdots=\|w^k\|=1} \|\omega(w^1, \ldots, w^k)\| < \infty;$$

we denote by $\mathcal{L}^k(\mathcal{W}_1 \times \cdots \times \mathcal{W}_k, \mathcal{W}_0)$ the Banach space of such bounded $k$-linear mappings. Let $\mathcal{V}_0, \ldots, \mathcal{V}_k, \mathcal{W}_0, \ldots, \mathcal{W}_k$ be Banach spaces over $\mathbb{C}$ equipped with admissible systems of rectangular matrix norms, and let $\Omega^{(j)} \subseteq \mathcal{V}_{j,\mathrm{nc}}$, $j = 0, \ldots, k$, be open nc sets. A higher order nc function $f \in \mathcal{T}^k(\Omega^{(0)}, \ldots, \Omega^{(k)}; \mathcal{W}_{0,\mathrm{nc}}, \ldots, \mathcal{W}_{k,\mathrm{nc}})$ is called *locally bounded* if for every $n_0, \ldots, n_k \in \mathbb{N}$ the function $f|_{\Omega^{(0)}_{n_0} \times \cdots \times \Omega^{(k)}_{n_k}}$ has values in $\mathcal{L}^k({\mathcal{W}_1}^{n_0 \times n_1} \times \cdots \times {\mathcal{W}_k}^{n_{k-1} \times n_k}, {\mathcal{W}_0}^{n_0 \times n_k})$ and is locally bounded, i.e., for every $Y^0 \in \Omega^{(0)}_{n_0}, \ldots, Y^k \in \Omega^{(k)}_{n_k}$ there exists $\delta > 0$ such that $\|f(X^0, \ldots, X^k)\|_{\mathcal{L}^k}$ is bounded for $X^0 \in B(Y^0, \delta), \ldots, X^k \in B(Y^k, \delta)$. Clearly, a locally bounded higher order nc function is $W$-locally bounded on slices.

A higher order nc function $f \in \mathcal{T}^k(\Omega^{(0)}, \ldots, \Omega^{(k)}; \mathcal{W}_{0,\mathrm{nc}}, \ldots, \mathcal{W}_{k,\mathrm{nc}})$ is called *G-differentiable* if for every $n_0, \ldots, n_k \in \mathbb{N}$ the function $f|_{\Omega^{(0)}_{n_0} \times \cdots \times \Omega^{(k)}_{n_k}}$ has values in $\mathcal{L}^k({\mathcal{W}_1}^{n_0 \times n_1} \times \cdots \times {\mathcal{W}_k}^{n_{k-1} \times n_k}, {\mathcal{W}_0}^{n_0 \times n_k})$ and is G-differentiable, i.e., for every $Y^0 \in \Omega^{(0)}_{n_0}, \ldots, Y^k \in \Omega^{(k)}_{n_k}$, $Z^0 \in {\mathcal{V}_0}^{n_0 \times n_0}, \ldots, Z^k \in \mathcal{V})k^{n_k \times n_k}$, the G-derivative

(7.48)

$$\delta f(Y^0, \ldots, Y^k)(Z^0, \ldots, Z^k) = \lim_{t \to 0} \frac{f(Y^0 + tZ^0, \ldots, Y^k + tZ^k) - f(Y^0, \ldots, Y^k)}{t} = \frac{d}{dt} f(Y^0 + tZ^0, \ldots, Y^k + tZ^k)\Big|_{t=0}$$

exists in the norm of $\mathcal{L}^k({\mathcal{W}_1}^{n_0 \times n_1} \times \cdots \times {\mathcal{W}_k}^{n_{k-1} \times n_k}, {\mathcal{W}_0}^{n_0 \times n_k})$. Clearly, a G-differentiable higher order nc function is $\mathrm{G}_W$-differentiable — compare (7.48) and (7.45).

A higher order nc function $f \in \mathcal{T}^k(\Omega^{(0)}, \ldots, \Omega^{(k)}; \mathcal{W}_{0,\mathrm{nc}}, \ldots, \mathcal{W}_{k,\mathrm{nc}})$ is called *Fréchet (F-) differentiable* if for every $n_0, \ldots, n_k \in \mathbb{N}$ the function $f|_{\Omega^{(0)}_{n_0} \times \cdots \times \Omega^{(k)}_{n_k}}$ has values in $\mathcal{L}^k({\mathcal{W}_1}^{n_0 \times n_1} \times \cdots \times {\mathcal{W}_k}^{n_{k-1} \times n_k}, {\mathcal{W}_0}^{n_0 \times n_k})$ and is F-differentiable, i.e., $f|_{\Omega^{(0)}_{n_0} \times \cdots \times \Omega^{(k)}_{n_k}}$ is G-differentiable and for any $Y^0 \in \Omega^{(0)}_{n_0}, \ldots, Y^k \in \Omega^{(k)}_{n_k}$ the linear operator

$$\delta f(Y^0, \ldots, Y^k)\colon {\mathcal{V}_0}^{n_0 \times n_0} \oplus \cdots \oplus {\mathcal{V}_k}^{n_k \times n_k} \to \mathcal{L}^k({\mathcal{W}_1}^{n_0 \times n_1} \times \cdots \times {\mathcal{W}_k}^{n_{k-1} \times n_k}, {\mathcal{W}_0}^{n_0 \times n_k})$$

is bounded.

A higher order nc function $f \in \mathcal{T}^k(\Omega^{(0)}, \ldots, \Omega^{(k)}; \mathcal{W}_{0,\mathrm{nc}}, \ldots, \mathcal{W}_{k,\mathrm{nc}})$ is called *analytic* if $f$ is locally bounded and G-differentiable. Clearly, an analytic higher order nc function is $W$-analytic. If $f \in \mathcal{T}^k(\Omega^{(0)}, \ldots, \Omega^{(k)}; \mathcal{W}_{0,\mathrm{nc}}, \ldots, \mathcal{W}_{k,\mathrm{nc}})$ is analytic, then by [**61**, Theorem 3.17.1] $f|_{\Omega^{(0)}_{n_0} \times \cdots \times \Omega^{(k)}_{n_k}}$ is also continuous and F-differentiable. It follows that an analytic higher order nc function $f$ is continuous (i.e., all its restrictions $f|_{\Omega^{(0)}_{n_0} \times \cdots \times \Omega^{(k)}_{n_k}}$ are continuous) and F-differentiable.

THEOREM 7.44. *If $f \in \mathcal{T}^k(\Omega^{(0)}, \ldots, \Omega^{(k)}; \mathcal{W}_{0,\mathrm{nc}}, \ldots, \mathcal{W}_{k,\mathrm{nc}})$ is locally bounded, then so is $\Delta_R f \in \mathcal{T}^{k+1}(\Omega^{(0)}, \ldots, \Omega^{(k)}, \Omega^{(k)}; \mathcal{W}_{0,\mathrm{nc}}, \ldots, \mathcal{W}_{k,\mathrm{nc}}, \mathcal{V}_{k,\mathrm{nc}})$.*

PROOF. Let $n_0, \dots, n_{k+1} \in \mathbb{N}$, and let $Y^0 \in \Omega^{(0)}_{n_0}, \dots, Y^k \in \Omega^{(k)}_{n_k}, Y^{k+1} \in \Omega^{(k)}_{n_{k+1}}$. Since $f$ is locally bounded, there exists $\delta > 0$ such that

$$\begin{aligned}\|f(X^0, \dots, X^{k-1}, X)\|_{\mathcal{L}^k} &= \sup_{\|W^1\|_{n_0,n_1} = \dots = \|W^{k-1}\|_{n_{k-2},n_{k-1}} = \|W\|_{n_{k-1},n_k+n_{k+1}} = 1} \|f(X^0, \dots, X^{k-1}, X) \\ &\qquad (W^1, \dots, W^{k-1}, W)\|_{n_0, n_k+n_{k+1}}\end{aligned}$$

is bounded for all $X^0 \in B(Y^0, \delta), \dots, X^{k-1} \in B(Y^{k-1}, \delta), X \in B(Y^k \oplus Y^{k+1}, \delta)$. Since

$$\begin{bmatrix} X^k & \frac{\delta}{2} W^{k+1} \\ 0 & X^{k+1} \end{bmatrix} \in B(Y^k \oplus Y^{k+1}, \delta)$$

whenever $X^k \in B(Y^k, \frac{\delta}{2})$, $X^{k+1} \in B(Y^{k+1}, \frac{\delta}{2})$, and $W^{k+1} \in \mathcal{V}_k{}^{n_k \times n_{k+1}}$ with $\|W^{k+1}\|_{n_k, n_{k+1}} = 1$, we obtain that

$$\left\| f\left( X^0, \dots, X^{k-1}, \begin{bmatrix} X^k & \frac{\delta}{2} W^{k+1} \\ 0 & X^{k+1} \end{bmatrix} \right) (W^1, \dots, W^{k-1}, \operatorname{row}[W^k, 0]) \right\|_{n_0, n_k+n_{k+1}}$$

is bounded for $X^0 \in B(Y^0, \frac{\delta}{2}), \dots, X^{k+1} \in B(Y^{k+1}, \frac{\delta}{2})$, $\|W^1\|_{n_0,n_1} = \dots = \|W^{k+1}\|_{n_k,n_{k+1}} = 1$. Here we used the boundedness of the block injection

$$W^k \in \mathcal{W}_k{}^{n_{k-1} \times n_k} \longmapsto \operatorname{row}[W^k, 0] \in \mathcal{W}_k{}^{n_{k-1} \times (n_k + n_{k+1})}.$$

Therefore,

$$\begin{aligned}\|\Delta_R f(X^0, \dots, X^{k+1})\|_{\mathcal{L}^{k+1}} &= \sup_{\|W^1\|_{n_0,n_1} = \dots = \|W^{k+1}\|_{n_k,n_{k+1}} = 1} \|\Delta_R f(X^0, \dots, X^{k+1})(W^1, \dots, W^{k+1})\|_{n_0, n_{k+1}} \\ &= \frac{2}{\delta} \sup_{\|W^1\|_{n_0,n_1} = \dots = \|W^{k+1}\|_{n_k,n_{k+1}} = 1} \left\| f\left( X^0, \dots, X^{k-1}, \begin{bmatrix} X^k & \frac{\delta}{2} W^{k+1} \\ 0 & X^{k+1} \end{bmatrix} \right) \right. \\ &\qquad \left. (W^1, \dots, W^{k-1}, \operatorname{row}[W^k, 0]) \begin{bmatrix} 0 \\ I_{n_{k+1}} \end{bmatrix} \right\|_{n_0, n_{k+1}}\end{aligned}$$

is bounded for $X^0 \in B(Y^0, \frac{\delta}{2}), \dots, X^{k+1} \in B(Y^{k+1}, \frac{\delta}{2})$. Here we used the boundedness of the block projection implemented by the multiplication with $\begin{bmatrix} 0 \\ I_{n_{k+1}} \end{bmatrix}$ on the right. □

REMARK 7.45. An analogue of Theorem 7.44 also holds, with an analogous proof, for ${}_j\Delta_R f$, see Remark 3.18.

THEOREM 7.46. *If $f \in \mathcal{T}^k(\Omega^{(0)}, \dots, \Omega^{(k)}; \mathcal{W}_{0,\mathrm{nc}}, \dots, \mathcal{W}_{k,\mathrm{nc}})$ is locally bounded, then $f$ is analytic.*

PROOF. We will first show that $f$ is *continuous on slices*, i.e.,

$$\lim_{t \to 0} \|f(Y^0 + tZ^0, \dots, Y^k + tZ^k) - f(Y^0, \dots, Y^k)\|_{\mathcal{L}^k} = 0$$

for all $n_0, \dots, n_k \in \mathbb{N}$, $Y^0 \in \Omega^{(0)}_{n_0}, \dots, Y^k \in \Omega^{(k)}_{n_k}$, $Z^0 \in \mathcal{V}_0{}^{n_0 \times n_0}, \dots, Z^k \in \mathcal{V}_k{}^{n_k \times n_k}$. Indeed, the convergence in (7.46) is uniform on the set of $W^1 \in \mathcal{W}_1{}^{n_0 \times n_1}, \dots, W^k \in \mathcal{W}_k{}^{n_{k-1} \times n_k}$ satisfying $\|W^1\|_{n_0,n_1} = \dots = \|W^k\|_{n_{k-1},n_k} = 1$ because the last sum in (7.46) is uniformly bounded on this set for small $t$, see Remark 7.45.

Since ${}_j\Delta_R f$ is continuous on slices by Remark 7.45 again and the paragraph above, (7.46) implies that $f$ is G-differentiable and $\delta f(Y^0, \dots, Y^k)(Z^0, \dots, Z^k)$ is calculated by (7.47) where the limit is uniform on the set of $W^1 \in \mathcal{W}_1{}^{n_0 \times n_1}$, $\dots$, $W^k \in \mathcal{W}_k{}^{n_{k-1} \times n_k}$ satisfying $\|W^1\|_{n_0,n_1} = \cdots = \|W^k\|_{n_{k-1},n_k} = 1$. We conclude that $f$ is analytic. $\square$

It follows from Theorem 7.46 that a higher order nc function $f$ is locally bounded if and only if $f$ is continuous if and only if $f$ is F-differentiable if and only if $f$ is analytic. We denote by $\mathcal{T}^k_{\rm an} = \mathcal{T}^k_{\rm an}(\Omega^{(0)}, \dots, \Omega^{(k)}; \mathcal{W}_{0,\rm nc}, \dots, \mathcal{W}_{k,\rm nc})$ the subclass of $\mathcal{T}^k(\Omega^{(0)}, \dots, \Omega^{(k)}; \mathcal{W}_{0,\rm nc}, \dots, \mathcal{W}_{k,\rm nc})$ consisting of nc functions of order $k$ that are analytic.

The following statement is an obvious corollary of Theorems 7.44 and 7.46.

COROLLARY 7.47. *Let* $f \in \mathcal{T}^k_{\rm an}(\Omega^{(0)}, \dots, \Omega^{(k)}; \mathcal{W}_{0,\rm nc}, \dots, \mathcal{W}_{k,\rm nc})$. *Then*

$$\Delta_R f \in \mathcal{T}^{k+1}_{\rm an}(\Omega^{(0)}, \dots, \Omega^{(k)}, \Omega^{(k)}; \mathcal{W}_{0,\rm nc}, \dots, \mathcal{W}_{k,\rm nc}, \mathcal{V}_{k,\rm nc}).$$

REMARK 7.48. An analogue of Corollary 7.47 also holds for ${}_j\Delta_R f$, see Remarks 3.18 and 7.45.

Let $\mathcal{W}_0$, $\dots$, $\mathcal{W}_k$ be Banach spaces equipped with admissible systems of rectangular matrix norms, and let $\omega\colon \mathcal{W}_1 \times \cdots \times \mathcal{W}_k \to \mathcal{W}_0$ be a $k$-linear mapping. For every $n_0$, $n_1$, $\dots$, $n_k \in \mathbb{N}$ we define a $k$-linear mapping $\omega^{(n_0,\dots,n_k)}\colon \mathcal{W}_1{}^{n_0\times n_1} \times \cdots \times \mathcal{W}_k{}^{n_{k-1}\times n_k} \to \mathcal{W}_0{}^{n_0 \times n_k}$ by

$$\omega^{(n_0,\dots,n_k)}(W^1, \dots, W^k) = (W^1 \odot \cdots \odot W^k)\omega, \tag{7.49}$$

where we use the notation (3.7) and $\omega$ acts on the matrix

$$W^1 \odot \cdots \odot W^k \in (\mathcal{W}_1 \otimes \cdots \otimes \mathcal{W}_k)^{n_0 \times n_k}$$

on the right entrywise; cf. (3.8). The $k$-linear mapping $\omega$ is called *completely bounded* (in the sense of Christensen and Sinclair, see, e.g., [**78**, Chapter 17]) if

$$\|\omega\|_{\mathcal{L}^k_{\rm cb}} := \sup_{n_0,\dots,n_k \in \mathbb{N}} \|\omega^{(n_0,\dots,n_k)}\|_{\mathcal{L}^k} < \infty.$$

We denote by $\mathcal{L}^k_{cb}(\mathcal{W}_1 \times \cdots \times \mathcal{W}_k, \mathcal{W}_0)$ the Banach space of such completely bounded $k$-linear mappings.

Recall [**78**, Page 185, Exercise 13.2] that an operator space $\mathcal{W}_{\rm nc}$ is equipped with a unique admissible system of rectangular matrix norms such that the analogues for rectangular matrices of (7.1) and of (7.2) (i.e., (7.39)) hold with all the constants equal to 1 independently of matrix sizes. We notice (see, e.g., [**78**, Chapter 17] or [**79**, Chapter 5]) that if $\mathcal{W}_0$, $\dots$, $\mathcal{W}_k$ are operator spaces, then $\|\omega\|_{\mathcal{L}^k_{\rm cb}}$ coincides with the completely bounded norm of $\omega$ viewed as a linear mapping $\mathcal{W}_1 \otimes \cdots \otimes \mathcal{W}_k \to \mathcal{W}_0$, where the vector spaces $(\mathcal{W}_1 \otimes \cdots \otimes \mathcal{W}_k)^{n_0 \times n_k}$, $n_0$, $n_k \in \mathbb{N}$, are endowed with the *Haagerup norm*, $\|\cdot\|_{H,n_0,n_k}$:

$$\begin{aligned} &\|W\|_{H,n_0,n_k} \\ &= \inf\{\|W^1\|_{n_0,n_1} \cdots \|W^k\|_{n_{k-1},n_k} \colon n_1, \dots, n_{k-1} \in \mathbb{N},\ W = W^1 \odot \cdots \odot W^k\}. \end{aligned} \tag{7.50}$$

PROPOSITION 7.49. *Let* $\mathcal{V}$ *and* $\mathcal{W}$ *be operator spaces, and let* $\omega\colon \mathcal{V}^k \to \mathcal{W}$ *be a* $k$*-linear mapping. Then*

$$\|\omega\|_{\mathcal{L}^k_{\rm cb}} = \sup\{\|W^{\odot k}\omega\|_n \colon n \in \mathbb{N}, W \in \mathcal{V}^{n\times n},\ \|W\|_n = 1\}. \tag{7.51}$$

PROOF. The inequality "$\geq$" is obvious. In order to show that "$\leq$" holds as well, let $n_0, \ldots, n_k \in \mathbb{N}$, $W^1 \in \mathcal{V}^{n_0 \times n_1}, \ldots, W^k \in \mathcal{V}^{n_{k-1} \times n_k}$ be arbitrary. Set

$$\widetilde{W} = \begin{bmatrix} 0_{n_0 \times n_0} & W^1 & 0_{n_0 \times n_2} & \cdots & 0_{n_0 \times n_k} \\ \vdots & \ddots & \ddots & \ddots & \vdots \\ \vdots & & \ddots & \ddots & 0_{n_{k-2} \times n_k} \\ \vdots & & & \ddots & W^k \\ 0_{n_k \times n_0} & \cdots & \cdots & \cdots & 0_{n_k \times n_k} \end{bmatrix} \in \mathcal{V}^{(n_0+\cdots+n_k)\times(n_0+\cdots+n_k)}.$$

Then

$$\begin{aligned} &\|(W^1 \odot \cdots \odot W^k)\omega\|_{n_0 \times n_k} \\ &= \left\| \begin{bmatrix} 0_{n_0 \times n_0} & \cdots & 0_{n_0 \times n_{k-1}} & (W^1 \odot \cdots \odot W^k)\omega \\ \vdots & \ddots & & 0_{n_1 \times n_k} \\ \vdots & & \ddots & \vdots \\ 0_{n_k \times n_0} & \cdots & \cdots & 0_{n_k \times n_k} \end{bmatrix} \right\|_{n_0+\cdots+n_k} \\ &\qquad = \|\widetilde{W}^{\odot k}\omega\|_{n_0+\cdots+n_k} \end{aligned}$$

is not greater than the right-hand side of (7.51). Taking the supremum of $\|(W^1 \odot \cdots \odot W^k)\omega\|_{n_0 \times n_k}$ over all $n_0, \ldots, n_k \in \mathbb{N}$, and all $W^1 \in \mathcal{V}^{n_0 \times n_1}, \ldots, W^k \in \mathcal{V}^{n_{k-1} \times n_k}$ of norm 1, we obtain the desired inequality "$\leq$" in (7.51). $\square$

REMARK 7.50. Proposition 7.49 tells us that the evaluation on the diagonal is an isometric isomorphism from the space $\mathcal{L}^k_{\mathrm{cb}}(\mathcal{V} \times \cdots \times \mathcal{V}, \mathcal{W})$ to the space of bounded nc homogeneous polynomials of degree $k$ on $\mathcal{V}_{\mathrm{nc}}$ with values in $\mathcal{W}_{\mathrm{nc}}$ and the supremum norm on $B_{\mathrm{nc}}(0,1)$, see the comments in the end of Chapter 6. This is a nc analogue of the well-known relation between bounded homogeneous polynomials and bounded symmetric multilinear forms on Banach spaces, where the isomorphism however is not isometric, see [**61**, Section 26.2] and [**72**, Section 2].

Let $\mathcal{V}_0, \ldots, \mathcal{V}_k, \mathcal{W}_0, \ldots, \mathcal{W}_k$ be operator spaces, and let $\Omega^{(j)} \subseteq \mathcal{V}_{j,\mathrm{nc}}$, $j = 0, \ldots, k$, be uniformly open nc sets. We call a higher order nc function $f \in \mathcal{T}^k(\Omega^{(0)}, \ldots, \Omega^{(k)}; \mathcal{W}_{0,\mathrm{nc}}, \ldots, \mathcal{W}_{k,\mathrm{nc}})$ *uniformly locally completely bounded* if for every $n_0, \ldots, n_k \in \mathbb{N}$ the function $f|_{\Omega^{(0)}_{n_0} \times \cdots \times \Omega^{(k)}_{n_k}}$ has values in

$$\mathcal{L}^k_{\mathrm{cb}}({\mathcal{W}_1}^{n_0 \times n_1} \times \cdots \times {\mathcal{W}_k}^{n_{k-1} \times n_k}, {\mathcal{W}_0}^{n_0 \times n_k})$$

and $f$ is locally bounded in the uniformly open topology, i.e., for every $Y^0 \in \Omega^{(0)}_{n_0}$, $\ldots$, $Y^k \in \Omega^{(k)}_{n_k}$ there exists $\delta > 0$ such that $\|f(X^0, \ldots, X^k)\|_{\mathcal{L}^k_{\mathrm{cb}}}$ is bounded for $X^0 \in B_{\mathrm{nc}}(Y^0, \delta)$, $\ldots$, $X^k \in B_{\mathrm{nc}}(Y^k, \delta)$. (Notice that $p \times q$ matrices over an operator space $\mathcal{W}$ form an operator space in its own right.)

For $\phi \in \mathcal{L}^k_{\rm cb}({\mathcal{W}_1}^{s_0 \times s_1} \times \cdots \times {\mathcal{W}_k}^{s_{k-1} \times s_k}, {\mathcal{W}_0}^{s_0 \times s_k})$ and $r > 0$, define a *nc ball centered at $\phi$ of radius $r$* as

$$
\begin{aligned}
&B_{\rm nc}(\phi, r)\\
&= \coprod_{m_0, \ldots, m_k = 1}^{\infty} \Big\{ \psi \in \mathcal{L}^k_{\rm cb}({\mathcal{W}_1}^{s_0 m_0 \times s_1 m_1} \times \cdots \times {\mathcal{W}_k}^{s_{k-1} m_{k-1} \times s_k m_k}, {\mathcal{W}_0}^{s_0 m_0 \times s_k m_k}):\\
&\qquad\qquad \|\psi - \phi^{(m_0, \ldots, m_k)}\|_{\mathcal{L}^k_{\rm cb}} < r \Big\}.
\end{aligned}
$$

Notice that here, for $W^1 \in {\mathcal{W}_1}^{s_0 m_0 \times s_1 m_1}, \ldots, W^k \in {\mathcal{W}_k}^{s_{k-1} m_{k-1} \times s_k m_k}$,

$$\phi^{(m_0, \ldots, m_k)}(W^1, \ldots, W^k) := (W^1_{s_0, s_2} \odot_{s_1} \cdots_{s_{k-2}, s_k} \odot_{s_{k-1}} W^k)\phi,$$

see (3.7) and (7.49). Analogously to Proposition 7.12, nc balls form a basis for a topology on

$$\coprod_{n_0, \ldots, n_k = 1}^{\infty} \mathcal{L}^k_{\rm cb}({\mathcal{W}_1}^{n_0 \times n_1} \times \cdots \times {\mathcal{W}_k}^{n_{k-1} \times n_k}, {\mathcal{W}_0}^{n_0 \times n_k}),$$

which will be called the *uniformly-open topology.* Analogously to Proposition 7.20, if a higher order nc function $f \in \mathcal{T}^k(\Omega^{(0)}, \ldots, \Omega^{(k)}; \mathcal{W}_{0,\rm nc}, \ldots, \mathcal{W}_{k,\rm nc})$ is continuous with respect to the product topology on

$$\underbrace{\mathcal{V}_{0,\rm nc} \times \cdots \times \mathcal{V}_{k,\rm nc}}_{k+1 \text{ times}}$$

(of the uniformly-open topologies on $\mathcal{V}_{j,\rm nc}$) and the uniformly-open topology on

$$\coprod_{n_0, \ldots, n_k = 1}^{\infty} \mathcal{L}^k_{\rm cb}({\mathcal{W}_1}^{n_0 \times n_1} \times \cdots \times {\mathcal{W}_k}^{n_{k-1} \times n_k}, {\mathcal{W}_0}^{n_0 \times n_k}),$$

then $f$ is uniformly locally completely bounded.

A higher order nc function $f \in \mathcal{T}^k(\Omega^{(0)}, \ldots, \Omega^{(k)}; \mathcal{W}_{0,\rm nc}, \ldots, \mathcal{W}_{k,\rm nc})$ is called *completely bounded Gâteaux ($G_{\rm cb}$-) differentiable* if for every $n_0, \ldots, n_k \in \mathbb{N}$ the function $f|_{\Omega^{(0)}_{n_0} \times \cdots \times \Omega^{(k)}_{n_k}}$ has values in $\mathcal{L}^k_{\rm cb}({\mathcal{W}_1}^{n_0 \times n_1} \times \cdots \times {\mathcal{W}_k}^{n_{k-1} \times n_k}, {\mathcal{W}_0}^{n_0 \times n_k})$ and is G-differentiable, i.e., for every $Y^0 \in \Omega^{(0)}_{n_0}, \ldots, Y^k \in \Omega^{(k)}_{n_k}$, $Z^0 \in {\mathcal{V}_0}^{n_0 \times n_0}, \ldots, Z^k \in {\mathcal{V}_k}^{n_k \times n_k}$, the G-derivative (7.48) exists in the norm of

$$\mathcal{L}^k_{\rm cb}({\mathcal{W}_1}^{n_0 \times n_1} \times \cdots \times {\mathcal{W}_k}^{n_{k-1} \times n_k}, {\mathcal{W}_0}^{n_0 \times n_k}).$$

A higher order nc function $f \in \mathcal{T}^k(\Omega; \mathcal{W}_{0,\rm nc}, \ldots, \mathcal{W}_{k,\rm nc})$ is called *uniformly completely bounded (uniformly cb-) analytic* if $f$ is uniformly locally completely bounded and $G_{\rm cb}$-differentiable.

THEOREM 7.51. *If $f \in \mathcal{T}^k(\Omega^{(0)}, \ldots, \Omega^{(k)}; \mathcal{W}_{0,\rm nc}, \ldots, \mathcal{W}_{k,\rm nc})$ is uniformly locally completely bounded, then so is*

$$\Delta_R f \in \mathcal{T}^{k+1}(\Omega^{(0)}, \ldots, \Omega^{(k)}, \Omega^{(k)}; \mathcal{W}_{0,\rm nc}, \ldots, \mathcal{W}_{k,\rm nc}, \mathcal{V}_{k,\rm nc}).$$

PROOF. Let $s_0, \ldots, s_{k+1} \in \mathbb{N}$, and let $Y^0 \in \Omega^{(0)}_{s_0}, \ldots, Y^k \in \Omega^{(k)}_{s_k}$, $Y^{k+1} \in \Omega^{(k)}_{s_{k+1}}$. Since $f$ is uniformly locally completely bounded, there exists $\delta > 0$ such that

$$\|(f(X^0, \ldots, X^{k-1}, X))^{(n_0, \ldots, n_{k-1}, n)}(W^1, \ldots, W^{k-1}, W)\|_{s_0 m_0 n_0, (s_k + s_{k+1}) mn} \tag{7.52}$$

is bounded for all $m_0, \ldots, m_{k-1}, m, n_0, \ldots, n_{k-1}, n \in \mathbb{N}$, $X^0 \in B_{\rm nc}(Y^0, \delta)_{s_0 m_0}, \ldots, X^{k-1} \in B_{\rm nc}(Y^{k-1}, \delta)_{s_{k-1} m_{k-1}}$, $X \in B_{\rm nc}(Y^k \oplus Y^{k+1}, \delta)_{(s_k + s_{k+1}) m}$, $W^1 \in$

${\mathcal{W}_1}^{s_0m_0n_0\times s_1m_1n_1}$, …, $W^{k-1}\in(\mathcal{W}_{k-1})^{s_{k-2}m_{k-2}n_{k-2}\times s_{k-1}m_{k-1}n_{k-1}}$, and $W\in{\mathcal{W}_k}^{s_{k-1}m_{k-1}n_{k-1}\times(s_k+s_{k+1})mn}$, with

$$\begin{aligned}\|W^1\|_{s_0m_0n_0,s_1m_1n_1}=\cdots=\|W^{k-1}\|_{s_{k-2}m_{k-2}n_{k-2},s_{k-1}m_{k-1}n_{k-1}}\\ =\|W\|_{s_{k-1}m_{k-1}n_{k-1},(s_k+s_{k+1})mn}=1.\end{aligned}$$

By definition, (7.52) is equal to

$$\begin{aligned}&\|(W^1_{s_0m_0,s_2m_2}\odot_{s_1m_1}\cdots_{s_{k-3}m_{k-3},s_{k-1}m_{k-1}}\odot_{s_{k-2}m_{k-2}}W^{k-1}\\ &\quad{}_{s_{k-2}m_{k-2},(s_k+s_{k+1})m}\odot_{s_{k-1}m_{k-1}}W)f(X^0,\dots,X^{k-1},X)\|_{s_0m_0n_0,(s_k+s_{k+1})mn}.\end{aligned}$$

Using (3.8), we rewrite this as

$$\Big\|f\Big(\bigoplus_{\alpha=1}^{n_0}X^0,\dots,\bigoplus_{\alpha=1}^{n_{k-1}}X^{k-1},\bigoplus_{\alpha=1}^{n}X\Big)(W^1,\dots,W^{k-1},W)\Big\|_{s_0m_0n_0,(s_k+s_{k+1})mn}.$$

Since

$$\begin{bmatrix}\bigoplus_{\alpha=1}^{n_k}X^k & \frac{\delta}{2}W^{k+1} & 0\\ 0 & \bigoplus_{\alpha=1}^{n_{k+1}}X^{k+1} & 0\\ 0 & 0 & \bigoplus_{\alpha=1}^{|m_{k+1}n_{k+1}-m_kn_k|}Y\end{bmatrix}\in B_{\rm nc}(Y^k\oplus Y^{k+1},\delta)_{(s_k+s_{k+1})\max\{m_kn_k,m_{k+1}n_{k+1}\}}$$

after a permutation of block rows and columns whenever $X^k\in B_{\rm nc}(Y^k,\frac{\delta}{2})_{s_km_k}$, $X^{k+1}\in B_{\rm nc}(Y^{k+1},\frac{\delta}{2})_{s_{k+1}m_{k+1}}$, and $W^{k+1}\in\mathcal{V}^{s_km_kn_k\times s_{k+1}m_{k+1}n_{k+1}}$ with $\|W^{k+1}\|_{s_km_kn_k,s_{k+1}m_{k+1}n_{k+1}}=1$, where

$$Y=\begin{cases}Y^k & \text{if } m_kn_k<m_{k+1}n_{k+1},\\ Y^{k+1} & \text{if } m_kn_k>m_{k+1}n_{k+1},\end{cases}\tag{7.53}$$

we obtain that

$$\begin{aligned}&\left\|f\left(\bigoplus_{\alpha=1}^{n_0}X^0,\dots,\bigoplus_{\alpha=1}^{n_{k-1}}X^{k-1},\begin{bmatrix}\bigoplus_{\alpha=1}^{n_k}X^k & \frac{\delta}{2}W^{k+1} & 0\\ 0 & \bigoplus_{\alpha=1}^{n_{k+1}}X^{k+1} & 0\\ 0 & 0 & \bigoplus_{\alpha=1}^{|m_{k+1}n_{k+1}-m_kn_k|}Y\end{bmatrix}\right)\right.\\ &\qquad\left.(W^1,\dots,W^{k-1},\operatorname{row}[W^k,0,0])\right\|_{s_0m_0n_0,(s_k+s_{k+1})\max\{m_kn_k,m_{k+1}n_{k+1}\}}\end{aligned}$$

is bounded for all $m_0$, …, $m_{k+1}$, $n_0$, …, $n_{k+1}\in\mathbb{N}$, $X^0\in B_{\rm nc}(Y^0,\frac{\delta}{2})_{s_0m_0}$, …, $X^{k+1}\in B_{\rm nc}(Y^{k+1},\frac{\delta}{2})_{s_{k+1}m_{k+1}}$, and $W^1\in{\mathcal{W}_1}^{s_0m_0n_0\times s_1m_1n_1}$, …, $W^{k+1}\in(\mathcal{W}_{k+1})^{s_km_kn_k\times s_{k+1}m_{k+1}n_{k+1}}$, with

$$\|W^1\|_{s_0m_0n_0,s_1m_1n_1}=\cdots=\|W^{k+1}\|_{s_km_kn_k,s_{k+1}m_{k+1}n_{k+1}}=1.$$

Here we used the isometricity of the block injection

$$W^k \in {\mathcal{W}_k}^{s_{k-1}m_{k-1}n_{k-1}\times s_k m_k n_k} \longmapsto \mathrm{row}[W^k,0,0] \in {\mathcal{W}_k}^{s_{k-1}m_{k-1}n_{k-1}\times(s_k+s_{k+1})\max\{m_k n_k, m_{k+1}n_{k+1}\}}.$$

Therefore,

$$\begin{aligned}&\Big\|\Delta_R f\Big(\bigoplus_{\alpha=1}^{n_0} X^0,\ldots,\bigoplus_{\alpha=1}^{n_{k+1}} X^{k+1}\Big)(W^1,\ldots,W^{k+1})\Big\|_{s_0m_0n_0,s_{k+1}m_{k+1}n_{k+1}}\\ &=\frac{2}{\delta}\Bigg\|f\left(\bigoplus_{\alpha=1}^{n_0} X^0,\ldots,\bigoplus_{\alpha=1}^{n_{k-1}} X^{k-1},\begin{bmatrix}\bigoplus_{\alpha=1}^{n_k} X^k & \frac{\delta}{2}W^{k+1} & 0\\ 0 & \bigoplus_{\alpha=1}^{n_{k+1}} X^{k+1} & 0\\ 0 & 0 & \bigoplus_{\alpha=1}^{|m_{k+1}n_{k+1}-m_kn_k|} Y\end{bmatrix}\right)\\ &\qquad (W^1,\ldots,W^{k-1},\mathrm{row}[W^k,0,0])\begin{bmatrix}0\\ I_{s_{k+1}m_{k+1}n_{k+1}}\\ 0\end{bmatrix}\Bigg\|_{s_0m_0n_0,s_{k+1}m_{k+1}n_{k+1}}\end{aligned}$$

is bounded for all $m_0, \ldots, m_{k+1}, n_0, \ldots, n_{k+1} \in \mathbb{N}$, $X^0 \in B_{\mathrm{nc}}(Y^0,\frac{\delta}{2})_{s_0m_0}$, $\ldots$, $X^{k+1} \in B_{\mathrm{nc}}(Y^{k+1},\frac{\delta}{2})_{s_{k+1}m_{k+1}}$, and $W^1 \in {\mathcal{W}_1}^{s_0m_0n_0\times s_1m_1n_1}$, $\ldots$, $W^{k+1} \in (\mathcal{W}_{k+1})^{s_km_kn_k\times s_{k+1}m_{k+1}n_{k+1}}$, with

$$\|W^1\|_{s_0m_0n_0,s_1m_1n_1} = \cdots = \|W^{k+1}\|_{s_km_kn_k,s_{k+1}m_{k+1}n_{k+1}} = 1.$$

Here we used the contractivity of the block projection implemented by the multiplication with

$$\begin{bmatrix}0\\ I_{s_{k+1}m_{k+1}n_{k+1}}\\ 0\end{bmatrix}$$

on the right.

Applying the same argument as in the beginning of the proof, for $\Delta_R f$ in the place of $f$, we obtain

$$\begin{aligned}&\|(\Delta_R f(X^0,\ldots,X^{k+1}))^{(n_0,\ldots,n_{k+1})}(W^1,\ldots,W^{k+1})\|_{s_0m_0n_0,s_{k+1}m_{k+1}n_{k+1}}\\ &\quad=\Big\|\Delta_R f\Big(\bigoplus_{\alpha=1}^{n_0} X^0,\ldots,\bigoplus_{\alpha=1}^{n_{k+1}} X^{k+1}\Big)(W^1,\ldots,W^{k+1})\Big\|_{s_0m_0n_0,s_{k+1}m_{k+1}n_{k+1}},\end{aligned}$$

which by the result of the preceding paragraph implies the uniform local complete boundedness of $\Delta_R f$. □

REMARK 7.52. An analogue of Theorem 7.51 also holds, with an analogous proof, for ${}_j\Delta_R f$, see Remark 3.18.

THEOREM 7.53. *If* $f \in \mathcal{T}^k(\Omega^{(0)},\ldots,\Omega^{(k)};\mathcal{W}_{0,\mathrm{nc}},\ldots,\mathcal{W}_{k,\mathrm{nc}})$ *is uniformly locally completely bounded, then* $f$ *is continuous with respect to the uniformly-open topologies and is uniformly cb analytic.*

PROOF. Let $s_0, \ldots, s_{k+1} \in \mathbb{N}$, and let $Y^0 \in \Omega^{(0)}_{s_0}, \ldots, Y^k \in \Omega^{(k)}_{s_k}$, $Y^{k+1} \in \Omega^{(k)}_{s_{k+1}}$. Since $f$ is uniformly locally completely bounded, according to Remark 7.52 so are ${}_j\Delta_R f$, $j = 0, \ldots, k$. Therefore, there exist $\delta > 0$ and $C > 0$ such that

$$\|({}_j\Delta_R f(X^0, \ldots, X^{k+1}))^{(n_0,\ldots,n_{k+1})}(W^1, \ldots, W^{k+1})\|_{s_0m_0n_0, s_{k+1}m_{k+1}n_{k+1}}$$
$$= \Big\|{}_j\Delta_R f\Big(\bigoplus_{\alpha=1}^{n_0} X^0, \ldots, \bigoplus_{\alpha=1}^{n_{k+1}} X^{k+1}\Big)(W^1, \ldots, W^{k+1})\Big\|_{s_0m_0n_0, s_{k+1}m_{k+1}n_{k+1}} \le C$$

for $j = 0, \ldots, k$, and for all $m_0, \ldots, m_{k+1}, n_0, \ldots, n_{k+1} \in \mathbb{N}$, $X^0 \in B_{\rm nc}(Y^0, \delta)_{s_0m_0}$, $\ldots$, $X^{k+1} \in B_{\rm nc}(Y^{k+1}, \delta)_{s_{k+1}m_{k+1}}$, and $W^1 \in \mathcal{W}_1{}^{s_0m_0n_0 \times s_1m_1n_1}$, $\ldots$, $W^{k+1} \in (\mathcal{W}_{k+1})^{s_km_kn_k \times s_{k+1}m_{k+1}n_{k+1}}$, with

$$\|W^1\|_{s_0m_0n_0, s_1m_1n_1} = \cdots = \|W^{k+1}\|_{s_km_kn_k, s_{k+1}m_{k+1}n_{k+1}} = 1.$$

We have by (3.43) (cf. (7.46))

$$\Big\|f\Big(\bigoplus_{\alpha=1}^{n_0} X^0, \ldots, \bigoplus_{\alpha=1}^{n_k} X^k\Big)(W^1, \ldots, W^k)$$
$$- f\Big(\bigoplus_{\alpha=1}^{m_0n_0} Y^0, \ldots, \bigoplus_{\alpha=1}^{m_kn_k} Y^k\Big)(W^1, \ldots, W^k)\Big\|_{s_0m_0n_0, s_km_kn_k}$$
$$= \Big\|\sum_{j=0}^{k}\Big(f\Big(\bigoplus_{\alpha=1}^{n_0} X^0, \ldots, \bigoplus_{\alpha=1}^{n_j} X^j, \bigoplus_{\alpha=1}^{m_{j+1}n_{j+1}} Y^{j+1}, \ldots, \bigoplus_{\alpha=1}^{m_kn_k} Y^k\Big)(W^1, \ldots, W^k)$$
$$- f\Big(\bigoplus_{\alpha=1}^{n_0} X^0, \ldots, \bigoplus_{\alpha=1}^{n_{j-1}} X^{j-1}, \bigoplus_{\alpha=1}^{m_jn_j} Y^j, \ldots, \bigoplus_{\alpha=1}^{m_kn_k} Y^k\Big)(W^1, \ldots, W^k)\Big)\Big\|_{s_0m_0n_0, s_km_kn_k}$$
$$= \Big\|\sum_{j=0}^{k} {}_j\Delta_R f\Big(\bigoplus_{\alpha=1}^{n_0} X^0, \ldots, \bigoplus_{\alpha=1}^{n_j} X^j, \bigoplus_{\alpha=1}^{m_jn_j} Y^j, \ldots, \bigoplus_{\alpha=1}^{m_kn_k} Y^k\Big)$$
$$\Big(W^1, \ldots, W^{j-1}, \bigoplus_{\alpha=1}^{n_j} X^j - \bigoplus_{\alpha=1}^{m_jn_j} Y^j, W^j, \ldots, W^k\Big)\Big\|_{s_0m_0n_0, s_km_kn_k}$$
$$\le C \sum_{j=0}^{k} \Big\|X^j - \bigoplus_{\alpha=1}^{m_j} Y^j\|_{s_jm_j}$$

for all $m_0, \ldots, m_k, n_0, \ldots, n_k \in \mathbb{N}$, $X^0 \in B_{\rm nc}(Y^0, \delta)_{s_0m_0}, \ldots, X^k \in B_{\rm nc}(Y^k, \delta)_{s_km_k}$, and $W^1 \in \mathcal{W}_1{}^{s_0m_0n_0 \times s_1m_1n_1}$, $\ldots$, $W^k \in (\mathcal{W}_k)^{s_{k-1}m_{k-1}n_{k-1} \times s_km_kn_k}$, with

$$\|W^1\|_{s_0m_0n_0, s_1m_1n_1} = \cdots = \|W^k\|_{s_{k-1}m_{k-1}n_{k-1}, s_km_kn_k} = 1.$$

Given $\epsilon > 0$, we set $\eta = \min\{\delta, \frac{\epsilon}{C(k+1)}\}$ and obtain that

$$\Big\|f\Big(\bigoplus_{\alpha=1}^{n_0} X^0, \ldots, \bigoplus_{\alpha=1}^{n_k} X^k\Big)(W^1, \ldots, W^k)$$
$$- f\Big(\bigoplus_{\alpha=1}^{m_0n_0} Y^0, \ldots, \bigoplus_{\alpha=1}^{m_kn_k} Y^k\Big)(W^1, \ldots, W^k)\Big\|_{s_0m_0n_0, s_km_kn_k} < \epsilon$$

for all $m_0, \dots, m_k, n_0, \dots, n_k \in \mathbb{N}$, $X^0 \in B_{\rm nc}(Y^0,\eta)_{s_0m_0}, \dots, X^k \in B_{\rm nc}(Y^k,\eta)_{s_km_k}$, and $W^1 \in \mathcal{W}_1{}^{s_0m_0n_0\times s_1m_1n_1}, \dots, W^k \in \mathcal{W}_k{}^{s_{k-1}m_{k-1}n_{k-1}\times s_km_kn_k}$, with

$$\|W^1\|_{s_0m_0n_0,s_1m_1n_1} = \cdots = \|W^k\|_{s_{k-1}m_{k-1}n_{k-1},s_km_kn_k} = 1.$$

Thus $f$ is continuous with respect to the uniformly-open topologies.

By Remark 7.52, ${}_j\Delta_R f$ is uniformly locally completely bounded for $j = 0, \dots, k$. Applying the first part of this theorem to ${}_j\Delta_R f$, we obtain that ${}_j\Delta_R f$ is continuous with respect to the uniformly-open topologies. Let $s_0, \dots, s_{k+1} \in \mathbb{N}$, and let $Y^0 \in \Omega^{(0)}_{s_0}, \dots, Y^k \in \Omega^{(k)}_{s_k}$. By (7.46) we have

$$\begin{aligned}(7.54)\quad & f\Big(\bigoplus_{\alpha=1}^{n_0} Y^0 + t\bigoplus_{\alpha=1}^{n_0} Z^0, \dots, \bigoplus_{\alpha=1}^{n_k} Y^k + t\bigoplus_{\alpha=1}^{n_k} Z^k\Big)(W^1,\dots,W^k)\\ &\qquad - f\Big(\bigoplus_{\alpha=1}^{n_0} Y^0, \dots, \bigoplus_{\alpha=1}^{n_k} Y^k\Big)(W^1,\dots,W^k)\\ &= t\sum_{j=0}^{k} {}_j\Delta_R f\Big(\bigoplus_{\alpha=1}^{n_0} Y^0 + t\bigoplus_{\alpha=1}^{n_0} Z^0, \dots, \bigoplus_{\alpha=1}^{n_j} Y^j + t\bigoplus_{\alpha=1}^{n_j} Z^j, \bigoplus_{\alpha=1}^{n_j} Y^j, \dots, \bigoplus_{\alpha=1}^{n_k} Y^k\Big)\\ &\qquad \Big(W^1,\dots,W^j,\bigoplus_{\alpha=1}^{n_j} Z^j, W^{j+1},\dots,W^k\Big)\end{aligned}$$

for any $n_0, \dots, n_k \in \mathbb{N}$, $Z^0 \in \mathcal{V}_0{}^{s_0\times s_0} \dots, Z^k \in \mathcal{V}_k{}^{s_k\times s_k}$, and $W^1 \in \mathcal{W}_1{}^{s_0n_0\times s_1n_1}, \dots, W^k \in \mathcal{W}_k{}^{s_{k-1}n_{k-1}\times s_kn_k}$, with

$$\|W^1\|_{s_0n_0,s_1n_1} = \cdots = \|W^k\|_{s_{k-1}n_{k-1},s_kn_k} = 1,$$

and $t$ small enough. Dividing both parts of (7.54) by $t$ and letting $t \to 0$, we obtain (cf. (7.47))

$$\begin{aligned}&\frac{d}{dt} f\Big(\bigoplus_{\alpha=1}^{n_0} Y^0 + t\bigoplus_{\alpha=1}^{n_0} Z^0, \dots, \bigoplus_{\alpha=1}^{n_k} Y^k + t\bigoplus_{\alpha=1}^{n_k} Z^k\Big)(W^1,\dots,W^k)\Big|_{t=0}\\ &= \sum_{j=0}^{k} {}_j\Delta_R f\Big(\bigoplus_{\alpha=1}^{n_0} Y^0, \dots, \bigoplus_{\alpha=1}^{n_{j-1}} Y^{j-1}, \bigoplus_{\alpha=1}^{n_j} Y^j, \bigoplus_{\alpha=1}^{n_j} Y^j, \bigoplus_{\alpha=1}^{n_{j+1}} Y^{j+1}, \dots, \bigoplus_{\alpha=1}^{n_k} Y^k\Big)\\ &\qquad \Big(W^1,\dots,W^{j-1},\bigoplus_{\alpha=1}^{n_j} Z^j, W^{j+1},\dots,W^k\Big)\end{aligned}$$

where the limit is uniform in $n_0, \dots, n_k \in \mathbb{N}$, and in $W^1 \in \mathcal{W}_1{}^{s_0n_0\times s_1n_1}, \dots, W^k \in \mathcal{W}_k{}^{s_{k-1}n_{k-1}\times s_kn_k}$, with

$$\|W^1\|_{s_0n_0,s_1n_1} = \cdots = \|W^k\|_{s_{k-1}n_{k-1},s_kn_k} = 1.$$

Therefore, the G-derivative

$$\delta f(Y^0,\dots,Y^k)(Z^0,\dots,Z^k) = \frac{d}{dt} f\left(Y^0 + tZ^0, \dots, Y^k + tZ^k\right)\Big|_{t=0}$$

(see (7.48)) exists in the norm of

$$\mathcal{L}^k_{\rm cb}(\mathcal{W}_1{}^{s_0\times s_1} \times \cdots \times \mathcal{W}_k{}^{s_{k-1}\times s_k}, \mathcal{W}_0{}^{s_0\times s_k}),$$

i.e., $f$ is $\mathrm{G_{cb}}$-differentiable. Since $f$ is also uniformly locally completely bounded, we conclude that $f$ is uniformly cb analytic. □

It follows from Theorem 7.53 that a higher order nc function $f$ is uniformly locally completely bounded if and only if $f$ is continuous with respect to the uniformly-open topologies if and only if $f$ is uniformly cb analytic. We denote by

$$\mathcal{T}^k_{\rm uan} = \mathcal{T}^k_{\rm uan}(\Omega^{(0)}, \ldots, \Omega^{(k)}; \mathcal{W}_{0,\rm nc}, \ldots, \mathcal{W}_{k,\rm nc})$$

the subclass of $\mathcal{T}^k(\Omega^{(0)}, \ldots, \Omega^{(k)}; \mathcal{W}_{0,\rm nc}, \ldots, \mathcal{W}_{k,\rm nc})$ consisting of nc functions of order $k$ that are uniformly cb analytic.

The following statement is an obvious corollary of Theorems 7.51 and 7.53.

COROLLARY 7.54. *Let* $f \in \mathcal{T}^k_{\rm uan}(\Omega^{(0)}, \ldots, \Omega^{(k)}; \mathcal{W}_{0,\rm nc}, \ldots, \mathcal{W}_{k,\rm nc})$. *Then*

$$\Delta_R f \in \mathcal{T}^{k+1}_{\rm uan}(\Omega^{(0)}, \ldots, \Omega^{(k)}, \Omega^{(k)}; \mathcal{W}_{0,\rm nc}, \ldots, \mathcal{W}_{k,\rm nc}, \mathcal{V}_{k,\rm nc}).$$

REMARK 7.55. An analogue of Corollary 7.54 also holds for ${}_j\Delta_R f$, see Remarks 3.18 and 7.52.

REMARK 7.56. The results of this section admit analogues in the real case — compare Remarks 7.11 and 7.35. Theorem 7.39, Remark 7.40, Theorem 7.41, Corollary 7.42, and Remark 7.43 are true for the case where $\mathcal{V}_0$, ..., $\mathcal{V}_k$, $\mathcal{W}_1$, ..., $\mathcal{W}_k$ are vector spaces over $\mathbb{R}$ and $\mathcal{W}_0$ is a Banach space over $\mathbb{R}$ equipped with an admissible system of rectangular matrix norms (all the notions are defined exactly as in the complex case). If $f \in \mathcal{T}^k(\Omega^{(0)}, \ldots, \Omega^{(k)}; \mathcal{W}_{0,\rm nc}, \ldots, \mathcal{W}_{k,\rm nc})$ is G$_W$-differentiable, it does not necessarily follow that $f$ is $W$-analytic on slices, nor that $f$ is infinitely many times differentiable on

$$(\mathcal{U}_0 \cap \Omega^{(0)}_{n_0}) \times \cdots \times (\mathcal{U}_k \cap \Omega^{(k)}_{n_k}) \times \mathcal{Y}_1 \times \cdots \times \mathcal{Y}_k \tag{7.55}$$

as a function of several real variables for every $n_0, \ldots, n_k \in \mathbb{N}$ and for all finite-dimensional subspaces $\mathcal{U}_0 \subseteq \mathcal{V}_0{}^{n_0\times n_0}$, ..., $\mathcal{U}_k \subseteq \mathcal{V}_k{}^{n_k\times n_k}$, $\mathcal{Y}_1 \subseteq \mathcal{W}^{n_0\times n_1}$, ..., $\mathcal{Y}_k \subseteq \mathcal{W}^{n_{k-1}\times n_k}$. Assume in addition that $f$ is *$W$-locally bounded on affine finite-dimensional subspaces*, i.e., that $f|_{\Omega^{(0)}_{n_0}\times\cdots\times\Omega^{(k)}_{n_k}}$ is locally bounded on affine finite-dimensional subspaces, similarly to Remark 7.11, for every fixed $W_1$, ..., $W_k$. Then $\Delta_R f \in \mathcal{T}^{k+1}(\Omega^{(0)}, \ldots, \Omega^{(k)}, \Omega^{(k)}; \mathcal{W}_{0,\rm nc}, \ldots, \mathcal{W}_{k,\rm nc}, \mathcal{V}_{k,\rm nc})$ is also $W$-locally bounded on affine finite-dimensional subspaces and $f$ is infinitely many times differentiable on the sets (7.55) — the proof is similar to the proofs of Theorems 7.39 and 7.41.

Theorem 7.44 and Remark 7.45 are true for the case where $\mathcal{V}_0$, ..., $\mathcal{V}_k$, $\mathcal{W}_0$, ..., $\mathcal{W}_k$ are Banach spaces over $\mathbb{R}$ equipped with an admissible system of rectangular matrix norms. The analogue of Theorem 7.41 is the following: if $f \in \mathcal{T}^k(\Omega^{(0)}, \ldots, \Omega^{(k)}; \mathcal{W}_{0,\rm nc}, \ldots, \mathcal{W}_{k,\rm nc})$ is locally bounded, then $f$ is $G$-differentiable, and furthermore, $f$ is infinitely many times differentiable on $(\mathcal{U}_0 \cap \Omega^{(0)}_{n_0}) \times \cdots \times (\mathcal{U}_k \cap \Omega^{(k)}_{n_k})$ as a function of several real variables for every $n_0, \ldots, n_k \in \mathbb{N}$ and for all finite-dimensional subspaces $\mathcal{U}_0 \subseteq \mathcal{V}_0{}^{n_0\times n_0}$, ..., $\mathcal{U}_k \subseteq \mathcal{V}_k{}^{n_k\times n_k}$, the derivative being taken in the norm of $\mathcal{L}^k(\mathcal{W}_1{}^{n_0\times n_1} \times \cdots \times \mathcal{W}_k{}^{n_{k-1}\times n_k}, \mathcal{W}_0{}^{n_0\times n_k})$.

Theorem 7.51, Remark 7.52, Theorem 7.53, Corollary 7.54, and Remark 7.55 are true for the case where $\mathcal{V}_0$, ..., $\mathcal{V}_k$, $\mathcal{W}_0$, ..., $\mathcal{W}_k$ are operator spaces over $\mathbb{R}$.

CHAPTER 8

# Convergence of nc power series

In this chapter, we shall study the convergence of power series of the form

$$\sum_{\ell=0}^{\infty}\Big(X-\bigoplus_{\alpha=1}^{m}Y\Big)^{\odot_s\ell}f_\ell \tag{8.1}$$

to a nc function. Here $\mathcal{V}$ and $\mathcal{W}$ are vector spaces over $\mathbb{C}$; $Y\in\mathcal{V}^{s\times s}$ is a given center; $X\in\mathcal{V}^{sm\times sm}$, $m=1,2,\ldots$; $f_\ell\colon(\mathcal{V}^{s\times s})^\ell\to\mathcal{W}^{s\times s}$, $\ell=0,1,\ldots$, is a given sequence of $\ell$-linear mappings, i.e., a linear mapping $f\colon\mathbf{T}(\mathcal{V}^{s\times s})\to\mathcal{W}^{s\times s}$; and conditions (4.4)–(4.7) are satisfied. Notice that for $s=1$ these conditions are trivially satisfied for any sequence $f_\ell$ of linear mappings. We shall be interested in the growth conditions on the coefficients $f_\ell$ under which the sum in (8.1) is an analytic or a uniformly analytic nc function. The results of this section are the converse of the results of Sections 7.1 and 7.3, much as the results of Section 5.2 is the converse of the results of Section 5.1 where we evaluate power series on nilpotent matrices, so that no questions of convergence arise.

## 8.1. Finitely open topology

Let $\mathcal{V}$ be a vector space over $\mathbb{C}$, let $\mathcal{W}$ be a Banach space over $\mathbb{C}$ equipped with an admissible system of matrix norms (see Section 7.1), and let $f\colon\mathbf{T}(\mathcal{V}^{s\times s})\to\mathcal{W}^{s\times s}$ be a linear mapping. Notice that, for $Z\in\mathcal{V}^{sm\times sm}$,

$$Z^{\odot_s\ell}f_\ell=f_\ell^{(m,\ldots,m)}(Z,\ldots,Z)$$

(see (7.49)) is a homogeneous polynomial of degree $\ell$ on $\mathcal{V}^{sm\times sm}$ with values in $\mathcal{W}^{sm\times sm}$ and the series (8.1) is a power series on $\mathcal{V}^{sm\times sm}$ centered at $\bigoplus_{\alpha=1}^{m}Y$ with values in $\mathcal{W}^{sm\times sm}$ as in [**61**, Chapter 26] and [**72**, Chapter 1].

If the series (8.1) converges on a finitely open nc set containing $Y$, then it follows from Theorem 7.9 and Remark 4.3 that conditions (4.4)–(4.7) are necessary for the sum of the series (8.1) to be a nc function. We will show in Theorem 8.2 below that these conditions are also sufficient, using the following lemma.

LEMMA 8.1. *Let $\mathcal{M}$ be a module over a ring $\mathcal{R}$, $A\in\mathcal{M}^{m\times m}$, $B\in\mathcal{M}^{m\times\tilde{m}}$, $\widetilde{A}\in\mathcal{M}^{\tilde{m}\times\tilde{m}}$. Then for any $\ell\in\mathbb{N}$,*

$$\begin{bmatrix}A & B\\ 0 & \widetilde{A}\end{bmatrix}^{\odot\ell}=\begin{bmatrix}A^{\odot\ell} & \sum_{k=0}^{\ell-1}A^{\odot k}\odot B\odot C^{\odot(\ell-1-k)}\\ 0 & C^{\odot\ell}\end{bmatrix}. \tag{8.2}$$

PROOF. A trivial induction. □

For all $m=1,2,\ldots$ and all $Z\in\mathcal{V}^{sm\times sm}$, we introduce the quantities

$$\mu(Z)=\limsup_{\ell\to\infty}\sqrt[\ell]{\|Z^{\odot_s\ell}f_\ell\|_{sm}},\qquad\rho(Z)=\mu(Z)^{-1}, \tag{8.3}$$

and

$$\rho_{\text{fin}}(Z) = \liminf_{W \overset{\text{fin}}{\to} Z} \rho(W), \tag{8.4}$$

where $W \overset{\text{fin}}{\to} Z$ means that the lower limit is taken in the finitely open topology on $\mathcal{V}^{sm \times sm}$. It is clear that the function $\rho(\cdot)$ is *positive-homogeneous of degree* $-1$, i.e., satisfies $\rho(\lambda Z) = \frac{1}{|\lambda|}\rho(Z)$ for every nonzero $\lambda \in \mathbb{C}$. Since the multiplication by $\lambda$ is continuous in the finitely open topology, $\rho_{\text{fin}}(\cdot)$ is positive-homogeneous of degree $-1$ as well.

THEOREM 8.2. *1. The convergence set, $\Upsilon_{\text{nc}} \subseteq \mathcal{V}_{\text{nc}}$, for the series (8.1) is a nc set which is complete circular about $Y$, so that the sum of the series (8.1), $f\colon \Upsilon_{\text{nc}} \to \mathcal{W}_{\text{nc}}$ is a well defined mapping. We have*

$$\begin{aligned} \coprod_{m=1}^{\infty} \Big\{X \in \mathcal{V}^{sm \times sm} \colon \rho\Big(X - \bigoplus_{\alpha=1}^{m} Y\Big) > 1\Big\} &\subseteq \Upsilon_{\text{nc}} \\ &\subseteq \coprod_{m=1}^{\infty} \Big\{X \in \mathcal{V}^{sm \times sm} \colon \rho\Big(X - \bigoplus_{\alpha=1}^{m} Y\Big) \geq 1\Big\}. \end{aligned} \tag{8.5}$$

*The set*

$$\Upsilon_{\text{nc,fin}} = \coprod_{m=1}^{\infty} \Big\{X \in \mathcal{V}^{sm \times sm} \colon \rho_{\text{fin}}\Big(X - \bigoplus_{\alpha=1}^{m} Y\Big) > 1\Big\} \tag{8.6}$$

*is the interior of $\Upsilon_{\text{nc}}$ in the finitely open topology on $\mathcal{V}_{\text{nc}}$.*

*2. For each $m$, $(\Upsilon_{\text{nc,fin}})_{sm}$ is nonempty if and only if it contains $\bigoplus_{\alpha=1}^{m} Y$. In this case,*

- *$\rho_{\text{fin}}(0_{sm \times sm}) = \infty$ and $\rho_{\text{fin}}(Z) > 0$ for every $Z \in \mathcal{V}^{sm \times sm}$,*
- *$(\Upsilon_{\text{nc,fin}})_{sm}$ is a complete circular set about $\bigoplus_{\alpha=1}^{m} Y$,*
- *the series (8.1) converges absolutely on $(\Upsilon_{\text{nc,fin}})_{sm}$, and*
- *$f|_{(\Upsilon_{\text{nc,fin}})_{sm}}$ is G-differentiable.*

*3. For every $m \in \mathbb{N}$ and $X \in (\Upsilon_{\text{nc,fin}})_{sm}$ there exists a finitely open complete circular set $\Upsilon_{\text{fin},X}$ about $\bigoplus_{\alpha=1}^{m} Y$ which contains $X$ and such that*

$$\sum_{\ell=0}^{\infty} \sup_{W \in \Upsilon_{\text{fin},X}} \Big\| \Big(W - \bigoplus_{\alpha=1}^{m} Y\Big)^{\odot_s \ell} f_\ell \Big\|_{sm} < \infty. \tag{8.7}$$

*4. For any finitely open nc set $\Gamma \subseteq \Upsilon_{\text{nc,fin}}$, $f|_\Gamma$ is a G-differentiable nc function. Let*

$$\Upsilon_{\text{nc,fin}-\text{G}} := \{X \in \Upsilon_{\text{nc,fin}} \colon \exists \Gamma, \text{ a finitely open nc set, such that } X \in \Gamma \subseteq \Upsilon_{\text{nc,fin}}\}. \tag{8.8}$$

*Then $\Upsilon_{\text{nc,fin}-\text{G}}$ is nonempty if and only if there exists a nonempty finitely open nc set $U \subseteq \mathcal{V}_{\text{nc}}$ such that $\inf_{Z \in U} \rho(Z) > 0$. We have $Y \in \Upsilon_{\text{nc,fin}-\text{G}}$ if and only if there exists a finitely open nc neighborhood $U$ of $Y$ such that $\inf_{X \in U} \rho(X - \bigoplus_{\alpha=1}^{m_X} Y) > 0$; here $m_X$ is the size of a block matrix $X$ with $s \times s$ blocks.*

REMARK 8.3. It follows that if $Y \in \Upsilon_{\text{nc,fin}-\text{G}}$, then the restriction of $f$ to a finitely open nc neighborhood of $Y$ is a G-differentiable nc function, and according to Theorem 7.9 one has

$$f_\ell = \Delta_R^\ell f(\underbrace{Y, \ldots, Y}_{\ell+1 \text{ times}}), \quad \ell = 0, 1, \ldots.$$

In particular, the coefficients $f_\ell$ (i.e., the original linear mapping $f$) are uniquely determined by the corresponding nc function, which is why we used the same notation for both the nc function and the linear mapping.

PROOF OF THEOREM 8.2. 1. The statement on $\Upsilon_{\rm nc}$ is straightforward. The inclusions in (8.5) follow from the classical Cauchy–Hadamard theorem since, for a fixed $Z$, $\rho(Z)$ is the radius of convergence of the series

$$\sum_{\ell=0}^{\infty} Z^{\odot_s \ell} f_\ell \lambda^\ell \tag{8.9}$$

as a series in $\lambda \in \mathbb{C}$. Next we observe that, by the construction, $\rho_{\rm fin}(Z)$ is a lower semicontinuous function of $Z$ in the finitely open topology on $\mathcal{V}_{\rm nc}$. This implies that $\Upsilon_{\rm nc,fin}$ is a finitely open set. Clearly, $\rho_{\rm fin}(Z) \le \rho(Z)$, therefore, again by the classical Cauchy–Hadamard theorem, the series (8.1) converges absolutely on $\Upsilon_{\rm nc,fin}$. On the other hand, if $Z \notin \Upsilon_{\rm nc,fin}$, i.e., if $\rho_{\rm fin}(Z) \le 1$, then by the positive-homogeneity of degree $-1$ of $\rho_{\rm fin}(\cdot)$ we have that $\rho_{\rm fin}((1+\epsilon)Z) \le \frac{1}{1+\epsilon}$ for an arbitrary $\epsilon > 0$. Therefore, in every finitely open neighborhood of $(1+\epsilon)Z$ there is a point $W$ such that $\rho(W) < 1$. This means that $(1+\epsilon)Z$ is not in the interior in the finitely open topology on $\mathcal{V}_{\rm nc}$ of the convergence set for the series (8.1), and then so is $Z$. Thus we obtain that $\Upsilon_{\rm nc,fin}$ is the interior of $\Upsilon_{\rm nc}$ in the finitely open topology on $\mathcal{V}_{\rm nc}$.

2. Let $m \in \mathbb{N}$ be fixed. If $\bigoplus_{\alpha=1}^m Y \in (\Upsilon_{\rm nc,fin})_{sm}$, then clearly $(\Upsilon_{\rm nc,fin})_{sm} \neq \emptyset$. Conversely, if $(\Upsilon_{\rm nc,fin})_{sm}$, which is the interior of $(\Upsilon_{\rm nc})_{sm}$ in the finitely open topology on $\mathcal{V}^{sm\times sm}$, is nonempty, then by [**61**, Theorem 26.5.9] (see also [**61**, Theorem 26.5.2]) $\bigoplus_{\alpha=1}^m Y$ is an interior point of $(\Upsilon_{\rm nc})_{sm}$ in this topology, i.e., $\rho_{\rm fin}(0_{sm\times sm}) > 1$. Moreover, by the positive-homogeneity of degree $-1$ and lower semicontinuity of $\rho_{\rm fin}(\cdot)$ we have that $\rho_{\rm fin}(Z) > 0$ for every $Z \in \mathcal{V}^{sm\times sm}$ and that $\rho_{\rm fin}(0_{sm\times sm}) = \infty$. From now on, assume $(\Upsilon_{\rm nc,fin})_{sm} \neq \emptyset$. By the positive-homogeneity of $\rho_{\rm fin}(\cdot)$ the set $(\Upsilon_{\rm nc,fin})_{sm}$ is a complete circular set about $\bigoplus_{\alpha=1}^m Y$, for every $m \in \mathbb{N}$ (see also [**61**, Theorem 26.5.9]). By [**61**, Theorem 26.5.6] (see also [**61**, Theorem 26.5.3 and Theorem 26.5.4]) the sum of the series (8.1) is G-differentiable on $(\Upsilon_{\rm nc,fin})_{sm}$. The absolute convergence of the series (8.1) on $(\Upsilon_{\rm nc,fin})_{sm}$ has been proved in the preceding paragraph. We also can prove this using Hartogs' theorem on the convergence in homogeneous polynomials [**97**, Theorem I.3.3].

3. This statement for any fixed $m$ and $X$ follows from [**61**, Theorem 26.3.8] (cf. Theorem 7.2(4)).

4. It is clear that $f((\Upsilon_{\rm nc})_{sm}) \subseteq \mathcal{W}^{sm\times sm}$. Let $X \in (\Upsilon_{\rm nc})_{sm}$, $\widetilde{X} \in (\Upsilon_{\rm nc})_{s\widetilde{m}}$, and $S \in \mathbb{C}^{s\widetilde{m}\times sm}$ be such that $SX = \widetilde{X}S$. Taking the limit as $N \to \infty$ in (5.12) (see Lemma 5.12) and using Lemma 8.1, we see that $Sf(X) = f(\widetilde{X})S$ if and only if the general term of the series (8.1) evaluated at $\begin{bmatrix} \widetilde{X} & S\bigoplus_{\nu=1}^m Y - \bigoplus_{\widetilde{\nu}=1}^{\widetilde{m}} YS \\ 0 & X \end{bmatrix}$ converges to 0. For $X, \widetilde{X} \in \Gamma$ where $\Gamma \subseteq \Upsilon_{\rm nc,fin}$ is a finitely open nc set, we have that $\begin{bmatrix} \widetilde{X} & \epsilon(S\bigoplus_{\nu=1}^m Y - \bigoplus_{\widetilde{\nu}=1}^{\widetilde{m}} YS) \\ 0 & X \end{bmatrix} \in \Gamma$ for $\epsilon > 0$ small enough, so that the series

(8.1) converges at this point and a fortiori its general term converges to 0; therefore $Sf(X) = f(\widetilde{X})S$. We conclude that $f|_\Gamma$ is a nc function. It is a G-differentiable nc function by part 2 of the theorem.

Next, if $\Upsilon_{\rm nc,fin-G} \neq \emptyset$, then there exists a nonempty finitely open nc set $\Gamma \subseteq \Upsilon_{\rm nc,fin}$, i.e., $\rho_{\rm fin}(W) > 1$ for every $W \in \Gamma$. We have

$$\inf_{W\in\Gamma} \rho(W) \geq \inf_{W\in\Gamma} \rho_{\rm fin}(W) \geq 1 > 0.$$

Conversely, if $U \subseteq \mathcal{V}_{\rm nc}$ is a nonempty finitely open nc set so that $\rho_0 := \inf_{W\in U} \rho(W) > 0$, then for any positive $\epsilon < \rho_0$ the set $(\rho_0 - \epsilon)U \subseteq \mathcal{V}_{\rm nc}$ is also a nonempty finitely open nc set and $\inf_{W\in(\rho_0-\epsilon)U} \rho(W) = \frac{\rho_0}{\rho_0-\epsilon}$. Therefore for any $Z \in (\rho_0 - \epsilon)U$ we have

$$\rho_{\rm fin}(Z) = \liminf_{\substack{{\rm fin}\\ W\to Z}} \rho(W) \geq \frac{\rho_0}{\rho_0 - \epsilon} > 1,$$

and thus the finitely open nc set $(\rho_0 - \epsilon)U$ is contained in $\Upsilon_{\rm nc,fin}$. We conclude that $\Upsilon_{\rm nc,fin-G} \neq \emptyset$.

If $Y \in \Upsilon_{\rm nc,fin-G}$, then there exists a finitely open nc neighborhood $\Gamma$ of $Y$ such that $\Gamma \subseteq \Upsilon_{\rm nc,fin}$. Then, as we have shown in the preceding paragraph, $\inf_{X\in\Gamma} \rho(X - \bigoplus_{\alpha=1}^{m_X} Y) > 0$. Conversely, if there exists a finitely open nc neighborhood $U$ of $Y$ such that $\rho_0 := \inf_{X\in U} \rho(X - \bigoplus_{\alpha=1}^{m_X} Y) > 0$, then arguing as in the preceding paragraph we obtain that, for $\epsilon > 0$ small enough, the set $\Gamma$ consisting of $X \in \Upsilon_{\rm nc}$ for which there is a $\widetilde{X} \in U$ satisfying $X - \bigoplus_{\alpha=1}^{m_X} Y = (\rho_0 - \epsilon)(\widetilde{X} - \bigoplus_{\alpha=1}^{m_X} Y)$, is a subset of $\Upsilon_{\rm nc,fin}$. Since $\Gamma$ is a finitely open nc neighborhood of $Y$, we conclude that $Y \in \Upsilon_{\rm nc,fin-G}$. □

REMARK 8.4. In the case $s = 1$, $f$ is a nc function on $\Upsilon_{\rm nc}$. This is clear from the form of the series (8.1). It also follows from the proof of part 4 of Theorem 8.2 since in this case $S \bigoplus_{\nu=1}^{m} Y - \bigoplus_{\tilde{\nu}=1}^{\tilde{m}} YS = 0$.

The following example demonstrates the possibility that $Y = 0_{1\times 1} \in \Upsilon_{\rm nc,fin-G} \subsetneq \Upsilon_{\rm nc,fin}$.

EXAMPLE 8.5. Consider the series

$$\sum_{\ell=0}^{\infty} Z_1^\ell (Z_1 Z_2 - Z_2 Z_1).$$

Here $\mathcal{V} = \mathbb{C}^2$, $\mathcal{W} = \mathbb{C}$, with the Euclidean topologies on $\mathcal{V}^{n\times n} = (\mathbb{C}^2)^{n\times n} \cong (\mathbb{C}^{n\times n})^2$ and $\mathcal{W}^{n\times n} = \mathbb{C}^{n\times n}$. If $Z = (Z_1, Z_2) \in (\mathbb{C}^{n\times n})^2$ satisfies $Z_1 Z_2 = Z_2 Z_1$ (in particular, it is always the case when $n = 1$), then $\rho(Z) = \infty$. If $z = (z_1, z_2) \in \mathcal{V}^{1\times 1} = \mathbb{C}^2$, then we have $\rho_{\rm fin}(z) = \rho(z) = \infty$ and therefore $(\Upsilon_{\rm nc,fin})_1 = \mathbb{C}^2$. If $Z = z \oplus z \in \mathcal{V}^{2\times 2}$ and $z_1 \neq 0$, then of course we have $\rho(Z) = \infty$, however $\rho_{\rm fin}(Z) = \frac{1}{|z_1|}$. To show the latter, we first calculate, for any fixed $\epsilon > 0$,

$$\mu\left(\begin{bmatrix} z_1 & \epsilon \\ 0 & z_1 \end{bmatrix}, \begin{bmatrix} z_2 & 0 \\ \epsilon & z_2 \end{bmatrix}\right) = \limsup_{\ell\to\infty} \sqrt[\ell+2]{\left\| \begin{bmatrix} z_1 & \epsilon \\ 0 & z_1 \end{bmatrix}^\ell \begin{bmatrix} \epsilon^2 & 0 \\ 0 & -\epsilon^2 \end{bmatrix} \right\|}$$

$$= \limsup_{\ell\to\infty} \sqrt[\ell+2]{\left\| \begin{bmatrix} \epsilon^2 z_1^\ell & -\ell\epsilon^3 z_1^{\ell-1} \\ 0 & -\epsilon^2 z_1^\ell \end{bmatrix} \right\|} = \limsup_{\ell\to\infty} \sqrt[\ell+2]{\epsilon^2 |z_1|^{\ell-1} \left\| \begin{bmatrix} z_1 & -\ell\epsilon \\ 0 & -z_1 \end{bmatrix} \right\|} = |z_1|.$$

Hence

$$\rho_{\mathrm{fin}}(Z) \leq \lim_{\epsilon\to 0} \rho\left(\begin{bmatrix} z_1 & \epsilon \\ 0 & z_1 \end{bmatrix}, \begin{bmatrix} z_2 & 0 \\ \epsilon & z_2 \end{bmatrix}\right) = \frac{1}{|z_1|}.$$

On the other hand, for any $M, N \in \mathbb{C}^{2\times 2}$ we have

$$\begin{aligned}\mu(z_1I_2 + M, z_2I_2 + N) &= \limsup_{\ell\to\infty} \sqrt[\ell+2]{\|(z_1I_2 + M)^\ell(MN - NM)\|} \\ &\leq \|z_1I_2 + M\| \lim_{\ell\to 0} \sqrt[\ell+2]{\|MN - NM\|} \leq \|z_1I_2 + M\|,\end{aligned}$$

and for $M$ small enough we have $\rho(z_1I_2 + M, z_2I_2 + N) \geq \frac{1}{\|z_1I_2+M\|}$. Hence

$$\rho_{\mathrm{fin}}(Z) \geq \frac{1}{|z_1|}.$$

We conclude that $\rho_{\mathrm{fin}}(Z) = \frac{1}{|z_1|}$. If $|z_1| > 1$, we have that $z = (z_1, z_2) \in \Upsilon_{\mathrm{nc,fin}}$, however $Z = z \oplus z \notin \Upsilon_{\mathrm{nc,fin}}$. Thus the set $\Upsilon_{\mathrm{nc,fin}}$ is not a nc set. On the other hand, the subset $\Gamma \subsetneq \Upsilon_{\mathrm{nc,fin}}$ consisting of pairs of matrices $Z = (Z_1, Z_2)$ satisfying $\|Z_1\| < 1$ (with the operator norm $\|\cdot\|$) is clearly a uniformly-open (and therefore finitely open) nc set, the series $\sum_{\ell=0}^\infty Z_1^\ell(Z_1Z_2 - Z_2Z_1)$ converges absolutely on $\Gamma$, and its sum $f\colon \Gamma \to \mathbb{C}_{\mathrm{nc}}$ is a G-differentiable nc function.

The next example shows that it is also possible that $(\Upsilon_{\mathrm{nc,fin}})_{1\cdot m} \neq \emptyset$, i.e., $\bigoplus_{\alpha=1}^m Y = 0_{1\cdot m\times 1\cdot m} \in \Upsilon_{\mathrm{nc,fin}}$ for all $m \in \mathbb{N}$, however $Y = 0_{1\times 1} \notin \Upsilon_{\mathrm{nc,fin-G}}$ (in fact, we have $0_{1\cdot m\times 1\cdot m} \notin \Upsilon_{\mathrm{nc,fin-G}}$ for all $m \in \mathbb{N}$).

EXAMPLE 8.6. Let $p_n$ be the homogeneous polynomial in two noncommuting variables of degree $\alpha_n = \frac{(n+1)(n+2)}{2}$, $n = 1, 2, \ldots$, from Example 6.3. We remind the reader that $p_n$ vanishes on all pairs of $n \times n$ matrices. Consider the series

$$\sum_{j=2}^{\infty}\sum_{k=1}^{j-1} (kZ_1)^{\alpha_j-\alpha_k-j+k} p_k(Z_1, Z_2).$$

We note that the total degrees of the homogeneous polynomials in this series strictly increase. Indeed, for any fixed $j$ we have

$$\deg\left((kx_1)^{\alpha_j-\alpha_k-j+k} p_k\right) = \alpha_j - j + k$$

increasing as $k$ changes from 1 to $j-1$. On the other hand, for any fixed $j$ we have

$$\alpha_j - j + (j-1) = \alpha_j - 1 < \alpha_{j+1} - j = \alpha_{j+1} - (j+1) + 1,$$

i.e., the degree increases when we pass from the last term in the $j$-th sum to the first term in the $j+1$-th sum. Similarly to Example 8.5, we have $\mathcal{V} = \mathbb{C}^2$, $\mathcal{W} = \mathbb{C}$, with the Euclidean topologies on $\mathcal{V}^{n\times n} = (\mathbb{C}^2)^{n\times n} \cong (\mathbb{C}^{n\times n})^2$ and $\mathcal{W}^{n\times n} = \mathbb{C}^{n\times n}$, and $(\Upsilon_{\mathrm{nc,fin}})_1 = \mathbb{C}^2$. Let $z = (z_1, z_2) \in \mathcal{V}^{1\times 1} = \mathbb{C}^2$ be such that $z_1 \neq 0$. Define $Z(k) = \bigoplus_{j=1}^k z \in \mathcal{V}^{k\times k}$. Arguing as in Example 8.5 and using the subsequence of homogeneous polynomials $(kx_1)^{\alpha_j-\alpha_k-j+k}p_k$, $j = 2, 3, \ldots$, we can easily show that $\rho_{\mathrm{fin}}(Z(k)) = \frac{1}{k|z_1|}$. Therefore, there is no finitely open nc neighborhood of $0_{m\times m}$, for any $m$, where the series converges (otherwise $|z_1| < \frac{1}{k}$ for all $k$, which is impossible).

## 8.2. Norm topology

Let $\mathcal{V}$ and $\mathcal{W}$ be Banach spaces over $\mathbb{C}$ equipped with admissible systems of matrix norms, and let $f_\ell\colon (\mathcal{V}^{s\times s})^\ell \to \mathcal{W}^{s\times s}$, $\ell = 0, 1, \ldots$, be a sequence of bounded $\ell$-linear mappings. It follows that, for all $m_0, \ldots, m_k \in \mathbb{N}$, $f_\ell^{(m_0,\ldots,m_k)}$ is also a sequence of bounded $\ell$-linear mappings. Therefore, for $Z \in \mathcal{V}^{sm\times sm}$,

$$Z^{\odot_s \ell} f_\ell = f_\ell^{(m,\ldots,m)}(Z, \ldots, Z)$$

is a bounded homogeneous polynomial of degree $\ell$ on $\mathcal{V}^{sm\times sm}$ with values in $\mathcal{W}^{sm\times sm}$ and the series (8.1) is a F-power series (a power series of bounded homogeneous polynomials) on $\mathcal{V}^{sm\times sm}$ centered at $\bigoplus_{\alpha=1}^m Y$ with values in $\mathcal{W}^{sm\times sm}$ as in [**61**, Chapter 26] and [**72**, Chapter 1].

We introduce the quantities

$$\rho_{\mathrm{norm}}(Z) = \liminf_{W\to Z} \rho(W) \tag{8.10}$$

(cf. (8.4)),

$$\mu_m = \limsup_{\ell\to\infty} \sqrt[\ell]{\sup_{\|Z\|_{sm}=1} \|Z^{\odot_s \ell} f_\ell\|_{sm}}, \qquad \rho_m = \mu_m^{-1}. \tag{8.11}$$

Analogously to $\rho_{\mathrm{fin}}(\cdot)$, the function $\rho_{\mathrm{norm}}(\cdot)$ is positive-homogeneous of degree $-1$. We notice that the expression under the $\ell$-th root is the norm of the homogeneous polynomial $Z^{\odot_s \ell} f_\ell$, see [**61**, Definition 26.2.4] and [**72**, Definition 2.1].

PROPOSITION 8.7. *The following inequalities hold:*

$$\frac{1}{e} \limsup_{\ell\to\infty} \sqrt[\ell]{\|f_\ell\|} \le \mu_1 \le K_1 \mu_m \le m K_1 K_2 \limsup_{\ell\to\infty} \sqrt[\ell]{\|f_\ell\|}, \tag{8.12}$$

*where*

$$K_1 = \min_{1\le i\le m} \|\iota_{ii}^s\|, \quad K_2 = \max_{1\le i,j\le m} \|\pi_{ij}^s\|;$$

*block injections, $\iota_{ij}^s$, and block projections, $\pi_{ij}^s$, are defined in* (7.5) *and* (7.6).

Notice that we also have estimates for $K_1$ and $K_2$ from the proof of Proposition 7.1 via the constants in (7.1) and (7.2) which determine the admissible system of norms on $\mathcal{V}^{n\times n}$ and $\mathcal{W}^{n\times n}$, $n = 1, 2, \ldots$:

$$K_1 \le C_2(sm) C_1'(s, sm - s), \quad K_2 \le C_1(s, sm - s) C_2(sm),$$

and that in the case where $\mathcal{V}$ and $\mathcal{W}$ are operator spaces one has $K_1 = K_2 = 1$.

PROOF OF PROPOSITION 8.7. The leftmost inequality in (8.12) follows from the estimate

$$\|f_\ell\| = \sup_{\|Z_1\|_s=\cdots=\|Z_\ell\|_s=1} \|(Z_1 \otimes \cdots \otimes Z_\ell) f_\ell\|_s \le \frac{\ell^\ell}{\ell!} \sup_{\|Z\|_s=1} \|Z^{\otimes \ell} f_\ell\|_s$$

(see [**72**, Theorem 2.2]) by taking the $\ell$-th root of both parts, then $\limsup_{\ell\to\infty}$, and then applying the Stirling formula. The second inequality in (8.12) follows from the estimate

$$\sup_{\|Z\|_s=1} \|Z^{\otimes \ell} f_\ell\|_s = \sup_{\|Z\|_s=1} \|\pi_{ii}^s \left((\iota_{ii}^s Z)^{\odot_s \ell} f_\ell\right)\|_s \le \|\pi_{ii}^s\| \|\iota_{ii}^s\|^\ell \sup_{\|W\|_{sm}=1} \|W^{\odot_s \ell} f_\ell\|_{sm}$$

by taking the $\ell$-th root of both parts, then $\limsup_{\ell\to\infty}$, and then $\min_{1\le i\le m}$. The last inequality in (8.12) follows from the estimate

$$\begin{aligned}\sup_{\|Z\|_{sm}=1}\|Z^{\odot_s \ell}f_\ell\|_{sm} &= \sup_{\|Z\|_{sm}=1}\Big\|\sum_{1\le i,j\le m}\iota^s_{ij}(Z^{\odot_s\ell}f_\ell)_{ij}\Big\|_{sm}\\ &\le \sum_{1\le i,j\le m}\|\iota^s_{ij}\|\sup_{\|Z\|_{sm}=1}\|(Z^{\odot_s\ell}f_\ell)_{ij}\|_s\\ = \sum_{1\le i,k_1,\ldots,k_{\ell-1},j\le m}\|\iota^s_{ij}\| &\sup_{\|Z\|_{sm}=1}\|(\pi^s_{ik_1}Z\otimes\pi^s_{k_1k_2}Z\otimes\cdots\otimes\pi^s_{k_{\ell-2}k_{\ell-1}}Z\otimes\pi^s_{k_{\ell-1}j}Z)f_\ell\|_s\\ \le \sum_{1\le i,k_1,\ldots,k_{\ell-1},j\le m} &\|\iota^s_{ij}\|\|\pi^s_{ik_1}\|\|\pi^s_{k_1k_2}\|\cdots\|\pi^s_{k_{\ell-2}k_{\ell-1}}\|\|\pi^s_{k_{\ell-1}j}\|\|f_\ell\|_s\\ &\le m^{\ell+1}K_2^\ell\sup_{1\le i,j\le m}\|\iota^s_{ij}\|\,\|f_\ell\|\end{aligned}$$

by taking the $\ell$-th root, then $\limsup_{\ell\to\infty}$, and then multiplying by $K_1$. □

We recall that $\Upsilon_{\rm nc}\subseteq\mathcal{V}_{\rm nc}$ denotes the convergence set for the series (8.1), and $f$ denote the sum of the series.

THEOREM 8.8. *1. The set*

$$\Upsilon_{\rm nc,norm}=\coprod_{m=1}^\infty\Big\{X\in\mathcal{V}^{sm\times sm}\colon \rho_{\rm norm}\Big(X-\bigoplus_{\alpha=1}^m Y\Big)>1\Big\}\tag{8.13}$$

*is contained in $\Upsilon_{\rm nc,fin}$ and is the interior of $\Upsilon_{\rm nc}$ in the norm topology on $\mathcal{V}_{\rm nc}$, i.e., in the topology on $\mathcal{V}_{\rm nc}$ defined in the paragraph next to Remark 7.3.*

*2. For each $m$, $(\Upsilon_{\rm nc,norm})_{sm}$ is nonempty if and only if it contains $\bigoplus_{\alpha=1}^m Y$. In this case,*

- *$\rho_{\rm norm}(0_{sm\times sm})=\infty$ and $\rho_{\rm norm}(Z)>0$ for every $Z\in\mathcal{V}^{sm\times sm}$,*
- *$(\Upsilon_{\rm nc,norm})_{sm}$ is a complete circular set about $\bigoplus_{\alpha=1}^m Y$, and*
- *$f|_{(\Upsilon_{\rm nc,norm})_{sm}}$ is analytic.*

*3. For every $m\in\mathbb{N}$ and $X\in(\Upsilon_{\rm nc,norm})_{sm}$ there exists an open complete circular set $\Upsilon_{{\rm norm},X}$ about $\bigoplus_{\alpha=1}^m Y$ which contains $X$ and such that*

$$\sum_{\ell=0}^\infty\sup_{W\in\Upsilon_{{\rm norm},X}}\Big\|\Big(W-\bigoplus_{\alpha=1}^m Y\Big)^{\odot_s\ell}f_\ell\Big\|_{sm}<\infty.\tag{8.14}$$

*4. For any open nc set $\Gamma\subseteq\Upsilon_{\rm nc,norm}$, $f|_\Gamma$ is an analytic nc function. Let*

$$\Upsilon_{\rm nc,norm-an}:=\{X\in\Upsilon_{\rm nc,norm}\colon\exists\Gamma,\text{ an open nc set, such that }X\in\Gamma\subseteq\Upsilon_{\rm nc,norm}\}.\tag{8.15}$$

*Then $\Upsilon_{\rm nc,norm-an}$ is nonempty if and only if there exists a nonempty open nc set $U\subseteq\mathcal{V}_{\rm nc}$ such that $\inf_{Z\in U}\rho(Z)>0$. We have $Y\in\Upsilon_{\rm nc,norm-an}$ if and only if there exists an open nc neighborhood of $Y$ such that $\inf_{X\in U}\rho(X-\bigoplus_{\alpha=1}^{m_X}Y)>0$; here $m_X$ is the size of a block matrix $X$ with $s\times s$ blocks.*

*5. The series (8.1) converges uniformly on every ball $B(\bigoplus_{\alpha=1}^m Y,\delta)$ with $\delta<\rho_m$, moreover*

$$\sum_{\ell=0}^\infty\sup_{X\in B(\bigoplus_{\alpha=1}^m Y,\delta)}\Big\|\Big(X-\bigoplus_{\alpha=1}^m Y\Big)^{\odot_s\ell}f_\ell\Big\|_{sm}<\infty.\tag{8.16}$$

*The series* (8.1) *fails to converge uniformly on every ball* $B(\bigoplus_{\alpha=1}^{m} Y, \delta)$ *with* $\delta > \rho_m$. *The set* $(\Upsilon_{\rm nc,norm})_{sm}$ *is nonempty if and only if* $\rho_m > 0$ *if and only if* $\inf\{\rho(W) \colon W \in \mathcal{V}^{sm\times sm}, \|W\|_{sm} = 1\} > 0$.

PROOF. 1. We first observe that the finitely open topology on $\mathcal{V}_{\rm nc}$ is stronger than the norm topology, hence $\rho_{\rm norm}(Z) \le \rho_{\rm fin}(Z)$ for any $Z$, and $\Upsilon_{\rm nc,norm} \subseteq \Upsilon_{\rm nc,fin}$. Next we observe that the function $\rho_{\rm norm}(\cdot)$ is lower semicontinuous in the norm topology and recall that $\rho_{\rm norm}(\cdot)$ is also positive-homogeneous of degree $-1$. The rest of the proof of part 1 is analogous to the proof of part 1 of Theorem 8.2.

2. Let $m \in \mathbb{N}$ be fixed. If $\bigoplus_{\alpha=1}^{m} Y \in (\Upsilon_{\rm nc,norm})_{sm}$, then clearly $(\Upsilon_{\rm nc,norm})_{sm} \neq \emptyset$. Conversely, if $(\Upsilon_{\rm nc,norm})_{sm}$, which is the interior of $(\Upsilon_{\rm nc})_{sm}$ in the norm topology on $\mathcal{V}^{sm\times sm}$, is nonempty, then by [**61**, Theorem 26.6.1] $\bigoplus_{\alpha=1}^{m} Y$ is an interior point of $(\Upsilon_{\rm nc})_{sm}$ in this topology and $(\Upsilon_{\rm nc,norm})_{sm}$ is a complete circular set about $\bigoplus_{\alpha=1}^{m} Y$. Then $\rho_{\rm norm}(0_{sm\times sm}) > 1$. Moreover, by the positive-homogeneity of degree $-1$ and lower semicontinuiuty of $\rho_{\rm norm}(\cdot)$ we have that $\rho_{\rm norm}(Z) > 0$ for every $Z \in \mathcal{V}^{sm\times sm}$ and that $\rho_{\rm norm}(0_{sm\times sm}) = \infty$. The analyticity of $f|(\Upsilon_{\rm norm})_{sm}$ follows from [**61**, Theorem 26.6.4].

3. This statement for any fixed $m$ and $X$ follows from [**61**, Theorem 26.6.5] (cf. Theorem 7.4).

4. Since an open set is finitely open, $f|\Gamma$ is a nc function by Theorem 8.2, part 4. It is an analytic nc function by part 2 of the present theorem. The remaining statements can be proved analogously to the proof of part 4 of Theorem 8.2.

5. Let $\delta < \rho_m$. Then for any $\epsilon > 0$ there exists $L > 0$ such that

$$\sup_{\|W\|_{sm}=1} \|W^{\odot_s \ell} f_\ell\|_{sm} < (\mu_m + \epsilon)^\ell$$

for all $\ell > L$. Choosing $\epsilon < \delta^{-1} - \mu_m$, we obtain that, for all $\ell > L$,

$$\sup_{W\in B(0_{sm\times sm},\delta)} \|W^{\odot_s \ell} f_\ell\|_{sm} < \delta^\ell(\mu_m + \epsilon)^\ell.$$

Since $\delta(\mu_m + \epsilon) < 1$, the series (8.1) converges absolutely and uniformly on the ball $B(0_{sm\times sm}, \delta)$, and

$$\sum_{\ell=0}^{\infty} \sup_{W\in B(0_{sm\times sm},\delta)} \|W^{\odot_s \ell} f_\ell\|_{sm} < \sum_{\ell=0}^{L} \sup_{W\in B(0_{sm\times sm},\delta)} \|W^{\odot_s \ell} f_\ell\|_{sm} + \sum_{\ell=L+1}^{\infty} \delta^\ell(\mu_m + \epsilon)^\ell < \infty,$$

i.e., (8.16) holds by setting $X = W + \bigoplus_{\alpha=1}^{m} Y$.

Now let $\delta > \rho_m$ be fixed. Choose any $\epsilon$ with $0 < \epsilon < \mu_m - \delta^{-1}$. Then for any $L > 0$ there exists $\ell > L$ such that

$$\sup_{\|W\|_{sm}=1} \|W^{\odot_s \ell} f_\ell\|_{sm} > (\mu_m - \epsilon)^\ell.$$

For any such $\ell$ there exists $Z \in \mathcal{V}^{sm\times sm}$ with $\|Z\|_{sm} = (\mu_m - \epsilon)^{-1} < \delta$ such that

$$\|Z^{\odot_s \ell} f_\ell\|_{sm} > (\mu_m - \epsilon)^\ell \|Z\|_{sm}^\ell = 1.$$

Hence the sequence $Z^{\odot_s \ell} f_\ell$, $\ell = 1, 2, \ldots$, does not converge to $0_{sm\times sm}$ uniformly on the ball $B(0_{sm\times sm}, \delta)$. We conclude that the series (8.1) does not converge uniformly on $B(\bigoplus_{\alpha=1}^{m} Y, \delta)$.

If $(\Upsilon_{\rm nc,norm})_{sm} \neq \emptyset$, i.e., the interior of $\Upsilon_{\rm nc}$ in the norm topology is nonempty, then by [**61**, Theorem 26.6.2] the series (8.1) converges uniformly on $B(\bigoplus_{\alpha=1}^m Y, \delta)$ for sufficiently small $\delta$, hence $\rho_m > 0$. Since $\rho_m \leq \rho(W)$ for every $W \in \mathcal{V}^{sm\times sm}$ with $\|W\|_{sm} = 1$, we have that $\rho_m > 0$ implies $\inf\{\rho(W) \colon W \in \mathcal{V}^{sm\times sm}, \|W\|_{sm} = 1\} > 0$. Finally, if $\alpha = \inf\{\rho(W) \colon W \in \mathcal{V}^{sm\times sm}, \|W\|_{sm} = 1\} > 0$, then by positive-homogeneity of degree $-1$ of both $\rho(\cdot)$ and $\rho_{\rm norm}(\cdot)$ we have that $\rho_{\rm norm}(Z) \geq \alpha$ for any $Z \in \mathcal{V}^{sm\times sm}$ with $\|Z\|_{sm} = 1$ and that $\rho_{\rm norm}(0_{sm\times sm}) = \infty$. The latter implies by part 2 that $(\Upsilon_{\rm nc,norm})_{sm} \neq \emptyset$. This completes the proof of the last statement of the theorem. □

REMARK 8.9. Since the finitely open and the norm topologies coincide for finite-dimensional spaces, Examples 8.5 and 8.6 also show that $\Upsilon_{\rm nc,norm}$ is not necessarily a nc set, and that a (norm-open) nc neighborhood of $Y$, which is contained in $\Upsilon_{\rm nc,norm}$, may or may not exist.

## 8.3. Uniformly-open topology

Let $\mathcal{V}$ and $\mathcal{W}$ be operator spaces over $\mathbb{C}$, and let $f_\ell \colon (\mathcal{V}^{s\times s})^\ell \to \mathcal{W}^{s\times s}$, $\ell = 0, 1, \ldots$, be a sequence of completely bounded $\ell$-linear mappings. We introduce the quantities

$$\rho_{\rm unif}(Z) = \liminf_{W \overset{\rm unif}{\to} Z} \rho(W) \tag{8.17}$$

(cf. (8.4) and (8.10)), where $W \overset{\rm unif}{\to} Z$ means that the lower limit is taken in the uniformly-open topology on $\mathcal{V}_{\rm nc}$,

$$\begin{aligned} \mu_{\rm unif-normal}(Z) &= \lim_{\delta\to 0} \limsup_{\ell\to\infty} \sqrt[\ell]{\sup_{W\in B_{\rm nc}(Z,\delta)} \|W^{\odot_s \ell} f_\ell\|}, \\ \rho_{\rm unif-normal}(Z) &= \mu_{\rm unif-normal}(Z)^{-1}, \end{aligned} \tag{8.18}$$

where $\|W^{\odot_s \ell} f_\ell\| = \|W^{\odot_s \ell} f_\ell\|_{sm}$ for $W \in \mathcal{V}^{sm\times sm}$, and

$$\mu_{\rm cb} = \limsup_{\ell\to\infty} \sqrt[\ell]{\|f_\ell\|_{\mathcal{L}^\ell_{\rm cb}}}, \qquad \rho_{\rm cb} = \mu_{\rm cb}^{-1}. \tag{8.19}$$

We notice that the limit $\lim\limits_{\delta\to 0}$ in (8.18) exists as a limit of a monotone function (decreasing as $\delta \downarrow 0$) bounded from below by 0. Analogously to $\rho_{\rm fin}(\cdot)$ and $\rho_{\rm norm}(\cdot)$, the functions $\rho_{\rm unif}(\cdot)$ and $\rho_{\rm unif-normal}(\cdot)$ are positive-homogeneous of degree $-1$; cf. Remark 7.13.

We will need the following lemma [**61**, Lemma 26.5.1]; we provide its proof for completeness.

LEMMA 8.10. *Let $\mathcal{V}$ be a vector space over $\mathbb{C}$, $\mathcal{W}$ a Banach space over $\mathbb{C}$, and let $\Gamma \subseteq \mathcal{V}$ be a complete circular set about 0. Suppose that $\omega \colon \mathcal{V}^k \to \mathcal{W}$ is a $k$-linear mapping and that $X \in \mathcal{V}$. Then*

$$\sup_{W\in \bigcup\limits_{\zeta\in\overline{\mathbb{D}}} (\zeta X+\Gamma)} \|\omega(W,\ldots,W)\| = \sup_{W\in X+\Gamma} \|\omega(W,\ldots,W)\|,$$

*where $\overline{\mathbb{D}} = \{\zeta \in \mathbb{C} \colon |\zeta| \leq 1\}$ is the closed unit disk, $X + \Gamma := \{X + Z \colon Z \in \Gamma\}$ is a complete circular set about $X$ and $\bigcup\limits_{\zeta\in\overline{\mathbb{D}}} (\zeta X + \Gamma)$ is a complete circular set about 0.*

PROOF. Fix an arbitrary $Z_0 \in \Gamma$. Consider the function

$$\phi\colon \overline{\mathbb{D}} \to \mathcal{W}, \quad \phi\colon \zeta \mapsto \omega(\zeta X + Z_0, \ldots, \zeta X + Z_0).$$

Clearly, $\phi$ is a polynomial and hence $\phi$ is analytic. By the maximum principle, there exists $\zeta_0$ with $|\zeta_0| = 1$ such that $\|\phi(\zeta_0)\| = \max_{\zeta \in \overline{\mathbb{D}}} \|\phi(\zeta)\|$. By the homogeneity (of degree $k$) of $\omega$,

$$\begin{aligned}\|\omega(\zeta_0 X + Z_0, \ldots, \zeta_0 X + Z_0)\| &= \|\omega(X + \zeta_0^{-1} Z_0, \ldots, X + \zeta_0^{-1} Z_0)\| \\ &\le \sup_{Z \in \Gamma} \|\omega(X + Z, \ldots, X + Z)\|\end{aligned}$$

since $\Gamma$ is completely circular about 0. This proves the lemma. $\square$

THEOREM 8.11. *1. The set*

$$\Upsilon_{\rm nc,unif} = \coprod_{m=1}^{\infty} \Big\{X \in \mathcal{V}^{sm \times sm} \colon \rho_{\rm unif}\Big(X - \bigoplus_{\alpha=1}^{m} Y\Big) > 1\Big\} \tag{8.20}$$

*is contained in $\Upsilon_{\rm nc,norm} \subseteq \Upsilon_{\rm nc,fin}$ and is the interior of $\Upsilon_{\rm nc}$ in the uniformly-open topology on $\mathcal{V}_{\rm nc}$.*

*2. The set*

$$\Upsilon_{\rm nc,unif-normal} = \coprod_{m=1}^{\infty} \Big\{X \in \mathcal{V}^{sm \times sm} \colon \rho_{\rm unif-normal}\Big(X - \bigoplus_{\alpha=1}^{m} Y\Big) > 1\Big\} \tag{8.21}$$

*is a uniformly-open subset of $\Upsilon_{\rm nc,unif} \subseteq \Upsilon_{\rm nc,norm} \subseteq \Upsilon_{\rm nc,fin}$ and*

$$\Upsilon_{\rm nc,unif-normal} = \Big\{X \in \Upsilon_{\rm nc,unif} \colon \exists \delta > 0$$

$$\text{such that} \sup_{\ell \in \mathbb{N}} \sup_{W \in B_{\rm nc}(X,\delta)} \Big\|\Big(W - \bigoplus_{\alpha=1}^{m_W} Y\Big)^{\odot_s \ell} f_\ell\Big\|_{sm_W} < \infty\Big\} \tag{8.22}$$

$$= \Big\{X \in \Upsilon_{\rm nc,unif} \colon \exists \delta > 0$$

$$\text{such that} \sum_{\ell=0}^{\infty} \sup_{W \in B_{\rm nc}(X,\delta)} \Big\|\Big(W - \bigoplus_{\alpha=1}^{m_W} Y\Big)^{\odot_s \ell} f_\ell\Big\|_{sm_W} < \infty\Big\}; \tag{8.23}$$

*here $m_W$ is the size of a block matrix $W$ with $s \times s$ blocks. $\Upsilon_{\rm nc,unif-normal}$ is nonempty if and only if it contains $Y$. In this case,*

- *$\rho_{\rm unif-normal}(0_{s\times s}) = \infty$ and $\rho_{\rm unif-normal}(Z) > 0$ for every $Z \in (\mathcal{V}^{s\times s})_{\rm nc}$,*
- *$\Upsilon_{\rm nc,unif-normal}$ is complete circular about $Y$ (i.e., $(\Upsilon_{\rm nc,unif-normal})_{sm}$ is a complete circular set about $\bigoplus_{\alpha=1}^{m} Y$ for every $m \in \mathbb{N}$).*

*3. For every $m \in \mathbb{N}$ and $X \in (\Upsilon_{\rm nc,unif-normal})_{sm}$ there exists a uniformly-open complete circular set $\Upsilon_{{\rm nc},X}$ about $\bigoplus_{\alpha=1}^{m} Y$, which contains $X$ and such that*

$$\sum_{\ell=0}^{\infty} \sup_{W \in \Upsilon_{{\rm nc},X}} \Big\|\Big(W - \bigoplus_{\alpha=1}^{m_W} Y\Big)^{\odot_s \ell} f_\ell\Big\|_{sm_W} < \infty. \tag{8.24}$$

*4. For any uniformly-open nc set $\Gamma \subseteq \Upsilon_{\rm nc,unif-normal}$, $f|_\Gamma$ is a uniformly analytic nc function.*

*5. The series* (8.1) *converges uniformly on every nc ball $B_{\rm nc}(Y,\delta)$ with $\delta < \rho_{\rm cb}$, moreover*

$$\sum_{\ell=0}^{\infty} \sup_{W \in B_{\rm nc}(Y,\delta)} \Big\|\Big(W - \bigoplus_{\alpha=1}^{m_W} Y\Big)^{\odot_s \ell} f_\ell\Big\|_{sm_W} < \infty. \tag{8.25}$$

*The series* (8.1) *fails to converge uniformly on every nc ball* $B_{\rm nc}(Y,\delta)$ *with* $\delta > \rho_{\rm cb}$. *The set* $\Upsilon_{\rm nc,unif-normal}$ *is nonempty if and only if* $\rho_{\rm cb} > 0$.

REMARK 8.12. It is an open question whether $\Upsilon_{\rm nc,unif}$ is nonempty if and only if $\rho_{\rm unif}(0_{s\times s}) > 0$. If $\rho_{\rm unif}(0_{s\times s}) > 0$, then similarly to Theorems 8.2 and 8.8, $\rho_{\rm unif}(0_{s\times s}) = \infty$ and $\Upsilon_{\rm nc,unif}$ is a complete circular set about $Y$.

PROOF OF THEOREM 8.11. 1. We first observe that the norm topology on $\mathcal{V}_{\rm nc}$ is stronger than the uniformly-open topology, hence $\rho_{\rm unif}(Z) \le \rho_{\rm norm}(Z)$ for any $Z$ and $\Upsilon_{\rm nc,unif} \subseteq \Upsilon_{\rm nc,norm}$. The proof of part 1 is analogous to the proof of part 1 of Theorems 8.2 and 8.8.

2 & 3. We first show that $\rho_{\rm unif-normal}(\cdot)$ is lower semicontinuous in the uniformly-open topology on $\mathcal{V}_{\rm nc}$. Let

$$\kappa := \limsup_{W \overset{\rm unif}{\to} Z} \mu_{\rm unif-normal}(W) = \limsup_{W \overset{\rm unif}{\to} Z} \lim_{\delta\to 0} \limsup_{\ell\to\infty} \sqrt[\ell]{\sup_{\widetilde{W}\in B_{\rm nc}(W,\delta)} \|\widetilde{W}^{\odot_s \ell} f_\ell\|}.$$

Then for an arbitrary $\epsilon > 0$ there is a sequence $W_n \overset{\rm unif}{\to} Z$ such that

$$\lim_{\delta\to 0} \limsup_{\ell\to\infty} \sqrt[\ell]{\sup_{\widetilde{W}\in B_{\rm nc}(W_n,\delta)} \|\widetilde{W}^{\odot_s \ell} f_\ell\|} > \kappa - \epsilon, \quad n = 1, 2, \ldots.$$

Clearly, there exists $N \in \mathbb{N}$ such that $W_n \in B_{\rm nc}(Z,\delta)$ for all $n \ge N$. Therefore,

$$\begin{aligned}\mu_{\rm unif-normal}(Z) &= \lim_{\delta\to 0} \limsup_{\ell\to\infty} \sqrt[\ell]{\sup_{\widetilde{W}\in B_{\rm nc}(Z,2\delta)} \|\widetilde{W}^{\odot_s \ell} f_\ell\|} \\ &\ge \lim_{\delta\to 0} \limsup_{\ell\to\infty} \sqrt[\ell]{\sup_{\widetilde{W}\in B_{\rm nc}(W_n,\delta)} \|\widetilde{W}^{\odot_s \ell} f_\ell\|} > \kappa - \epsilon, \quad n \ge N.\end{aligned}$$

Since $\epsilon$ can be chosen arbitrarily small, we obtain that

$$\mu_{\rm unif-normal}(Z) \ge \kappa = \limsup_{W \overset{\rm unif}{\to} Z} \mu_{\rm unif-normal}(W),$$

i.e., $\mu_{\rm unif-normal}(\cdot)$ is upper semicontinuous, and thus $\rho_{\rm unif-normal}(\cdot)$ is lower semicontinuous, in the uniformly-open topology on $\mathcal{V}_{\rm nc}$. This implies also that the set $\Upsilon_{\rm nc,unif-normal}$ is uniformly-open.

Since clearly $\rho_{\rm unif-normal}(Z) \le \rho(Z)$ for any $Z$, by the lower semicontinuity of $\rho_{\rm unif-normal}(\cdot)$ we obtain that $\rho_{\rm unif-normal}(Z) \le \rho_{\rm unif}(Z)$ for any $Z$ and hence $\Upsilon_{\rm nc,unif-normal} \subseteq \Upsilon_{\rm nc,unif}$.

Let $X \in \Upsilon_{\rm nc,unif-normal}$. Then

$$\mu_{\rm unif-normal}\Big(X - \bigoplus_{\alpha=1}^{m_X} Y\Big) = \lim_{\delta\to 0} \limsup_{\ell\to\infty} \sqrt[\ell]{\sup_{W\in B_{\rm nc}(X,\delta)} \Big\|\Big(W - \bigoplus_{\alpha=1}^{m_W} Y\Big)^{\odot_s \ell} f_\ell\Big\|_{sm_W}} < 1.$$

Hence, there exists $\delta > 0$ such that

$$\limsup_{\ell\to\infty} \sqrt[\ell]{\sup_{W\in B_{\rm nc}(X,\delta)} \Big\|\Big(W - \bigoplus_{\alpha=1}^{m_W} Y\Big)^{\odot_s \ell} f_\ell\Big\|_{sm_W}} < 1.$$

This implies that the sequence

$$\sup_{W\in B_{\rm nc}(X,\delta)} \Big\| \Big(W-\bigoplus_{\alpha=1}^{m_W} Y\Big)^{\odot_s \ell} f_\ell \Big\|_{sm_W}, \quad \ell=1,2,\ldots,$$

is bounded, moreover there exists $C>0$ and $\eta$: $0<\eta<1$ such that

$$\sup_{W\in B_{\rm nc}(X,\delta)} \Big\| \Big(W-\bigoplus_{\alpha=1}^{m_W} Y\Big)^{\odot_s \ell} f_\ell \Big\|_{sm_W} \le C\eta^\ell, \quad \ell=1,2,\ldots.$$

Consequently,

$$\sum_{\ell=0}^{\infty} \sup_{W\in B_{\rm nc}(X,\delta)} \Big\| \Big(W-\bigoplus_{\alpha=1}^{m_W} Y\Big)^{\odot_s \ell} f_\ell \Big\|_{sm_W} < \infty.$$

Conversely, if $X\in \Upsilon_{\rm nc,unif}$ is such that for some $\delta>0$ the series

$$\sum_{\ell=0}^{\infty} \sup_{W\in B_{\rm nc}(X,\delta)} \left\| \left(W-\bigoplus_{\alpha=1}^{m_W} Y\right)^{\odot_s \ell} f_\ell \right\|_{sm_W}$$

converges, then the series

$$\sum_{\ell=0}^{\infty} \lambda^\ell \sup_{W\in B_{\rm nc}(X,\delta)} \Big\| \Big(W-\bigoplus_{\alpha=1}^{m_W} Y\Big)^{\odot_s \ell} f_\ell \Big\|_{sm_W}$$

as a series in $\lambda$ converges uniformly and absolutely on the closed unit disk $\overline{\mathbb{D}}$, and thus its radius of convergence

$$\left(\limsup_{\ell\to\infty} \sqrt[\ell]{\sup_{W\in B_{\rm nc}(X,\delta)} \Big\| \Big(W-\bigoplus_{\alpha=1}^{m_W} Y\Big)^{\odot_s \ell} f_\ell \Big\|_{sm_W}}\right)^{-1} > 1.$$

Since the left-hand side is increasing as $\delta\downarrow 0$, we have that

$$\begin{aligned}\rho_{\rm unif-normal}&\Big(X-\bigoplus_{\alpha=1}^{m_X} Y\Big)\\ &= \left(\lim_{\delta\to 0}\limsup_{\ell\to\infty} \sqrt[\ell]{\sup_{W\in B_{\rm nc}(X,\delta)} \Big\| \Big(W-\bigoplus_{\alpha=1}^{m_W} Y\Big)^{\odot_s \ell} f_\ell \Big\|_{sm_W}}\right)^{-1} > 1,\end{aligned}$$

i.e., $X\in\Upsilon_{\rm nc,unif-normal}$. This proves equalities (8.22) and (8.23).

Next, if $Y\in\Upsilon_{\rm nc,unif-normal}$, then clearly $\Upsilon_{\rm nc,unif-normal}\neq\emptyset$. Conversely, assume that $X\in(\Upsilon_{\rm nc,unif-normal})_{sm}$ for some $m\in\mathbb{N}$. Then by (8.23) there exists a nc ball $B_{\rm nc}(X,\delta)$ such that

$$\sum_{\ell=0}^{\infty} \sup_{W\in B_{\rm nc}(X,\delta)} \Big\| \Big(W-\bigoplus_{\alpha=1}^{m_W} Y\Big)^{\odot_s \ell} f_\ell \Big\|_{sm_W} < \infty.$$

By Lemma 8.10 this inequality holds with the suprema over $B_{\rm nc}(X,\delta)$ replaced by the suprema over

$$\Upsilon_{{\rm nc},X} := \bigcup_{\zeta\in\overline{\mathbb{D}}} B_{\rm nc}\left(\bigoplus_{\alpha=1}^{m} Y+\zeta\Big(X-\bigoplus_{\alpha=1}^{m} Y\Big),\delta\right).$$

This implies part 3 of the theorem, with $\Upsilon_{\text{nc},X}$ as above. In particular, we obtain that $\bigoplus_{\alpha=1}^m Y \in \Upsilon_{\text{nc,unif-normal}}$. We claim $\rho_{\text{unif-normal}}(\bigoplus_{j=1}^m W) = \rho_{\text{unif-normal}}(W)$ for every $k \in \mathbb{N}$ and $W \in \mathcal{V}^{sk\times sk}$ which yields $Y \in \Upsilon_{\text{nc,unif-normal}}$. Indeed, this follows from the equality

$$\Big\| \Big(\bigoplus_{j=1}^m W\Big)^{\odot_s \ell} f_\ell \Big\|_{skm} = \|W^{\odot_s \ell} f_\ell\|_{sk}$$

and the definition of $\rho_{\text{unif-normal}}(\cdot)$.

The remaining statements of part 2 (i.e., the ones under the bullets) follow from the positive-homogeneity and lower semicontinuity of $\rho_{\text{unif-normal}}(\cdot)$.

4. Since a uniformly-open nc set is a finitely open nc set, $f|_\Gamma$ is a G-differentiable nc function by Theorem 8.2, part 4. It follows from (8.23) that $f|_\Gamma$ is locally uniformly bounded. We conclude that $f|_\Gamma$ is a uniformly analytic nc function.

5. If $\rho_{\text{cb}} > \delta > 0$, then we have

$$\mu_{\text{cb}} = \limsup_{\ell\to\infty} \sqrt[\ell]{\|f_\ell\|_{\mathcal{L}^\ell_{\text{cb}}}} < \infty$$

and there exists $C > 0$ such that

$$\|f_\ell\|_{\mathcal{L}^\ell_{\text{cb}}} \le C\mu_{\text{cb}}^\ell, \quad \ell = 0, 1, \dots.$$

Therefore

$$\begin{aligned}\sum_{\ell=0}^\infty \sup_{W\in B_{\text{nc}}(0_{s\times s},\delta)} \|W^{\odot_s \ell} f_\ell\| &= \sum_{\ell=0}^\infty \sup_{\|W\|=\delta} \|W^{\odot_s \ell} f_\ell\| \\ &= \sum_{\ell=0}^\infty \delta^\ell \sup_{\|W\|=1} \|W^{\odot_s \ell} f_\ell\| = \sum_{\ell=0}^\infty \delta^\ell \|f_\ell\|_{\mathcal{L}^\ell_{\text{cb}}} \le C\sum_{\ell=0}^\infty \delta^\ell \mu_{\text{cb}}^\ell < \infty,\end{aligned}$$

which is equivalent to (8.25); here we used Proposition 7.49 and the inequality $\delta\mu_{\text{cb}} < 1$. In particular, the series (8.1) converges absolutely and uniformly on $B_{\text{nc}}(Y,\delta)$, so that $Y \in \Upsilon_{\text{nc,unif-normal}}$ and thus $\Upsilon_{\text{nc,unif-normal}} \ne \emptyset$.

Now suppose $\delta > \rho_{\text{cb}}$. Then for any $\epsilon$: $0 < \epsilon < \mu_{\text{cb}} - \delta^{-1}$ and $L > 0$ there exists $\ell > L$ such that $\|f_\ell\|_{\mathcal{L}^\ell_{\text{cb}}} > (\mu_{\text{cb}} - \epsilon)^\ell$. For any such $\ell$ there exists $W \in (\mathcal{V}^{s\times s})_{\text{nc}}$ with $\|W\| = (\mu_{\text{cb}} - \epsilon)^{-1} < \delta$ such that

$$\|W^{\odot_s \ell} f_\ell\| > (\mu_{\text{cb}} - \epsilon)^\ell \|W\| = 1.$$

Hence the sequence $\|W^{\odot_s \ell} f_\ell\|$, $\ell = 0, 1, \dots$, does not converge to 0 uniformly on $B_{\text{nc}}(0_{s\times s},\delta)$, and the series (8.1) does not converge uniformly on $B_{\text{nc}}(Y,\delta)$.

Finally, if $\Upsilon_{\text{nc,unif-normal}} \ne \emptyset$, then by part 2 we have that $Y \in \Upsilon_{\text{nc,unif-normal}}$. Therefore, again by part 2, there exists $\delta > 0$ such that the series (8.1) converges absolutely and uniformly on $B_{\text{nc}}(Y,\delta)$. This implies that $\rho_{\text{cb}} \ge \delta > 0$. The proof is complete. □

The following example demonstrates a possibility that $0 = \rho_{\text{unif}}(Z) < \rho_{\text{norm}}(Z)$ for some $Z$ and thus $\Upsilon_{\text{nc,unif}} \subsetneq \Upsilon_{\text{nc,norm}}$. We also have in this example that $0 = \rho_{\text{cb}} < \rho_m$ for every fixed $m \in \mathbb{N}$.

EXAMPLE 8.13. It was shown in [**77**] (see also [**79**, Chapter 3]) that for every $p = 3, 4, \dots$ there exist $k_p \in \mathbb{N}$ (with $k_p$ increasing as $p$ increases) and a linear mapping $\phi_p\colon \mathbb{C}^p \to \mathbb{C}^{k_p\times k_p}$ such that $\max_{\zeta\in\overline{\mathbb{D}^p}} \|\phi_p(\zeta)\| = 1$ and

$$\sup_{n\in\mathbb{N}, Z_{p1},\dots,Z_{pp}\in\mathbb{C}^{n\times n}\colon \|Z_{p1}\|=\cdots=\|Z_{pp}\|=1} \|(\mathrm{id}_{\mathbb{C}^{n\times n}}\otimes\phi_p)(Z_{p1},\dots,Z_{pp})\| = \|(\mathrm{id}_{\mathbb{C}^{k_p\times k_p}}\otimes\phi_p)(Z^0_{p1},\dots,Z^0_{pp})\| = \sqrt{\frac{p}{2}} \tag{8.26}$$

for some $Z^0_{p1}, \dots, Z^0_{pp} \in \mathbb{C}^{k_p\times k_p}$ of norm 1, where $\|\cdot\|$ denotes the operator norm with respect to the standard Euclidean norms. Moreover,

$$\|(\mathrm{id}_{\mathbb{C}^{k_p\times k_p}}\otimes\phi_p)(Z^0_{p1},\dots,Z^0_{pp})^\ell\| = \left(\sqrt{\frac{p}{2}}\right)^\ell, \quad \ell = 1,2,\dots. \tag{8.27}$$

Let $\mathcal{V}$ be a Banach space consisting of bounded double-indexed sequences of complex numbers $z = (z_{pq})_{p=3,4,\dots;\, q=1,\dots,p}$ with the norm $\|z\|_\infty = \sup_{p,q}|z_{pq}|$. One can identify $n\times n$ matrices over $\mathcal{V}$ with sequences of $n\times n$ matrices over $\mathbb{C}$, $Z = (Z_{pq})_{p=3,4,\dots;\, q=1,\dots,p}$. Then multiplication of such a $Z$ on the right (on the left) with a $n\times n$ matrix over $\mathbb{C}$ is defined as simultaneous right (left) multiplication with this matrix for all components $Z_{pq}$. It is easy to see that the norms on $\mathcal{V}^{n\times n}$ defined by $\|Z\|_{\infty,n} = \sup_{p,q}\|Z_{pq}\|$ for $n = 1,2,\dots$ determine an operator space structure on $\mathcal{V}$. Let $\mathcal{W}$ be a Banach space consisting of block-diagonal infinite matrices (infinite direct sums) $W = \bigoplus_{p=3}^\infty W_p$ with $W_p \in \mathbb{C}^{k_p\times k_p}$ and with the norm $\|W\|_\infty = \sup_p \|W_p\|$. One can identify $n\times n$ matrices over $\mathcal{W}$ with direct sums of $n\times n$ matrices over $\mathbb{C}^{k_p\times k_p}$, $W = \bigoplus_{p=3}^\infty W_p$. Multiplication of such a $W$ on the right (on the left) with a $n\times n$ matrix $S$ over $\mathbb{C}$ is then reduced to the simultaneous right (left) multiplication with this matrix for all diagonal (block) entries $W_p$ as follows: $(WS)_p = W_p(S\otimes I_{k_p})$ (resp., $(SW)_p = (S\otimes I_{k_p})W_p$. It is easy to see that the norms on $\mathcal{W}^{n\times n}$ defined by $\|W\|_{\infty,n} = \sup_p\|W_p\|$ for $n = 1,2,\dots$ determine an operator space structure on $\mathcal{W}$. For each $\ell = 3,4,\dots$ we define a linear mapping $\phi^{(\ell)}\colon \mathcal{V}\to\mathcal{W}$ by

$$\phi^{(\ell)}(z) = \phi_3(z_{31}, z_{32}, z_{33}) \oplus \cdots \oplus \phi_\ell(z_{\ell 1},\dots,z_{\ell\ell}) \oplus 0_{k_{\ell+1}\times k_{\ell+1}} \oplus 0_{k_{\ell+2}\times k_{\ell+2}} \oplus \cdots$$

and a $\ell$-linear mapping $f_\ell\colon \mathcal{V}^\ell \to \mathcal{W}$ by

$$f_\ell(z^1,\dots,z^\ell) = \phi^{(\ell)}(z^1)\cdots\phi^{(\ell)}(z^\ell),$$

so that

$$f_\ell(z,\dots,z) = \phi^{(\ell)}(z)^\ell.$$

The mappings $\phi^{(\ell)}$ and $f_\ell$ are extended to $n\times n$ matrices over $\mathcal{V}$ in the standard way. Then by Proposition 7.49 and the equalities (8.26), (8.27) we have

$$\|f_\ell\|_{\mathcal{L}^\ell_{\mathrm{cb}}} = \sup_{n\in\mathbb{N}, Z\in\mathcal{V}^{n\times n}\colon \|Z\|_{\infty,n}=1} \|Z^{\odot\ell} f_\ell\|_{\infty,n} = \sup_{n\in\mathbb{N}, Z\in\mathcal{V}^{n\times n}\colon \|Z\|_{\infty,n}=1} \|f_\ell(Z,\dots,Z)\|_{\infty,n}$$

$$= \sup_{n\in\mathbb{N}, Z\in\mathcal{V}^{n\times n}:\ \|Z\|_{\infty,n}=1} \|\phi^{(\ell)}(Z)^{\ell}\|_{\infty,n}$$

$$= \sup_{n\in\mathbb{N}, Z_{\ell 1},\ldots,Z_{\ell\ell}\in\mathbb{C}^{n\times n}:\ \|Z_{\ell 1}\|=\cdots=\|Z_{\ell\ell}\|=1} \|(\mathrm{id}_{\mathbb{C}^{n\times n}}\otimes\phi_\ell)(Z_{\ell 1},\ldots,Z_{\ell\ell})^{\ell}\|$$

$$= \left(\sqrt{\frac{\ell}{2}}\right)^{\ell},$$

hence the $\ell$-linear mappings $f_\ell$, $\ell = 3, 4, \ldots$, are completely bounded. Now, consider the power series

$$\sum_{\ell=3}^{\infty} Z^{\odot \ell} f_\ell.$$

We have for every fixed $m \in \mathbb{N}$ and $Z = (Z_{pq})_{p=3,4,\ldots;q=1,\ldots,p} \in \mathcal{V}^{m\times m}$ with $\|Z\|_{\infty,m} = 1$ that

$$\mu(Z) = \limsup_{\ell\to\infty} \sqrt[\ell]{\|Z^{\odot\ell} f_\ell\|_{\infty,m}} \le \limsup_{\ell\to\infty} \|\phi^{(\ell)}(Z)\|$$
$$\le m^2 \limsup_{\ell\to\infty} \max_{p=3,\ldots,\ell} \max_{1\le i,j\le m} \|\phi_p(Z_{p1}^{ij},\ldots,Z_{pp}^{ij})\| \le m^2,$$

where $Z_{pq}^{ij}$ is the $(i,j)$-th entry of the matrix $Z_{pq}$. Therefore $\rho(Z) \ge m^{-2}$ for every $Z \in \mathcal{V}^{m\times m}$ with $\|Z\|_{\infty,m} = 1$. It is easy to see that by the homogeneity of degree $-1$ we also have $\rho_{\mathrm{norm}}(Z) \ge m^{-2}$ for every $Z \in \mathcal{V}^{m\times m}$ with $\|Z\|_{\infty,m} = 1$. This implies that $\rho_{\mathrm{norm}}(Z) > 0$ for every $Z \in \mathcal{V}_{\mathrm{nc}}$ and hence $\rho_{\mathrm{norm}}(0_{m\times m}) = \infty$ for every $m \in \mathbb{N}$. We also obtain that $\rho_m \ge m^{-2}$ for every $m \in \mathbb{N}$.

Define the sequence $Z^j = (Z^j_{pq})_{p=3,4,\ldots;q=1,\ldots,p} \in \mathcal{V}^{k_j\times k_j}$, $j = 3, 4, \ldots$, by

$$Z^j_{pq} = \begin{cases} 0_{k_j\times k_j} & \text{for } p \ne j, \\ \frac{1}{\sqrt[4]{j}} Z^0_{jq} & \text{for } p = j,\ 1 \le q \le j. \end{cases}$$

Then

$$\lim_{j\to\infty} \|Z^j\|_{\infty,k_j} = \lim_{j\to\infty} \frac{1}{\sqrt[4]{j}} = 0,$$

however

$$\lim_{j\to\infty} \mu(Z^j) = \lim_{j\to\infty} \limsup_{\ell\to\infty} \sqrt[\ell]{\|(Z^j)^{\odot\ell} f_\ell\|_{\infty,k_j}}$$

$$= \lim_{j\to\infty} \frac{1}{\sqrt[4]{j}} \limsup_{\ell\to\infty} \sqrt[\ell]{\|(\mathrm{id}_{\mathbb{C}^{k_j\times k_j}}\otimes\phi_j)(Z^0_{j1},\ldots,Z^0_{jj})^{\ell}\|_{\infty,k_j}} = \lim_{j\to\infty} \frac{1}{\sqrt[4]{j}}\sqrt{\frac{j}{2}} = \infty.$$

Thus

$$\rho_{\mathrm{unif}}(0_{1\times 1}) = \lim_{j\to\infty} \rho(Z^j) = 0.$$

Note that we can show analogously that $\rho_{\mathrm{unif}}(Z) = 0$ when $Z \ne 0$. E.g., for $z = (z_{pq})_{p=3,4,\ldots;q=1,\ldots,p}$ with any $z_{3,q} \ne 0$, $q = 1, 2, 3$, and $z_{pq} = 0$ for all $p > 3$ and all $q$, we can define a sequence $z \otimes I_{k_j} + Z^j$ with $Z^j$ as above and verify that

$$\rho_{\mathrm{unif}}(z) = \lim_{j\to\infty} \rho(z \otimes I_{k_j} + Z^j) = 0.$$

Finally, we observe that

$$\rho_{\mathrm{cb}} = \left(\limsup_{\ell\to\infty} \sqrt[\ell]{\|f_\ell\|_{\mathcal{L}^{\ell}_{\mathrm{cb}}}}\right)^{-1} = \left(\limsup_{\ell\to\infty} \sqrt{\frac{\ell}{2}}\right)^{-1} = 0.$$

Thus all the statements formulated in the paragraph preceding this example are true.

In the next example we demonstrate the possibility that $0=\rho_{\text{unif}-\text{normal}}(0_{1\times 1})<\rho_{\text{unif}}(0_{1\times 1})$ and thus $\emptyset = \Upsilon_{\text{nc,unif}-\text{normal}} \subsetneq \Upsilon_{\text{nc,unif}}$. The sum of the nc power series in this example is an entire (i.e., analytic everywhere and bounded on every ball of $n \times n$ matrices, $n = 1, 2, \ldots$) nc function which is not uniformly analytic at $0_{1\times 1}$, that is, not uniformly analytic in any uniformly-open nc neighborhood of $0_{1\times 1}$.

EXAMPLE 8.14. Let $\mathcal{V} = \mathbb{C}^2$ with the norm $\|(z_1, z_2)\| = \max\{|z_1|, |z_2|\}$ and $\mathcal{W} = \mathbb{C}$. We identify $n \times n$ matrices over $\mathcal{V} = \mathbb{C}^2$ with pairs of matrices over $\mathbb{C}$. It is clear that the norms on $\mathcal{V}^{n\times n}$, $n = 1, 2, \ldots$, defined by $\|(Z_1, Z_2)\|_n = \max\{\|Z_1\|, \|Z_2\|\}$ and componentwise multiplication of pairs $(Z_1, Z_2)$ of matrices over $\mathbb{C}$ on the left (on the right) with a matrix over $\mathbb{C}$ determine the operator space structure on $\mathcal{V}$. The canonical (i.e., with $\|1\|_1 = 1$) operator space structure on $\mathcal{W} = \mathbb{C}$ is uniquely determined. Consider the power series

$$\sum_{k=1}^{\infty} Z^{\odot \alpha_k} f_{\alpha_k} = \sum_{k=1}^{\infty} c_k p_k(Z_1, Z_2) = \sum_{k=1}^{\infty} c_k \sum_{\pi \in \mathcal{S}_{k+1}} \operatorname{sign}(\pi) Z_1^{\pi(1)-1} Z_2 \cdots Z_1^{\pi(k+1)-1} Z_2$$

with positive constants $c_k$ which will be defined later and $\alpha_k = \deg p_k = \frac{(k+1)(k+2)}{2}$ (see Example 6.3 for more details on the polynomials $p_k$). The $\alpha_k$-linear mappings $f_{\alpha_k}$ are completely bounded. Indeed, by Proposition 7.49 we have

$$\|f_{\alpha_k}\|_{\mathcal{L}_{\text{cb}}^{\alpha_k}} = \sup_{n\in\mathbb{N},\ Z=(Z_1,Z_2)\in(\mathbb{C}^{n\times n})^2:\ \max\{\|Z_1\|,\|Z_2\|\}=1} c_k \|p_k(Z_1, Z_2)\|_n \le c_k (k+1)! < \infty.$$

Since the polynomials $p_k$ vanish identically on pairs of $k\times k$ matrices, for every fixed $n \in \mathbb{N}$ and any $Z = (Z_1, Z_2) \in (\mathbb{C}^{n\times n})^2$ the series contains only the powers smaller than $\alpha_n$ and thus converges uniformly on $(\mathbb{C}^{n\times n})^2$. Therefore, $\rho(Z) = \infty$ for every $Z \in \mathcal{V}_{\text{nc}}$. We also have $\rho_{\text{unif}}(Z) = \infty$ for every $Z \in \mathcal{V}_{\text{nc}}$ and $\Upsilon_{\text{nc,unif}} = \mathcal{V}_{\text{nc}}$. On the other hand, any nonzero polynomial $p$ in two noncommuting indeterminates is not identically zero on pairs of square matrices of sufficiently large size due to the absence of polynomial identities for matrices of all sizes — see [**91**, Example 1.4.4 and Theorem 1.4.5, page 22]. Therefore

$$\sup_{n\in\mathbb{N},\ Z=(Z_1,Z_2)\in(\mathbb{C}^{n\times n})^2:\ \max\{\|Z_1\|,\|Z_2\|\}=1} \|p_k(Z_1, Z_2)\|_n \neq 0.$$

Define

$$c_k = k^{\alpha_k} \Big( \sup_{n\in\mathbb{N},\ Z=(Z_1,Z_2)\in(\mathbb{C}^{n\times n})^2:\ \max\{\|Z_1\|,\|Z_2\|\}=1} \|p_k(Z_1, Z_2)\|_n \Big)^{-1}.$$

Then

$$\|f_{\alpha_k}\|_{\mathcal{L}_{\text{cb}}^{\alpha_k}} = k^{\alpha_k}.$$

For any $\delta > 0$ we have

$$\begin{aligned}\sup_{\ell\in\mathbb{N}} \sup_{W\in B_{\text{nc}}(0_{1\times 1},\delta)} \|W^{\odot \ell} f_\ell\| &= \sup_{k\in\mathbb{N}} c_k \delta^{\alpha_k} k^{\alpha_k} \sup_{n\in\mathbb{N},\ Z=(Z_1,Z_2)\in(\mathbb{C}^{n\times n})^2:\ \max\{\|Z_1\|,\|Z_2\|\}=1} \|p_k(Z_1, Z_2)\|_n \\ &= \sup_{k\in\mathbb{N}} (\delta k)^{\alpha_k} = \infty.\end{aligned}$$

Then by part 2 of Theorem 8.11 we have that $0_{1\times 1} \notin \Upsilon_{\mathrm{nc,unif-normal}}$ and thus $\Upsilon_{\mathrm{nc,unif-normal}} = \emptyset$. Alternatively, we can compute

$$\mu_{\mathrm{unif-normal}}(0_{1\times 1}) = \lim_{\delta\to 0} \limsup_{\ell\to 0} \sqrt[\ell]{\sup_{W\in B_{\mathrm{nc}}(0_{1\times 1},\delta)} \|W^{\odot \ell} f_\ell\|} = \lim_{\delta\to 0} \lim_{k\to\infty} \delta k = \infty,$$

and thus $\rho_{\mathrm{unif-normal}}(0_{1\times 1}) = 0$. By Remark 8.3 and Theorem 7.21, the sum of the series, $f(Z) = \sum_{k=1}^{\infty} Z^{\odot \alpha_k} f_{\alpha_k}$ is not uniformly analytic at $0_{1\times 1}$.

REMARK 8.15. It is not too difficult to modify Example 8.14 by changing the coefficients $c_k$ so that $\Upsilon_{\mathrm{nc,unif-normal}} \subsetneq \Upsilon_{\mathrm{nc,unif}} = (\mathbb{C}^2)_{\mathrm{nc}}$ contains a nc ball centered at $0_{1\times 1}$ of a given finite radius.

We consider now the case where $\mathcal{V} = \mathbb{C}^d$ with some operator space structure and we expand each term of the power series (8.1) obtaining a series along the free monoid $\mathcal{G}_d$. Our results will be the converse of Theorem 7.29. In other words, we shall consider the convergence of power series of the form

$$\sum_{w\in\mathcal{G}_d} \Big(X - \bigoplus_{\alpha=1}^{m} Y\Big)^{\odot_s w} f_w. \tag{8.28}$$

Here $Y = (Y_1, \ldots, Y_d) \in (\mathbb{C}^{s\times s})^d$ is a given center; $X = (X_1, \ldots, X_d) \in (\mathbb{C}^{sm\times sm})^d$, $m = 1, 2, \ldots$; $Z^{\odot_s w}$ for $w = g_{i_1} \cdots g_{i_\ell}$ and $Z = (Z_1, \ldots, Z_d) \in (\mathbb{C}^{sm\times sm})^d$ is defined in (4.8); $f_w \colon (\mathbb{C}^{s\times s})^\ell \to \mathcal{W}^{s\times s}$, $w \in \mathcal{G}_d$, $\ell = |w|$, is a given sequence of $\ell$-linear mappings (here $\mathbb{C}$ and, hence, $\mathbb{C}^{s\times s}$ are equipped with the canonical operator space structure); and conditions (4.14)–(4.17) are satisfied. The series (8.28) can be alternatively written as

$$\sum_{w\in\mathcal{G}_d} \left(\Big(\bigoplus_{\alpha=1}^{m} A_{w,(0)}\Big) \otimes \Big(\bigoplus_{\alpha=1}^{m} A_{w,(1)}\Big) \otimes \cdots \otimes \Big(\bigoplus_{\alpha=1}^{m} A_{w,(|w|)}\Big)\right) \star Z^{[w]}.$$

Here

$$f_w = A_{w,(0)} \otimes A_{w,(1)} \otimes \cdots \otimes A_{w,(|w|)} \in \mathcal{W}^{s\times s} \otimes \underbrace{\mathbb{C}^{s\times s} \otimes \cdots \otimes \mathbb{C}^{s\times s}}_{|w| \text{ times}},$$

with the tensor product interpretation for the multilinear mappings $f_w$ (Remark 3.4), the sumless Sweedler notation (4.19), and the pseudo-power notation (4.21). In the case $s = 1$, the series (8.28) becomes

$$\sum_{w\in\mathcal{G}_d} Z^w f_w,$$

with $f_w \in \mathcal{W}$; more precisely, we identify the $\ell$-linear mapping $f_w \colon \mathbb{C}^\ell \to \mathcal{W}$ with the vector $f_w(1, \ldots, 1) \in \mathcal{W}$, and

$$\|Z^w f_w\|_m = \|Z^w\| \|f_w\|_1 \tag{8.29}$$

for every $m \in \mathbb{N}$ and $Z \in (\mathbb{C}^{m\times m})^d$ (see [**79**, Proposition 1.10(ii)]), so that

$$\|f_w\|_{\mathcal{L}^\ell_{\mathrm{cb}}} = \|f_w\|_1. \tag{8.30}$$

We associate with the series (8.28) the series (8.1) where the $\ell$-linear mappings $f_\ell \colon ((\mathbb{C}^{s\times s})^d)^\ell \to \mathcal{W}^{s\times s}$ are defined by (5.5), so that (5.6) and (5.7) hold. Similarly to Remark 4.5, conditions (4.14)–(4.17) on the linear mappings $f_w$ are easily seen to be equivalent to conditions (4.4)–(4.7) on the linear mappings $f_\ell$. Notice that the completely bounded norms of $f_\ell$ and of $f_w$ for all $w$ with $|w| = \ell$ are related by

$$\|f_\ell\|_{\mathcal{L}^\ell_{\rm cb}} \le \sum_{|w|=\ell} \mathbf{C}_1^w \|f_w\|_{\mathcal{L}^\ell_{\rm cb}}, \tag{8.31}$$

where $\mathbf{C}_1 = (\|\pi_1\|_{\rm cb}, \ldots, \|\pi_d\|_{\rm cb})$ for the $i$-th coordinate mapping $\pi_i \colon \mathbb{C}^d \to \mathbb{C}$, and

$$\|f_w\|_{\mathcal{L}^\ell_{\rm cb}} \le \mathbf{C}_2^w \|f_\ell\|_{\mathcal{L}^\ell_{\rm cb}}, \tag{8.32}$$

where $\mathbf{C}_2 = (\|e_1\|_1, \ldots, \|e_d\|_1)$.

Let $\mathcal{X}$ be a vector space over $\mathbb{C}$ with a direct sum decomposition $\mathcal{X} = \bigoplus_{j=1}^d \mathcal{X}_j$, and let $Y = (Y_1, \ldots, Y_d) \in \mathcal{X}$. A set $\Upsilon \subseteq \mathcal{X}$ is called *complete Reinhardt-like about* $Y$ if for every $X = (X_1, \ldots, X_d) \in \Upsilon$ and every $t = (t_1, \ldots, t_d) \in \mathbb{C}^d$ with $|t_j| \le 1$ for all $j$, one has $Y + t\circ(X - Y) \in \Upsilon$, where $t\circ(X-Y) := (t_1(X_1 - Y_1), \ldots, t_d(X_d - Y_d))$. Clearly, a complete Reinhardt-like set is complete circular. In the case of $d = 1$ we recover the definition of complete circular set. In the case where $\mathcal{X} = \mathbb{C}^d$ and $\mathcal{X}_j = \mathbb{C}e_j$, $j = 1, \ldots, d$, we recover the usual definition of complete Reinhardt set [**97**, Page 8].

The following is an analogue of Lemma 8.10 for complete Reinhardt-like sets.

LEMMA 8.16. *Let $\mathcal{X}$ be a vector space over $\mathbb{C}$ with the direct sum decomposition $\mathcal{X} = \bigoplus_{j=1}^d \mathcal{X}_j$, let $\mathcal{W}$ be a Banach space over $\mathbb{C}$, and let $\Gamma \subseteq \mathcal{X}$ be a complete Reinhardt-like set about* 0. *Suppose that $\omega \colon \mathcal{X}^k \to \mathcal{W}$ is a $k$-linear mapping whose restriction to the set*

$$\operatorname{diag}[\mathcal{X}^k] = \{(\underbrace{W, \ldots, W}_{k \text{ times}}) \colon W = (W_1, \ldots, W_d) \in \mathcal{X}\}$$

*is homogeneous in $W_j$ of degree $k_j$ (so that $k_1 + \cdots + k_d = k$), and suppose that $X = (X_1, \ldots, X_d) \in \mathcal{X}$. Then*

$$\sup_{W \in \bigcup_{t \in \overline{\mathbb{D}^d}} (t\circ X + \Gamma)} \|\omega(W, \ldots, W)\| = \sup_{W \in X + \Gamma} \|\omega(W, \ldots, W)\|,$$

*where $\overline{\mathbb{D}^d} = \overline{\mathbb{D}}^d = \{t = (t_1, \ldots, t_d) \in \mathbb{C}^d \colon |t_j| \le 1\}$ is the closed unit polydisk, $X + \Gamma := \{X + Z \colon Z \in \Gamma\}$ is a complete Reinhardt-like set about $X$ and $\bigcup_{t \in \overline{\mathbb{D}^d}} (t\circ X + \Gamma)$ is a complete Reinhardt-like set about* 0.

PROOF. Fix an arbitrary $Z^0 \in \Gamma$. Consider the function

$$\phi \colon \overline{\mathbb{D}^d} \to \mathcal{W}, \quad \phi \colon t \mapsto \omega(t \circ X + Z^0, \ldots, t \circ X + Z^0).$$

Clearly, $\phi$ is a polynomial in $t_1, \ldots, t_d$ and hence $\phi$ is analytic. By the maximum principle, there exists $t^0$ with $|t_j^0| = 1$, $j = 1, \ldots, d$, such that $\|\phi(t^0)\| = \max_{t \in \overline{\mathbb{D}^d}} \|\phi(t)\|$. By the simultaneous homogeneity of $\omega$,

$$\begin{aligned} \|\omega(t^0 \circ X + Z^0, \ldots, t^0 \circ X + Z^0)\| &= \|\omega(X + (t^0)^{-1} \circ Z^0, \ldots, X + (t^0)^{-1} \circ Z^0)\| \\ &\le \sup_{Z \in \Gamma} \|\omega(X + Z, \ldots, X + Z)\|, \end{aligned}$$

where $(t^0)^{-1} = ((t_1^0)^{-1}, \ldots, (t_d^0)^{-1})$, since $\Gamma$ is complete Reinhardt-like about 0. This proves the lemma. ◻

We shall be mostly interested in the case where $\mathcal{X} = (\mathbb{C}^{n\times n})^d$ and $\mathcal{X}_j = \mathbb{C}^{n\times n}e_j$. A nc set $\Upsilon_{\rm nc} \subseteq (\mathbb{C}^d)_{\rm nc}$ is called *complete Reinhardt-like about* $Y \in (\mathbb{C}^{s\times s})^d$ if $(\Upsilon_{\rm nc})_{sm} \subseteq (\mathbb{C}^{sm\times sm})^d$ is complete Reinhardt-like about $\bigoplus_{\alpha=1}^m Y$ for every $m\in\mathbb{N}$.

For $Y = (Y_1,\ldots,Y_d) \in (\mathbb{C}^{s\times s})^d$ and $\mathbf{r} = (r_1,\ldots,r_d)$ a $d$-tuple of positive extended real numbers (i.e., $0 < r_j \le \infty$, $j = 1, \ldots, d$), define the *nc polydisk centered at* $Y$ *of multiradius* $\mathbf{r}$

$$(\mathbb{D}^d)_{\rm nc}(Y,\mathbf{r})$$
$$= \Big\{Z = (Z_1,\ldots,Z_d) \in \coprod_{m=1}^{\infty} (\mathbb{C}^{sm\times sm})^d : \Big\| Z_j - \bigoplus_{\alpha=1}^{m} Y_j\Big\|_{sm} < r_j,\ j = 1,\ldots,d\Big\}.$$

Clearly, $(\mathbb{D}^d)_{\rm nc}(Y,\mathbf{r})$ is a complete Reinhardt-like nc set about $Y$.

For $\mathbf{r} = (r_1,\ldots,r_d)$ a $d$-tuple of nonnegative real numbers, we define

$$\mu(\mathbf{r}) = \limsup_{\ell\to\infty} \sqrt[\ell]{\sum_{w\in\mathcal{G}_d\colon |w|=\ell} \mathbf{r}^w \|f_w\|_{\mathcal{L}^\ell_{\rm cb}}}. \tag{8.33}$$

LEMMA 8.17. *The function $\mu(\cdot)$ is continuous on $d$-tuples of strictly positive real numbers.*

PROOF. Let $\mathbf{r}^0 = (r_1^0,\ldots,r_d^0) \in \mathbb{R}_+^d$ with $r_j^0 > 0$ for all $j = 1,\ldots,d$. Let $\epsilon > 0$. Then for all $d$-tuples $\mathbf{r}$ of positive real numbers satisfying $1-\epsilon < \frac{r_j^0}{r_j} < 1+\epsilon$ for all $j = 1,\ldots,d$ we have

$$(1-\epsilon)\mu(\mathbf{r}) = \mu((1-\epsilon)\mathbf{r}) \le \mu(\mathbf{r}^0) \le \mu((1+\epsilon)\mathbf{r}) = (1+\epsilon)\mu(\mathbf{r}),$$

since the function $\mu(\cdot)$ is nondecreasing in each argument and homogeneous of degree one. Therefore $\lim_{\mathbf{r}\to\mathbf{r}^0}\mu(\mathbf{r}) = \mu(\mathbf{r}^0)$, i.e., $\mu(\cdot)$ is continuous at $\mathbf{r}^0$. $\square$

We call nonnegative extended real numbers $\rho_1, \ldots, \rho_d$ *associated radii of normal convergence of the series* (8.28) if $\mu(\mathbf{r}) \le 1$ for every $\mathbf{r} = (r_1,\ldots,r_d)$ with $r_j \le \rho_j$, $j = 1,\ldots,d$, and for every $d$-tuple $\mathbf{r}' = (r_1',\ldots,r_d')$ of nonnegative extended real numbers with $r_j' \ge \rho_j$ so that strict inequality occurs for at least one $j$, there exists a $d$-tuple $\mathbf{r} = (r_1,\ldots,r_d)$ of nonnegative real numbers $r_j \le r_j'$, $j = 1,\ldots,d$, such that $\mu(\mathbf{r}) > 1$.

THEOREM 8.18. *1. The set of absolute convergence of the series* (8.28),

$$\Upsilon_{\text{free,abs}} = \coprod_{m=1}^{\infty} \Big\{X \in (\mathbb{C}^{sm\times sm})^d : \sum_{w\in\mathcal{G}_d} \Big\|\Big(X - \bigoplus_{\alpha=1}^{m} Y\Big)^{\odot_s w} f_w\Big\|_{sm} < \infty\Big\} \tag{8.34}$$
$$\subseteq \Upsilon_{\rm nc} \subseteq (\mathbb{C}^d)_{\rm nc},$$

*is a nc set which is complete Reinhardt-like about $Y$. Here $\Upsilon_{\rm nc}$ is the convergence set (see Theorem 8.2) of the corresponding series* (8.1); *see* (5.5) *and* (5.6).

*2. The set*

$$\Upsilon_{\text{free,unif}-\text{normal}} = \Big\{X \in \Upsilon_{\text{free,abs}} \colon \exists\delta > 0 \text{ such that} \tag{8.35}$$
$$\sum_{w\in\mathcal{G}_d} \sup_{W\in B_{\rm nc}(X,\delta)} \Big\|\Big(W - \bigoplus_{\alpha=1}^{m_W} Y\Big)^{\odot_s w} f_w\Big\|_{sm_W} < \infty\Big\} \subseteq \Upsilon_{\text{nc,unif}-\text{normal}}$$

*(see Theorem 8.11 for the definition of the set $\Upsilon_{\mathrm{nc,unif-normal}}$; here $m_W$ is the size of a block matrix $W$ with $s \times s$ blocks) is a uniformly-open subset of $\Upsilon_{\mathrm{free,abs}}$ which is nonempty if and only if it contains $Y$. In this case, $\Upsilon_{\mathrm{free,unif-normal}}$ is complete Reinhardt-like about $Y$ (i.e., $(\Upsilon_{\mathrm{free,unif-normal}})_{sm}$ is a complete Reinhardt-like set about $\bigoplus_{\alpha=1}^{m} Y$ for every $m \in \mathbb{N}$).*

*3. For every $m \in \mathbb{N}$ and $X \in (\Upsilon_{\mathrm{free,unif-normal}})_{sm}$ there exists a uniformly-open complete Reinhardt-like set $\Upsilon_{\mathrm{free},X}$ about $\bigoplus_{\alpha=1}^{m} Y$, which contains $X$ and such that*

$$\sum_{w \in \mathcal{G}_d} \sup_{W \in \Upsilon_{\mathrm{free},X}} \left\| \left( W - \bigoplus_{\alpha=1}^{m_W} Y \right)^{\odot_s w} f_w \right\|_{sm_W} < \infty. \tag{8.36}$$

*4. Let $\boldsymbol{\rho} = (\rho_1, \ldots, \rho_d)$ be a $d$-tuple of associated radii of normal convergence of the series (8.28). Then the series converges absolutely on every nc polydisk $(\mathbb{D}^d)_{\mathrm{nc}}(Y, \mathbf{r})$ with $r_j < \rho_j$ for all $j = 1, \ldots, d$, moreover*

$$\sum_{w \in \mathcal{G}_d} \sup_{W \in (\mathbb{D}^d)_{\mathrm{nc}}(Y, \mathbf{r})} \left\| \left( W - \bigoplus_{\alpha=1}^{m_W} Y \right)^{\odot_s w} f_w \right\|_{sm_W} < \infty. \tag{8.37}$$

*For every $\mathbf{r}$ with $r_j \geq \rho_j$ for all $j = 1, \ldots, d$, so that strict inequality occurs for at least one $j$, (8.37) fails. In the case $s = 1$, there exists $Z \in (\mathbb{D}^d)_{\mathrm{nc}}(Y, \mathbf{r})$ such that the series (8.28) does not converge absolutely at $Z$. The set $\Upsilon_{\mathrm{free,unif-normal}}$ is nonempty if and only if there exists a $d$-tuple of strictly positive associated radii of normal convergence of the series (8.28).*

Proof of Theorem 8.18. 1. The statements in this part are straightforward.

2 & 3. The inclusion in (8.35) and the fact that $\Upsilon_{\mathrm{free,unif-normal}}$ is a uniformly-open subset of $\Upsilon_{\mathrm{free,abs}}$ are obvious.

Clearly, if $Y \in \Upsilon_{\mathrm{free,unif-normal}}$, then $\Upsilon_{\mathrm{free,unif-normal}}$ is nonempty. Conversely, suppose there exist $m \in \mathbb{N}$ and $X \in (\Upsilon_{\mathrm{free,unif-normal}})_{sm}$. Then there exists $\delta > 0$ such that

$$\sum_{w \in \mathcal{G}_d} \sup_{W \in B_{\mathrm{nc}}(X,\delta)} \left\| \left( W - \bigoplus_{\alpha=1}^{m_W} Y \right)^{\odot_s w} f_w \right\|_{sm_W} < \infty. \tag{8.38}$$

There exists a nc polydisk $(\mathbb{D}^d)_{\mathrm{nc}}(X, \boldsymbol{\rho})$ contained in $B_{\mathrm{nc}}(X, \delta)$, e.g., one can set $\rho_j = \frac{\delta}{d\|e_j\|_1}$, $j = 1, \ldots, d$, and then for $W \in (\mathbb{D}^d)_{\mathrm{nc}}(X, \boldsymbol{\rho})_{smk}$ one has

$$\left\| W - \bigoplus_{\alpha=1}^{k} X \right\|_{smk} \leq \sum_{j=1}^{d} \left\| W_j - \bigoplus_{\alpha=1}^{k} X_j \right\|_{smk} \|e_j\|_1 < \delta,$$

i.e., $W \in B_{\mathrm{nc}}(X, \delta)_{smk}$. Since $\Gamma = (\mathbb{D}^d)_{\mathrm{nc}} \left( \bigoplus_{\alpha=1}^{m} Y, \boldsymbol{\rho} \right)$ is a complete Reinhardt-like nc set about $\bigoplus_{\alpha=1}^{m} Y$, it follows from Lemma 8.16 that the inequality in (8.38) holds with the suprema over $B_{\mathrm{nc}}(X, \delta)$ replaced by the suprema over

$$\Upsilon_{\mathrm{free,X}} := \bigcup_{t \in \overline{\mathbb{D}^d}} (\mathbb{D}^d)_{\mathrm{nc}} \left( \bigoplus_{\alpha=1}^{m} Y + t \circ \left( X - \bigoplus_{\alpha=1}^{m} Y \right), \boldsymbol{\rho} \right).$$

This implies part 3 of the theorem with $\Upsilon_{\mathrm{free,X}}$ as above. In particular, $\bigoplus_{\alpha=1}^{m} Y \in \Upsilon_{\mathrm{free,unif-normal}}$ and

$$\sum_{w \in \mathcal{G}_d} \sup_{W \in B_{\mathrm{nc}}(\bigoplus_{\alpha=1}^{m} Y, \delta)} \left\| \left( W - \bigoplus_{\alpha=1}^{m_W} Y \right)^{\odot_s w} f_w \right\|_{sm_W} < \infty.$$

By picking up only $W \in B_{\rm nc}\left(\bigoplus_{\alpha=1}^{m} Y, \delta\right)$ of the form $W = \bigoplus_{\alpha=1}^{m} V$, where $V \in B_{\rm nc}(Y, \delta)$, we obtain that

$$\sum_{w \in \mathcal{G}_d} \sup_{V \in B_{\rm nc}(Y,\delta)} \left\| \left(V - \bigoplus_{\alpha=1}^{m_V} Y\right)^{\odot_s w} f_w \right\|_{sm_V} < \infty,$$

i.e., $Y \in \Upsilon_{\rm free,unif-normal}$.

Since $Y \in \Upsilon_{\rm free,unif-normal}$ and since the union of complete Reinhardt-like sets about $\bigoplus_{\alpha=1}^{m} Y$ for all $m \in \mathbb{N}$ is a complete Reinhardt-like set about $Y$, the set

$$\Upsilon_{\rm free,unif-normal} = \bigcup_{X \in \Upsilon_{\rm free,unif-normal}} \Upsilon_{\rm free,X}$$

is a complete Reinhardt-like set about $Y$. This completes the proof of part 2.

4. Let $\boldsymbol{\rho} = (\rho_1, \ldots, \rho_d)$ be a $d$-tuple of associated radii of absolute convergence of the series (8.28), and let $\mathbf{r} = (r_1, \ldots, r_d)$ be a $d$-tuple of positive real numbers such that $r_j < \rho_j$. Then there exists $\epsilon > 0$ small enough, so that $\widetilde{r}_j = r_j(1+\epsilon) < \rho_j$, $j = 1, \ldots, d$. Then for $\widetilde{\mathbf{r}} = (\widetilde{r}_1, \ldots, \widetilde{r}_d)$ we have $\mu(\widetilde{\mathbf{r}}) \le 1$. Since the function $\mu(\cdot)$ is homogeneous of degree 1, we have

$$\mu(\mathbf{r}) = \frac{\mu(\widetilde{\mathbf{r}})}{1+\epsilon} \le \frac{1}{1+\epsilon}.$$

Hence there exists $\ell_0 \in \mathbb{N}$ such that

$$\sqrt[\ell]{\sum_{|w|=\ell} \mathbf{r}^w \|f_w\|_{\mathcal{L}^{\ell}_{\rm cb}}} < \frac{1+0.5\epsilon}{1+\epsilon} < 1$$

for all $\ell > \ell_0$. This implies that the series (8.28) converges absolutely on the nc polydisk $(\mathbb{D}^d)_{\rm nc}(Y, \mathbf{r})$, moreover

$$\begin{aligned}\sum_{w \in \mathcal{G}_d} \sup_{W \in (\mathbb{D}^d)_{\rm nc}(Y,\mathbf{r})} \left\| \left(W - \bigoplus_{\alpha=1}^{m_W} Y\right)^{\odot_s w} f_w \right\|_{sm_W} &= \sum_{\ell=0}^{\infty} \sum_{|w|=\ell} \mathbf{r}^w \|f_w\|_{\mathcal{L}^{\ell}_{\rm cb}} \\ &< \sum_{\ell=0}^{\ell_0} \sum_{|w|=\ell} \mathbf{r}^w \|f_w\|_{\mathcal{L}^{\ell}_{\rm cb}} + \sum_{\ell=\ell_0+1}^{\infty} \left(\frac{1+0.5\epsilon}{1+\epsilon}\right)^{\ell} < \infty.\end{aligned}$$

Now suppose that $\mathbf{r} = (r_1, \ldots, r_d)$ is a $d$-tuple of positive extended real numbers with $r_j \ge \rho_j$ for all $j$ and $r_i > \rho_i$ for some $i$. Then there exists a $d$-tuple $\widetilde{\mathbf{r}} = (\widetilde{r}_1, \ldots, \widetilde{r}_d)$ of positive real numbers with $\widetilde{r}_j < r_j$ such that $\mu(\widetilde{\mathbf{r}}) > 1$. Then there exists $\epsilon > 0$ such that $\mu(\widetilde{\mathbf{r}}) > 1 + \epsilon$. Hence

$$\sum_{|w|=\ell} \widetilde{\mathbf{r}}^w \|f_w\|_{\mathcal{L}^{\ell}_{\rm cb}} > (1+0.5\epsilon)^{\ell} > 1 + 0.5\ell\epsilon \tag{8.39}$$

for infinitely many $\ell \in \mathbb{N}$. Let $S \subseteq \mathbb{N}$ be the (infinite) set of all such $\ell$'s. For any $\ell \in S$ and $w \in \mathcal{G}_d$ with $|w| = \ell$, using an argument similar to the one in Proposition 7.49 we obtain that

$$\begin{aligned}\|f_w\|_{\mathcal{L}^{\ell}_{\rm cb}} = \sup\{&\|W^{\odot_s w} f_w\|_{sm} : \\ &m \in \mathbb{N},\ W \in (\mathbb{C}^{sm \times sm})^d,\ \|W_j\|_{sm} \le 1,\ j = 1, \ldots, d\}.\end{aligned} \tag{8.40}$$

Therefore, there exist $m_w \in \mathbb{N}$ and $W_w \in (\mathbb{D}^d)_{\rm nc}(0_{s\times s}, \widetilde{\mathbf{r}})_{sm_w}$ such that

$$\|W_w^{\odot_s w} f_w\|_{sm_w} > \widetilde{\mathbf{r}}^w \|f_w\|_{\mathcal{L}^{\ell}_{\rm cb}} - \frac{0.5\ell\epsilon}{d^{\ell}}.$$

Set $m(\ell) = \sum_{|w|=\ell} m_w$ and $W(\ell) = \bigoplus_{|w|=\ell}\Big(W_w + \bigoplus_{\alpha=1}^{m_w} Y\Big)$. Then $W(\ell) \in (\mathbb{D}^d)_{\rm nc}(Y, \widetilde{\mathbf{r}})_{sm(\ell)}$ and

$$\sum_{|w|=\ell} \Big\| \Big(W(\ell) - \bigoplus_{\alpha=1}^{m(\ell)} Y\Big)^{\odot_s w} f_w\|_{sm(\ell)} > \sum_{|w|=\ell} \widetilde{\mathbf{r}}^w \|f_w\|_{\mathcal{L}^\ell_{\rm cb}} - 0.5\ell\epsilon > 1.$$

Thus for $\widetilde{\mathbf{r}}$, and hence for $\mathbf{r}$, (8.37) fails.

Now, let $s = 1$. Choosing $\widetilde{\mathbf{r}} \in (\mathbb{D}^d)_{\rm nc}(0_{1\times 1}, \mathbf{r})_1$ as in the preceding paragraph, we obtain from (8.29), (8.30), and (8.39), that

$$\sum_{w\in\mathcal{G}_d} \|\widetilde{\mathbf{r}}^w f_w\|_1 = \sum_{w\in\mathcal{G}_d} \widetilde{\mathbf{r}}^w \|f_w\|_{\mathcal{L}^{|w|}_{\rm cb}} = \infty,$$

i.e., the series (8.28) does not converge absolutely at $Z = \widetilde{\mathbf{r}} + Y$.

For the proof of the last statement, first suppose that $\boldsymbol{\rho} = (\rho_1, \ldots, \rho_d)$ is a $d$-tuple of strictly positive associated radii of normal convergence of the series (8.28). We have $B_{\rm nc}(Y, \delta) \subseteq (\mathbb{D}^d)_{\rm nc}(Y, (\delta, \ldots, \delta)) \subseteq (\mathbb{D}^d)_{\rm nc}(Y, \boldsymbol{\rho})$ for any positive real $\delta < \min_{1\le j\le d} \rho_j$, where for the first inclusion we used the fact that the block projections $\pi^s_{ij}$ are completely contractive. By the (already proven) first statement of part 4, (8.37) holds with $\mathbf{r} = (\delta, \ldots, \delta)$. Then

$$\sum_{w\in\mathcal{G}_d} \sup_{W\in B_{\rm nc}(Y,\delta)} \Big\| \Big(W - \bigoplus_{\alpha=1}^{m_W} Y\Big)^{\odot_s w} f_w\Big\|_{sm_W} < \infty, \tag{8.41}$$

i.e., $Y \in \Upsilon_{\rm free, unif-normal}$. Conversely, if $\Upsilon_{\rm free, unif-normal}$ is nonempty, then by part 2 of the theorem we have that $Y \in \Upsilon_{\rm free, unif-normal}$. The latter means that (8.41) holds for some $\delta > 0$. Then there exists a polydisk $(\mathbb{D}^d)_{\rm nc}(Y, \mathbf{r}^0)$ contained in $B_{\rm nc}(Y, \delta)$, say for $r^0_j = \frac{\delta}{d\|e_j\|_1}$, $j = 1, \ldots, d$, so that (8.37) holds with $\mathbf{r}^0$ in the place of $\mathbf{r}$. Clearly, in this case $\mu(\mathbf{r}^0) \le 1$. Let

$$\rho_1 = \sup\{r_1 \ge r^0_1 \colon \mu(r_1, r^0_2, \ldots, r^0_d) \le 1\}.$$

Then for any positive real number $r_1 \le \rho_1$ we have $\mu(r_1, r^0_2, \ldots, r^0_d) \le 1$ (in the case of $\rho_1 < \infty$ this is a consequence of Lemma 8.17). Let

$$\rho_2 = \sup\{r_2 \ge r^0_2 \colon \mu(r_1, r_2, r^0_3, \ldots, r^0_d) \le 1 \text{ for every } r_1 < \rho_1\}.$$

Then for any positive real numbers $r_1 \le \rho_1$ and $r_2 \le \rho_2$ we have $\mu(r_1, r_2, r^0_3, \ldots, r^0_d) \le 1$ (here we use Lemma 8.17 again in the case of $\rho_1 < \infty$ and/or $\rho_2 < \infty$). Continuing this construction in an obvious way, we define $\rho_3, \ldots, \rho_{d-1}$, and finally,

$$\rho_d = \sup\{r_d \ge r^0_d \colon \mu(r_1, \ldots, r_{d-1}, r_d) \le 1 \text{ for every } r_1 < \rho_1, \ldots, r_{d-1} < \rho_{d-1}\}.$$

Then for any $d$-tuple $\mathbf{r} = (r_1, \ldots, r_d)$ of positive real numbers $r_j \le \rho_j$ we have $\mu(\mathbf{r}) \le 1$ (we use Lemma 8.17 again when $\rho_j$ are finite). It is clear that $\boldsymbol{\rho} = (\rho_1, \ldots, \rho_d)$ is a $d$-tuple of strictly positive associated radii of normal convergence of the series (8.28). □

REMARK 8.19. It follows from the homogeneity and monotonicity of $\mu(\cdot)$ that we have the following trichotomy:

(1) $\mu(\mathbf{r}) = \infty$ for some, and hence for all, $d$-tuples $\mathbf{r}$ of strictly positive real numbers; thus there is no strictly positive associated radii of normal convergence of the series (8.28).

(2) $0 < \mu(\mathbf{r}) < \infty$ for some, and hence for all, $d$-tuples $\mathbf{r}$ of strictly positive real numbers; the argument in the last paragraph of the proof of Theorem 8.18 shows that there exists a $d$-tuple of strictly positive associated radii of normal convergence of (8.28) not all of which equal $\infty$.

(3) $\mu(\mathbf{r}) = 0$ for some, and hence for all, $d$-tuples $\mathbf{r}$ of strictly positive real numbers; thus the only $d$-tuple of associated radii of normal convergence of (8.28) is $(\infty, \dots, \infty)$.

REMARK 8.20. There exists a $d$-tuple of strictly positive associated radii of normal convergence of the series (8.28) if and only if $\rho_{\rm cb} > 0$ — see (8.19). One direction is obvious since $\Upsilon_{\rm free,unif-normal} \subseteq \Upsilon_{\rm nc,unif-normal}$. The other direction follows by applying Theorem 7.29 (or Corollary 7.31) to the nc function $f$ which is uniformly analytic on $\Upsilon_{\rm nc,unif-normal}$. The statement also follows by direct estimates. We obtain from (8.31), (8.32), (8.33), and (8.19) that

$$\mu(\mathbf{r}) \le (\mathbf{r} \cdot \mathbf{C}_2)\mu_{\rm cb} = \Big(\sum_{j=1}^{d} r_j \|e_j\|_1\Big)\mu_{\rm cb}, \tag{8.42}$$

$$\mu_{\rm cb} \le \mu(\mathbf{C}_1). \tag{8.43}$$

These inequalities imply that $\mu_{\rm cb} < \infty$ if and only if $\mu(\mathbf{r}) < \infty$ for some $d$-tuple $\mathbf{r}$ of strictly positive real numbers. Clearly, $\mu_{\rm cb} < \infty$ is equivalent to $\rho_{\rm cb} > 0$, and the conclusion follows from Remark 8.19.

For power series in several complex variables, the classical Abel lemma states that if the terms of the series at a point $\mu = (\mu_1, \dots, \mu_d)$ are bounded, then the series converges absolutely and locally normally in the polydisk with multiradius $(|\mu_1|, \dots, |\mu_d|)$ — see [**97**, Page 19]. We proceed to establish a noncommutative counterpart of this statement.

For $Y = (Y_1, \dots, Y_d) \in (\mathbb{C}^{s\times s})^d$ and $\mathbf{r} = (r_1, \dots, r_d)$ a $d$-tuple of positive extended real numbers, define the *nc diamond centered at* $Y$ *of multiradius* $\mathbf{r}$

$$(\diamondsuit^d)_{\rm nc}(Y, \mathbf{r}) := \coprod_{m=1}^{\infty} \Big\{Z = (Z_1, \dots, Z_d) \in \big(\mathbb{C}^{sm\times sm}\big)^d : \sum_{j=1}^{d} \frac{\|Z_j - \bigoplus_{\alpha=1}^m Y_j\|}{r_j} < 1\Big\}.$$

Notice that $\diamondsuit_{\rm nc}(Y, r) = (\diamondsuit^d)_{\rm nc}(Y, \mathbf{r})$ with $\mathbf{r} = \Big(\frac{r}{\|e_1\|_1}, \dots, \frac{r}{\|e_d\|_1}\Big)$ — see (7.37).

THEOREM 8.21. *1. Suppose there exist* $\mu = (\mu_1, \dots, \mu_d) \in \mathbb{C}^d$ *with all* $\mu_j \neq 0$ *and* $C > 0$ *such that* $|\mu^w| \|f_w\|_{\mathcal{L}_{\rm cb}^{|w|}} \le C$ *for every* $w \in \mathcal{G}_d$ *(in the case* $s = 1$ *this means that the terms of the series* (8.28) *at* $Y + \mu$ *are bounded). Then the series* (8.28) *converges absolutely on every nc diamond* $(\diamondsuit^d)_{\rm nc}(Y, \mathbf{r})$ *with* $r_j < |\mu_j|$ *for all* $j = 1, \dots, d$. *Moreover,*

$$\sum_{w\in\mathcal{G}_d} \sup_{W\in(\diamondsuit^d)_{\rm nc}(Y,\mathbf{r})} \Big\|\Big(W - \bigoplus_{\alpha=1}^{m_W} Y\Big)^{\odot_s w} f_w\Big\|_{sm_W} < \infty; \tag{8.44}$$

*here* $m_W$ *is the size of a block matrix* $W$ *with* $s \times s$ *blocks.*

*2. Assume there exist* $m \in \mathbb{N}$ *and* $X = (X_1, \dots, X_d) \in \big(\mathbb{C}^{sm\times sm}\big)^d$ *such that the terms of the series* (8.28) *are bounded on some uniformly-open neighborhood of* $X$*, i.e., there exist* $\delta > 0$ *and* $C > 0$ *such that* $\|(W - \bigoplus_{\alpha=1}^{m_W} Y)^{\odot_s w} f_w\|_{sm_W} \le C$ *for every* $W \in B_{\rm nc}(X, \delta)$ *and* $w \in \mathcal{G}_d$. *Then the set* $\Upsilon_{\rm free,unif-normal}$ *is nonempty.*

PROOF. 1. We may assume that $0 < r_j < (1-\epsilon)|\mu_j|$ for some $\epsilon > 0$ and for all $j = 1, \ldots, d$. Then

$$\begin{aligned}
\sum_{w \in \mathcal{G}_d} \sup_{W \in (\diamondsuit^d)_{\rm nc}(0_{s\times s}, \mathbf{r})} \|W^{\odot_s w} f_w\| &= \sum_{\ell=0}^{\infty} \sum_{|w|=\ell} \sup_{W \in (\diamondsuit^d)_{\rm nc}(0_{s\times s}, \mathbf{r})} \|W^{\odot_s w} f_w\| \\
&= \sum_{\ell=0}^{\infty} \sum_{|w|=\ell} \sup_{W \in (\diamondsuit^d)_{\rm nc}(0_{s\times s}, \mathbf{r})} \left(\frac{\|W_1\|}{r_1}, \ldots, \frac{\|W_d\|}{r_d}\right)^w \mathbf{r}^w \|f_w\|_{\mathcal{L}^\ell_{\rm cb}} \\
&< \sum_{\ell=0}^{\infty} (1-\epsilon)^\ell \sum_{|w|=\ell} \sup_{W \in (\diamondsuit^d)_{\rm nc}(0_{s\times s}, \mathbf{r})} \left(\frac{\|W_1\|}{r_1}, \ldots, \frac{\|W_d\|}{r_d}\right)^w |\mu^w| \|f_w\|_{\mathcal{L}^\ell_{\rm cb}} \\
&\le C \sum_{\ell=0}^{\infty} (1-\epsilon)^\ell \sum_{|w|=\ell} \sup_{W \in (\diamondsuit^d)_{\rm nc}(0_{s\times s}, \mathbf{r})} \left(\frac{\|W_1\|}{r_1}, \ldots, \frac{\|W_d\|}{r_d}\right)^w.
\end{aligned}$$

For any fixed $\ell$ and any $w \in \mathcal{G}_d$ with $|w| = \ell$ there exists $W_w \in (\diamondsuit^d)_{\rm nc}(0_{s\times s}, \mathbf{r})$ such that

$$\sup_{W \in (\diamondsuit^d)_{\rm nc}(0_{s\times s}, \mathbf{r})} \left(\frac{\|W_1\|}{r_1}, \ldots, \frac{\|W_d\|}{r_d}\right)^w \le (1+\epsilon)^\ell \left(\frac{\|(W_w)_1\|}{r_1}, \ldots, \frac{\|(W_w)_d\|}{r_d}\right)^w.$$

Define $W(\ell) = \bigoplus_{|w|=\ell} W_w \in (\diamondsuit^d)_{\rm nc}(0_{s\times s}, \mathbf{r})$. Clearly,

$$\left(\frac{\|(W_w)_1\|}{r_1}, \ldots, \frac{\|(W_w)_d\|}{r_d}\right)^w \le \left(\frac{\|W(\ell)_1\|}{r_1}, \ldots, \frac{\|W(\ell)_d\|}{r_d}\right)^w.$$

Therefore,

$$\begin{aligned}
\sum_{w \in \mathcal{G}_d} & \sup_{W \in (\diamondsuit^d)_{\rm nc}(0_{s\times s}, \mathbf{r})} \|W^{\odot_s w} f_w\| \\
&\le C \sum_{\ell=0}^{\infty} (1-\epsilon^2)^\ell \sum_{|w|=\ell} \left(\frac{\|W(\ell)_1\|}{r_1}, \ldots, \frac{\|W(\ell)_d\|}{r_d}\right)^w \\
&= C \sum_{\ell=0}^{\infty} (1-\epsilon^2)^\ell \Big(\sum_{j=1}^{d} \frac{\|W(\ell)_j\|}{r_j}\Big)^\ell < C \sum_{\ell=0}^{\infty} (1-\epsilon^2)^\ell = \frac{C}{\epsilon^2} < \infty,
\end{aligned}$$

i.e., (8.44) holds. In the case $s = 1$ we have by (8.29) and (8.30) that

$$\|f_w(\mu_1, \ldots, \mu_d)\|_1 = |\mu^w| \|f_w\|_{\mathcal{L}^{|w|}_{\rm cb}} \le C, \quad w \in \mathcal{G}_d.$$

2. By Lemma 8.10, the inequality

$$\sup_{W \in B_{\rm nc}(X,\delta)} \left\| \Big(W - \bigoplus_{\alpha=1}^{m_W} Y\Big)^{\odot_s w} f_w \right\|_{s m_W} \le C, \quad w \in \mathcal{G}_d,$$

holds with the suprema over $B_{\rm nc}(X, \delta)$ replaced by the suprema over

$$\bigcup_{\zeta \in \overline{\mathbb{D}}} B_{\rm nc}\left(\bigoplus_{\alpha=1}^{m} Y + \zeta\Big(X - \bigoplus_{\alpha=1}^{m} Y\Big), \delta\right).$$

In particular, taking $\zeta = 0$ we obtain that

$$\sup_{W \in B_{\rm nc}(\bigoplus_{\alpha=1}^{m} Y, \delta)} \left\| \Big(W - \bigoplus_{\alpha=1}^{m_W} Y\Big)^{\odot_s w} f_w \right\|_{s m_W} \le C, \quad w \in \mathcal{G}_d.$$

By picking up only $W \in B_{\rm nc}\Big(\bigoplus_{\alpha=1}^{m} Y, \delta\Big)$ of the form $W = \bigoplus_{\alpha=1}^{m} V$, where $V \in B_{\rm nc}(Y, \delta)$, we obtain that

$$\sup_{V \in B_{\rm nc}(Y,\delta)} \Big\| \Big(V - \bigoplus_{\alpha=1}^{m_V}\Big)^{\odot_s w} f_w \Big\|_{sm_V} \le C, \quad w \in \mathcal{G}_d.$$

As in the proof of Theorem 8.18, we can find a nc polydisk $(\mathbb{D}^d)_{\rm nc}(Y, \boldsymbol{\rho})$ which is contained in $B_{\rm nc}(Y, \delta)$, with $\rho_j = \frac{\delta}{d\|e_j\|_1}$, $j = 1, \ldots, d$. Then

$$\begin{aligned}\boldsymbol{\rho}^w \|f_w\|_{\mathcal{L}_{\rm cb}^{|w|}} &= \sup_{W \in (\mathbb{D}^d)_{\rm nc}(Y,\boldsymbol{\rho})} \Big( \Big\| V_1 - \bigoplus_{\alpha=1}^{m_V} Y_1 \Big\|_{sm_V}, \ldots, \Big\| V_d - \bigoplus_{\alpha=1}^{m_V} Y_d \Big\|_{sm_V} \Big)^w \|f_w\|_{\mathcal{L}_{\rm cb}^{|w|}} \\ &= \sup_{V \in (\mathbb{D}^d)_{\rm nc}(Y,\boldsymbol{\rho})} \Big\| \Big(V - \bigoplus_{\alpha=1}^{m_V} Y\Big)^{\odot_s w} f_w \|_{sm_V} \le C, \quad w \in \mathcal{G}_d.\end{aligned}$$

By part 1 of the theorem for $\mu = \boldsymbol{\rho}$ and every $\mathbf{r} = (r_1, \ldots, r_d)$ with $0 < r_j < \mu_j$, $j = 1, \ldots, d$, (8.44) holds. Clearly, the nc diamond $(\diamondsuit^d)_{\rm nc}(Y, \mathbf{r})$ is open in the uniformly-open topology on $(\mathbb{C}^d)_{\rm nc}$. In particular, $Y$ is its interior point. Indeed, $B_{\rm nc}(Y, \frac{1}{d}\min_{1\le j\le d} r_j) \subseteq (\diamondsuit^d)_{\rm nc}(Y, \mathbf{r})$. Therefore, we obtain that

$$\sum_{w \in \mathcal{G}_d} \sup_{W \in B_{\rm nc}(Y, \frac{1}{d}\min_{1\le j\le d} r_j)} \Big\| \Big(W - \bigoplus_{\alpha=1}^{m_W} Y\Big)^{\odot_s w} f_w \Big\|_{sm_W} < \infty,$$

i.e., $Y \in \Upsilon_{\rm free, unif-normal}$. □

REMARK 8.22. It follows from Theorem 8.21, part 1, that one can formulate an analogue of Theorem 8.18, part 4, for nc diamonds rather than nc polydisks, replacing (8.33) by

$$\mu_\diamondsuit(\mathbf{r}) = \limsup_{\ell \to \infty} \sqrt[\ell]{\max_{w \in \mathcal{G}_d\colon |w| = \ell} \mathbf{r}^w \|f_w\|_{\mathcal{L}_{\rm cb}^{\ell}}} \tag{8.45}$$

and defining the corresponding associated radii of normal convergence.

For power series in several complex variables, the union of polydiscs of (absolute) convergence equals the interior of the set of locally normal convergence equals the interior of the set of absolute convergence equals the interior of the set of convergence. On the other hand, the interior of the set of convergence of a power series in several complex variables may be strictly contained in the interior of the set of convergence of the corresponding series of homogeneous polynomials — consider, e.g., the power series $\sum_{n=0}^{\infty} (z_1 + z_2)^n$.

In the noncommutative setting we have, for the series (8.28), the following sequence of inclusions:

(8.46)
$$\bigcup_{\boldsymbol{\rho}} (\mathbb{D}^d)_{\rm nc}(0_{s\times s}, \boldsymbol{\rho}) \subseteq \Upsilon_{\rm free, unif-normal} \subseteq {\rm Int}\,\Upsilon_{\rm free, abs} \subseteq {\rm Int}\,\Upsilon_{{\rm free},\prec} \subseteq \Upsilon_{\rm nc, unif}.$$

Here the union is taken over all $d$-tuples $\boldsymbol{\rho} = (\rho_1, \ldots, \rho_d)$ of strictly positive associated radii of normal convergence, Int denotes the interior in the uniformly-open topology, $\Upsilon_{{\rm free},\prec}$ is the convergence set of the series (8.28) with respect to some order $\prec$ on the free monoid $\mathcal{G}_d$ which is compatible with the word length (i.e., $w \prec w' \Longrightarrow |w| < |w'|$), and $\Upsilon_{\rm nc, unif} = {\rm Int}\,\Upsilon_{\rm nc}$ (see Theorem 8.11).

We will calculate various sets appearing in (8.46) for several examples. We need the following lemma (where $r_{\rm spec}(Z)$ denotes the spectral radius of a matrix $Z$).

LEMMA 8.23.

$$\lim_{\delta\to 0}\limsup_{\ell\to\infty}\sqrt[\ell]{\sup_{W\in B_{\rm nc}(Z,\delta)}\|W^\ell\|}=\lim_{\delta\to 0}\liminf_{\ell\to\infty}\sqrt[\ell]{\sup_{W\in B_{\rm nc}(Z,\delta)}\|W^\ell\|}=r_{\rm spec}(Z).$$

PROOF. Suppose $Z\in\mathbb{C}^{s\times s}$ and let $\epsilon>0$. Using the lower semicontinuity of the spectral radius for elements of $\mathcal{L}(\ell^2)$ [**61**, page 167], we obtain that there exists $\delta>0$ such that

$$r_{\rm spec}(X)<r_{\rm spec}\Big(\bigoplus_{\alpha=1}^{\infty}Z\Big)+\epsilon=r_{\rm spec}(Z)+\epsilon$$

for every $X\in\mathcal{L}(\ell^2)$ with $\|X-\bigoplus_{\alpha=1}^{\infty}Z\|<\delta$. It follows that $r_{\rm spec}(W)<r_{\rm spec}(Z)+\epsilon$ for every $W\in B_{\rm nc}(Z,\delta)$ (by taking $X=\bigoplus_{\beta=1}^{\infty}W$). Set

$$M_\ell:=\sup_{W\in B_{\rm nc}(Z,\delta)}\|W^\ell\|,\quad \ell\in\mathbb{N}.$$

Then for any fixed $\ell$ we have $M_\ell\le(\|Z\|+\delta)^\ell<\infty$. Next, for every fixed $\ell$ one can find $W_\ell\in B_{\rm nc}(Z,\delta)$ such that $\|W_\ell^\ell\|\ge\frac{M_\ell}{2}$. Set $X:=\bigoplus_{\ell=1}^{\infty}W_\ell\in\mathcal{L}(\ell^2)$. Since $r_{\rm spec}(X)=\sup_\ell r_{\rm spec}(W_\ell)$ and $r_{\rm spec}(W_\ell)<r_{\rm spec}(Z)+\epsilon$, we have

$$\begin{aligned}r_{\rm spec}(Z)+\epsilon\ge r_{\rm spec}(X)&=\lim_{\ell\to\infty}\sqrt[\ell]{\|X^\ell\|}\ge\limsup_{\ell\to\infty}\sqrt[\ell]{\|W_\ell^\ell\|}\ge\limsup_{\ell\to\infty}\sqrt[\ell]{\frac{M_\ell}{2}}\\ =\limsup_{\ell\to\infty}\sqrt[\ell]{M_\ell}&=\limsup_{\ell\to\infty}\sqrt[\ell]{\sup_{W\in B_{\rm nc}(Z,\delta)}\|W^\ell\|}\ge\lim_{\delta\to 0}\liminf_{\ell\to\infty}\sqrt[\ell]{\sup_{W\in B_{\rm nc}(Z,\delta)}\|W^\ell\|}\\ &\ge\lim_{\ell\to\infty}\sqrt[\ell]{\|Z^\ell\|}=r_{\rm spec}(Z).\end{aligned}$$

We conclude that

$$\lim_{\delta\to 0}\limsup_{\ell\to\infty}\sqrt[\ell]{\sup_{W\in B_{\rm nc}(Z,\delta)}\|W^\ell\|}=\lim_{\delta\to 0}\liminf_{\ell\to\infty}\sqrt[\ell]{\sup_{W\in B_{\rm nc}(Z,\delta)}\|W^\ell\|}=r_{\rm spec}(Z)$$

as required. □

The following two examples show that the first inclusion in (8.46) can be proper.

EXAMPLE 8.24. Let $d=1$ and $s=1$, i.e., (8.1) and (8.28) both collapse to

$$\sum_{\ell=0}^{\infty}Z^\ell f_\ell,\tag{8.47}$$

where $Z\in\mathbb{C}^{m\times m}$, $m=1,2,\ldots$, for a given sequence of vectors $f_\ell\in\mathcal{W}$. Let $\boldsymbol{\rho}=\Big(\limsup_{\ell\to\infty}\sqrt[\ell]{\|f_\ell\|}\Big)^{-1}$ be the (unique associated) radius of normal convergence; of course, $\boldsymbol{\rho}=\rho_{\rm cb}=\rho_m$ for all $m$. Then clearly

$$\rho(Z)=\frac{\boldsymbol{\rho}}{r_{\rm spec}(Z)},$$

with the convention that the fraction equals $\infty$ when the denominator is zero, irrespectively of the numerator. It follows from Lemma 8.23 that if $\boldsymbol{\rho}$ and $r_{\rm spec}(Z)$ are not both zero, then we have also

$$\rho_{\rm unif-normal}(Z) = \frac{\boldsymbol{\rho}}{r_{\rm spec}(Z)}.$$

Indeed,

$$\begin{aligned}\rho(Z) \geq \rho_{\rm unif-normal}(Z) &= \frac{1}{\lim\limits_{\delta\to 0}\limsup\limits_{\ell\to\infty}\sqrt[\ell]{\sup\limits_{W\in B_{\rm nc}(Z,\delta)}\|W^\ell f_\ell\|}}\\ &= \frac{1}{\lim\limits_{\delta\to 0}\limsup\limits_{\ell\to\infty}\sqrt[\ell]{\sup\limits_{W\in B_{\rm nc}(Z,\delta)}\|W^\ell\|\|f_\ell\|}}\\ \geq \frac{1}{\lim\limits_{\delta\to 0}\limsup\limits_{\ell\to\infty}\sqrt[\ell]{\sup\limits_{W\in B_{\rm nc}(Z,\delta)}\|W^\ell\|}\limsup\limits_{\ell\to\infty}\sqrt[\ell]{\|f_\ell\|}} &= \frac{\boldsymbol{\rho}}{r_{\rm spec}(Z)} = \rho(Z),\end{aligned}$$

which implies the required equality. It follows that

$$\Upsilon_{\rm nc,unif-normal} = \Upsilon_{\rm nc,unif} = \Upsilon_{\rm nc,norm} = \{Z\in\mathbb{C}_{\rm nc}\colon r_{\rm spec}(Z)<\boldsymbol{\rho}\}.$$

Of course, $\Upsilon_{\rm free,unif-normal} = \Upsilon_{\rm nc,unif-normal}$; except for the cases $\boldsymbol{\rho}=0$ and $\boldsymbol{\rho}=\infty$ this is a much bigger set than

$$(\mathbb{D}^1)_{\rm nc}(0_{1\times 1},\boldsymbol{\rho}) = \{Z\in\mathbb{C}_{\rm nc}\colon \|Z\|<\boldsymbol{\rho}\}.$$

EXAMPLE 8.25. Consider the series

$$\sum_{n=0}^{\infty}(Z_1Z_2)^n.$$

We have

$$\mu(\mathbf{r}) = \limsup_{n\to\infty}\sqrt[2n]{r_1^n r_2^n} = \sqrt{r_1 r_2}.$$

It follows that a pair $\boldsymbol{\rho}=(\rho_1,\rho_2)$ of strictly positive extended real numbers is a pair of associated radii of normal convergence of the series if and only if $\rho_1\rho_2=1$, and

$$\bigcup_{\boldsymbol{\rho}}(\mathbb{D}^2)_{\rm nc}(0_{1\times 1},\boldsymbol{\rho}) = \{(Z_1,Z_2)\in(\mathbb{C}^2)_{\rm nc}\colon \|Z_1Z_2\|<1\}.$$

It is clear that the convergence set of this series is

$$\Upsilon_{\rm nc} = \{(Z_1,Z_2)\in(\mathbb{C}^2)_{\rm nc}\colon r_{\rm spec}(Z_1Z_2)<1\},$$

and $\rho(Z) = (r_{\rm spec}(Z_1Z_2))^{-1}$. It follows from Lemma 8.23 that

$$\begin{aligned}\rho(Z)\geq\rho_{\rm unif-normal}(Z) &= \Big(\lim_{\delta\to 0}\limsup_{\ell\to\infty}\sqrt[\ell]{\sup_{(W_1,W_2)\in B_{\rm nc}(Z,\delta)}\|(W_1W_2)^\ell\|}\Big)^{-1}\\ &\geq \Big(\lim_{\delta\to 0}\limsup_{\ell\to\infty}\sqrt[\ell]{\sup_{(W)\in B_{\rm nc}(Z_1Z_2,\delta)}\|W^\ell\|}\Big)^{-1} = (r_{\rm spec}(Z_1Z_2))^{-1} = \rho(Z),\end{aligned}$$

and therefore

$$\Upsilon_{\rm nc,unif-normal} = \Upsilon_{\rm nc,unif} = \Upsilon_{\rm nc,norm} = \{(Z_1,Z_2)\in(\mathbb{C}^2)_{\rm nc}\colon r_{\rm spec}(Z_1Z_2)<1\}.$$

Since each homogeneous polynomial in the series has only one term, the convergence of this series along the free monoid is the same as its convergence as a series of homogeneous polynomials, so that $\Upsilon_{\mathrm{free,unif-normal}} = \Upsilon_{\mathrm{nc,unif-normal}}$. Similarly to the previous example, the second set in (8.46) is much bigger than the first one.

The following two examples show that the last inclusion in (8.46) may be proper. The first example also shows that, unlike in the classical Abel lemma, the convergence of the series (8.28) at a point $\mu = (\mu_1, \ldots, \mu_d) \in \mathbb{C}^d$ does not imply its convergence on the nc polydisk of multiradius $(|\mu_1|, \ldots, |\mu_d|)$. The second example also shows that the TT series along $\mathcal{G}_d$ of a uniformly analytic nc function does not necessarily converge on a nc polydisk when the nc function is bounded there — cf. Theorem 7.29 and Corollary 7.31.

EXAMPLE 8.26. Let $d > 1$ and let us consider "nc geometric series"

$$\sum_{w \in \mathcal{G}_d} Z^w, \tag{8.48}$$

where $Z \in \mathbb{C}^{m\times m}$, $m = 1, 2, \ldots$ (i.e., $s = 1$, $\mathcal{W} = \mathbb{C}$, and $f_w = 1$ for all $w \in \mathcal{G}_d$). Notice that

$$\sum_{w\in\mathcal{G}_d,\, |w|=\ell} Z^w = (Z_1 + \cdots + Z_d)^\ell,$$

therefore

$$\Upsilon_{\mathrm{nc}} = \left\{(Z_1, \ldots, Z_d) \in (\mathbb{C}^d)_{\mathrm{nc}} \colon r_{\mathrm{spec}}(Z_1 + \cdots + Z_d) < 1\right\}$$

and $\rho(Z) = (r_{\mathrm{spec}}(Z_1 + \cdots + Z_d))^{-1}$. Similarly to Example 8.25, it follows from Lemma 8.23 that

$$\begin{aligned}\Upsilon_{\mathrm{nc,unif-normal}} &= \Upsilon_{\mathrm{nc,unif}} = \Upsilon_{\mathrm{nc,norm}} \\ &= \left\{(Z_1, \ldots, Z_d) \in (\mathbb{C}^d)_{\mathrm{nc}} \colon r_{\mathrm{spec}}(Z_1 + \cdots + Z_d) < 1\right\}\end{aligned} \tag{8.49}$$

as well. Notice that the set $\Upsilon_{\mathrm{free,prec}}$ is properly contained in (8.49) for any order $\prec$ on $\mathcal{G}_d$ which is compatible with the word length — because $r_{\mathrm{spec}}(Z^w) < 1$ for every $w \in \mathcal{G}_d$ is a necessary condition for the convergence of the series (8.48), and there are plenty of $d$-tuples of matrices in (8.49) which violate this condition.

Next, for a $d$-tuple $\mathbf{r} = (r_1, \ldots, r_d)$ of nonnegative real numbers, $\mu(\mathbf{r}) = r_1 + \cdots + r_d$. Therefore a $d$-tuple $\boldsymbol{\rho} = (\rho_1, \ldots, \rho_d)$ of strictly positive extended real numbers is a $d$-tuple of associated radii of normal convergence if and only if $\rho_1 + \cdots + \rho_d = 1$, so that

$$\begin{aligned}\bigcup_{\boldsymbol{\rho}} (\mathbb{D}^d)_{\mathrm{nc}}(0_{1\times 1}, \boldsymbol{\rho}) &= \left\{Z \in (\mathbb{C}^d)_{\mathrm{nc}} \colon \|Z_1\| + \cdots + \|Z_d\| < 1\right\} \\ &= (\diamondsuit^d)_{\mathrm{nc}}(0_{1\times 1}, (1, \ldots, 1)).\end{aligned} \tag{8.50}$$

On the other hand, for a $d$-tuple $\mathbf{r} = (r_1, \ldots, r_d)$ of nonnegative real numbers, $\mu_\diamondsuit(\mathbf{r}) = \max_j r_j$. Therefore a $d$-tuple $\boldsymbol{\rho} = (\rho_1, \ldots, \rho_d)$ of strictly positive extended real numbers is a $d$-tuple of associated radii of normal convergence in the sense of diamonds — see Remark 8.22 — if and only if $\rho_1 = \cdots = \rho_d = 1$, so that the set (8.50) coincides with $\bigcup_{\boldsymbol{\rho}} (\diamondsuit^d)_{\mathrm{nc}}(0_{1\times 1}, \boldsymbol{\rho})$ where the union is taken over all $d$-tuples $\boldsymbol{\rho} = (\rho_1, \ldots, \rho_d)$ of of strictly positive associated radii of normal convergence in the sense of diamonds.

Clearly, $Z \in \Upsilon_{\mathrm{free},\prec}$ if and only if $Z \in \Upsilon_{\mathrm{nc}}$, i.e., $r_{\mathrm{spec}}(Z_1 + \cdots + Z_d) < 1$, and

$$\lim_{w_1 \preceq w_2, |w_1|=|w_2| \to \infty} \sum_{w_1 \preceq w \preceq w_2} Z^w = 0.$$

Let $d = 2$ and let $\zeta \in \mathbb{C}$, $\frac{1}{2} < |\zeta| < 1$. It folows that $(\zeta, -\zeta) \in \Upsilon_{\mathrm{free},\prec}$ if and only if

$$\lim_{w_1 \preceq w_2, |w_1|=|w_2| \to \infty} \zeta^{|w_1|} \sum_{w_1 \preceq w \preceq w_2} (-1)^{|w|_2} = 0, \tag{8.51}$$

where $|w|_2$ denotes the number of times the letter $g_2$ appears in the word $w$. Notice that the number of words of length $\ell$ with $|w|_2$ even is the same as the number of words of length $\ell$ with $|w|_2$ odd and equals $2^{\ell-1}$. Thus if we choose the order $\prec$ so that all the words of a given length with $|w|_2$ even preceed all the words of the same length with $|w|_2$ odd, then (8.51) fails. On the other hand, if we choose the order $\prec$ so that words with $|w|_2$ even and words with $|w|_2$ odd interlace, then (8.51) holds. Therefore, $(\zeta, -\zeta) \notin \Upsilon_{\mathrm{free},\prec}$ or $(\zeta, -\zeta) \in \Upsilon_{\mathrm{free},\prec}$, depending on the choice of $\prec$. Furthemore, in the case where $(\zeta, -\zeta) \in \Upsilon_{\mathrm{free},\prec}$ the series does not converge on $(\mathbb{D}^2)_{\mathrm{nc}}(0_{1\times 1}, (|\zeta|, |\zeta|))$, e.g., it does not converge at the point $(\epsilon\zeta, \epsilon\zeta)$ for $\frac{1}{2|\zeta|} < \epsilon < 1$.

EXAMPLE 8.27. Consider

$$f(Z) = (I - \mathcal{Z}(Z) \otimes A)^{-1} = (I - Z_1(P_1 A) - Z_2(P_2 A))^{-1},$$

where

$$A = \begin{bmatrix} a_{11} & a_{12} \\ a_{21} & a_{22} \end{bmatrix}, \quad P_1 = \begin{bmatrix} 1 & 0 \\ 0 & 0 \end{bmatrix}, \quad P_2 = \begin{bmatrix} 0 & 0 \\ 0 & 1 \end{bmatrix}$$

are $2 \times 2$ matrices over $\mathbb{C}$, and $\mathcal{Z}(Z) = \begin{bmatrix} Z_1 & 0 \\ 0 & Z_2 \end{bmatrix} = Z_1 \otimes P_1 + Z_2 \otimes P_2$ is a $2n \times 2n$ matrix over $\mathbb{C}$ for every $n \in \mathbb{N}$ and $Z \in \mathbb{C}^{n\times n}$. It is easy to see that $f$ is a uniformly analytic nc function on the uniformly-open nc set

$$\left\{Z \in (\mathbb{C}^2)_{\mathrm{nc}} \colon \det(I - \mathcal{Z}(Z) \otimes A) \neq 0\right\}.$$

From now on, we assume that $\|A\| < 1$, so that $f$ is bounded by $(1 - \|A\|)^{-1}$ on $(\mathbb{D}^2)_{\mathrm{nc}}(0_{1\times 1}, (1, \dots, 1))$. The TT series of $f$ at $0_{1\times 1}$ is

$$\sum_{\ell=0}^{\infty} (\mathcal{Z}(Z) \otimes A)^\ell.$$

For this series, it is clear that

$$\Upsilon_{\mathrm{nc}} = \left\{Z \in (\mathbb{C}^2)_{\mathrm{nc}} \colon r_{\mathrm{spec}}(\mathcal{Z}(Z) \otimes A) < 1\right\}$$

and $\rho(Z) = (r_{\mathrm{spec}}(\mathcal{Z}(Z) \otimes A))^{-1}$. Using Lemma 8.23, we obtain that

$$\Upsilon_{\mathrm{nc,unif-normal}} = \Upsilon_{\mathrm{nc,unif}} = \Upsilon_{\mathrm{nc,norm}} = \left\{Z \in (\mathbb{C}^2)_{\mathrm{nc}} \colon r_{\mathrm{spec}}(\mathcal{Z}(Z) \otimes A) < 1\right\}$$

as well.

We consider now the corresponding series along $\mathcal{G}_2$. The $(i, j)$-th entry of the sum is given by

$$\delta_{ij} + \sum_{\ell=0}^{\infty} \sum_{1 \le i_1, \dots, i_{\ell-1} \le 2} a_{ii_1} a_{i_1 i_2} \cdots a_{i_{\ell-1} j} Z_i Z_{i_1} \cdots Z_{i_{\ell-1}},$$

for all $i, j = 1, 2$. For $(z_1, z_2) \in \mathbb{C}^2$, the series converges absolutely if and only if the series

$$\sum_{\ell=0}^{\infty} \left( \begin{bmatrix} |z_1| & 0 \\ 0 & |z_2| \end{bmatrix} \begin{bmatrix} |a_{11}| & |a_{12}| \\ |a_{21}| & |a_{22}| \end{bmatrix} \right)^{\ell} \tag{8.52}$$

converges. We choose $A$ with $\|A\| < 1$ and $r_{\text{spec}} \begin{bmatrix} |a_{11}| & |a_{12}| \\ |a_{21}| & |a_{22}| \end{bmatrix} > 1$, e.g.,

$$A = \frac{1 - \epsilon}{\sqrt{2}} \begin{bmatrix} 1 & -1 \\ 1 & 1 \end{bmatrix},$$

with $\epsilon > 0$ small enough. Then for $|z_1|, |z_2| < 1$ sufficiently close to 1, the series (8.52) diverges.

CHAPTER 9

# Direct summands extensions of nc sets and nc functions

In this chapter, we consider a natural extension of a similarity invariant nc set $\Omega$ to a larger nc set, $\Omega_{\text{d.s.e.}}$ which contains all direct summands of matrices from $\Omega$, and the corresponding extension of nc functions on $\Omega$ to nc functions on $\Omega_{\text{d.s.e.}}$. Clearly, $\Omega_{\text{d.s.e.}}$ includes not only $\Omega$, but also the set $\operatorname{rad}\Omega$ which was introduced in Section 7.2—see the paragraph preceding Proposition 7.15.

Let $\mathcal{M}$ be a module over a commutative unital ring $\mathcal{R}$, and let $\Omega \subseteq \mathcal{M}_{\text{nc}}$ be a similarity invariant nc set. We define the *direct summands extension* of $\Omega$ by

$$\Omega_{\text{d.s.e.}} := \{X \in \mathcal{M}_{\text{nc}} \colon X \oplus Y \in \Omega \text{ for some } Y \in \mathcal{M}_{\text{nc}}\}. \tag{9.1}$$

Since $\Omega$ is invariant under similarities, in particular, when a similarity matrix is a permutation of rows (columns), $X \in \Omega_{\text{d.s.e.}}$ if and only if $Y^1 \oplus X \oplus Y^2 \in \Omega$ for some $Y^1$, $Y^2 \in \mathcal{M}_{\text{nc}}$.

PROPOSITION 9.1. *Let $\Omega \subseteq \mathcal{M}_{\text{nc}}$ be a similarity invariant nc set. Then*

1. *$\Omega_{\text{d.s.e.}}$ is a radical similarity invariant nc set;*
2. *If $\Omega$ is right (resp., left) admissible, then so is $\Omega_{\text{d.s.e.}}$.*
3. *If $\Omega$ is finitely open (resp., open, uniformly-open), then so is $\Omega_{\text{d.s.e.}}$; moreover, if $\Omega$ is uniformly-open, then $\Omega_{\text{d.s.e.}}$ is $\tau$-open (see Section 7.2 for the definition of the pseudometric $\tau$).*

PROOF. (1) If $X^1$, $X^2 \in \Omega_{\text{d.s.e.}}$, then $X^1 \oplus Y^1 \in \Omega$ and $X^2 \oplus Y^2 \in \Omega$ for some $Y^1, Y^2 \in \mathcal{M}_{\text{nc}}$. Since $\Omega$ is a nc set, $X^1 \oplus Y^1 \oplus X^2 \oplus Y^2 \in \Omega$. Since $\Omega$ is similarity invariant, and the matrix $X^1 \oplus Y^1 \oplus X^2 \oplus Y^2$ is similar, with the similarity matrix being a permutation, to $X^1 \oplus X^2 \oplus Y^1 \oplus Y^2$, the latter belongs to $\Omega$. Therefore, $X^1 \oplus X^2 \in \Omega_{\text{d.s.e.}}$, and we conclude that $\Omega_{\text{d.s.e.}}$ is a nc set.

Let $X \in (\Omega_{\text{d.s.e.}})_n$ and let $S \in \mathcal{R}^{n \times n}$ be invertible. Then $X \oplus Y \in \Omega_{n+m}$ for some $m \in \mathbb{N}$ and $Y \in \mathcal{M}^{m \times m}$, and since $\Omega$ is similarity invariant,

$$SXS^{-1} \oplus Y = (S \oplus I_m)(X \oplus Y)(S \oplus I_m)^{-1} \in \Omega_{n+m}.$$

Therefore, $SXS^{-1} \in (\Omega_{\text{d.s.e.}})_n$, and we conclude that $\Omega_{\text{d.s.e.}}$ is similarity invariant.

Suppose that $Y \in \operatorname{rad}\Omega_{\text{d.s.e.}}$. Then $X = \bigoplus_{\alpha=1}^{m} Y \in \Omega_{\text{d.s.e.}}$ for some $m \in \mathbb{N}$. Therefore, $X \oplus Z \in \Omega$ for some $Z \in \mathcal{M}_{\text{nc}}$, i.e.,

$$Y \oplus \Big(\bigoplus_{\alpha=1}^{m-1} Y\Big) \oplus Z \in \Omega.$$

Hence, $Y \in \Omega_{\text{d.s.e.}}$. This proves the inclusion $\operatorname{rad}\Omega_{\text{d.s.e.}} \subseteq \Omega_{\text{d.s.e.}}$. Together with the obvious inclusion $\operatorname{rad}\Omega_{\text{d.s.e.}} \supseteq \Omega_{\text{d.s.e.}}$, this implies that $\operatorname{rad}\Omega_{\text{d.s.e.}} = \Omega_{\text{d.s.e.}}$, i.e., the set $\Omega_{\text{d.s.e.}}$ is radical.

(2) Suppose that $\Omega$ is right admissible. Let $X^1 \in (\Omega_{\text{d.s.e.}})_{n_1}$, $X^2 \in (\Omega_{\text{d.s.e.}})_{n_2}$, and $Z \in \mathcal{M}^{n_1\times n_2}$. Then there exist $m_1, m_2 \in \mathbb{N}$ and $Y^1 \in \mathcal{M}^{m_1\times m_1}$, $Y^2 \in \mathcal{M}^{m_2\times m_2}$ such that $X^1 \oplus Y^1 \in \Omega_{n_1+m_1}$ and $X^2 \oplus Y^2 \in \Omega_{n_2+m_2}$. Since $\Omega$ is a similarity invariant right admissible nc set, by Proposition A.2,

$$\begin{bmatrix} X^1 & 0 & Z & 0 \\ 0 & Y^1 & 0 & 0 \\ 0 & 0 & X^2 & 0 \\ 0 & 0 & 0 & Y^2 \end{bmatrix} \in \Omega_{n_1+m_1+n_2+m_2}.$$

The latter matrix is similar, with the similarity matrix being a permutation, to

$$\begin{bmatrix} X^1 & Z & 0 & 0 \\ 0 & X^2 & 0 & 0 \\ 0 & 0 & Y^1 & 0 \\ 0 & 0 & 0 & Y^2 \end{bmatrix},$$

which therefore also belongs to $\Omega_{n_1+m_1+n_2+m_2} = \Omega_{n_1+n_2+m_1+m_2}$. Hence,

$$\begin{bmatrix} X^1 & Z \\ 0 & X^2 \end{bmatrix} \in (\Omega_{\text{d.s.e.}})_{n_1+n_2},$$

and we conclude that $\Omega_{\text{d.s.e.}}$ is right admissible.

The statement for a left admissible $\Omega$ is proved analogously.

(3) Let $\mathcal{M} = \mathcal{V}$ be a vector space over $\mathbb{C}$, and let the nc set $\Omega \subseteq \mathcal{V}_{\text{nc}}$ be finitely open. Let $X \in (\Omega_{\text{d.s.e.}})_n$. Then there exist $m \in \mathbb{N}$ and $Y \in \mathcal{V}^{m\times m}$ such that $X \oplus Y \in \Omega_{n+m}$. Let $\mathcal{U}$ be a finite-dimensional subspace of $\mathcal{V}^{n\times n}$ which contains $X$. Then the space $\mathcal{U} \oplus \text{span}\{Y\} := \{U \oplus \alpha Y : U \in \mathcal{U}, \alpha \in \mathbb{C}\}$ is finite-dimensional and contains $Y$. Since $\Omega_{n+m}$ is finitely open, there exists an open neighborhood $\Gamma$ of $X \oplus Y$ in $\mathcal{U} \oplus \text{span}\{Y\}$ which is contained in $\Omega_{n+m}$. Moreover, one can choose such a neighborhood of the form $\Gamma = \Phi \oplus \Psi$ where $\Phi$ is an open neighborhood of $X$ in $\mathcal{U}$, and $\Psi$ is an open neighborhood of $Y$ in $\text{span}\{Y\}$. Since for every $U \in \Phi$ and $W \in \Psi$ one has $U \oplus W \in \Gamma \subseteq \Omega_{n+m}$, we have that $\Phi \subseteq (\Omega_{\text{d.s.e.}})_n$. We conclude that $\Omega_{\text{d.s.e.}}$ is finitely open.

Let $\mathcal{M} = \mathcal{V}$ be a Banach space equipped with an admissible system of matrix norms over $\mathcal{V}$, and let the nc set $\Omega \subseteq \mathcal{V}_{\text{nc}}$ be open. Let $X \in (\Omega_{\text{d.s.e.}})_n$. Then there exist $m \in \mathbb{N}$ and $Y \in \mathcal{V}^{m\times m}$ such that $X \oplus Y \in \Omega_{n+m}$. Let $\epsilon > 0$ be such that $B(X \oplus Y, \epsilon) \subseteq \Omega_{n+m}$. Set $\delta := C_1'(n,m)^{-1}\epsilon$ (see (7.1) where $C_1'(n,m)$ is introduced). Then for every $U \in B(X, \delta)$ one has $U \oplus Y \in B(X \oplus Y, \epsilon) \subseteq \Omega_{n+m}$, hence $U \in (\Omega_{\text{d.s.e.}})_n$. We conclude that $\Omega_{\text{d.s.e.}}$ is open.

Let $\mathcal{M} = \mathcal{V}$ be an operator space, and let the nc set $\Omega \subseteq \mathcal{V}_{\text{nc}}$ be uniformly-open. Let $X \in (\Omega_{\text{d.s.e.}})_n$. Then there exist $m \in \mathbb{N}$ and $Y \in \mathcal{V}^{m\times m}$ such that $X \oplus Y \in \Omega_{n+m}$. Let $\epsilon > 0$ be such that $B_{\text{nc}}(X \oplus Y, \epsilon) \subseteq \Omega$. Let $Z \in B_{\text{nc}}(X, \epsilon)_{nk}$, and let $U \in \mathcal{R}^{(n+m)k\times(n+m)k}$ be a permutation matrix such that

$$U\Big(\bigoplus_{\alpha=1}^{k}(X \oplus Y)\Big)U^{-1} = \Big(\bigoplus_{\alpha=1}^{k} X\Big) \oplus \Big(\bigoplus_{\beta=1}^{k} Y\Big). \tag{9.2}$$

Since $\Omega$ is a similarity invariant nc set, this matrix belongs to $\Omega_{(n+m)k}$. Since $U$ is unitary,

$$U^{-1}\Big(Z \oplus \bigoplus_{\beta=1}^{k} Y\Big)U \in B_{\text{nc}}(X \oplus Y, \epsilon) \subseteq \Omega_{(n+m)k}, \tag{9.3}$$

and since $\Omega$ is similarity invariant, $Z \oplus \bigoplus_{\beta=1}^{k} Y \in \Omega_{(n+m)k}$. Therefore, $Z \in (\Omega_{\text{d.s.e.}})_{nk}$. This proves that $B_{\text{nc}}(X, \epsilon) \subseteq \Omega_{\text{d.s.e.}}$, and we conclude that $\Omega_{\text{d.s.e.}}$ is uniformly-open. □

We define next a direct summands extension $f_{\text{d.s.e.}}$ of a nc function $f$.

PROPOSITION 9.2. *Let $\mathcal{M}$, $\mathcal{N}$ be modules over a unital commutative ring $\mathcal{R}$, let $\Omega \subseteq \mathcal{M}_{\text{nc}}$ be a similarity invariant nc set, and let $f\colon \Omega \to \mathcal{N}_{\text{nc}}$ be a nc function. Then:*

(1) *There exists a uniquely determined* direct summands extension *of $f$, i.e., a nc function $f_{\text{d.s.e.}}\colon \Omega_{\text{d.s.e.}} \to \mathcal{N}_{\text{nc}}$ such that $f_{\text{d.s.e.}}|_{\Omega} = f$.*
(2) *If $\mathcal{R} = \mathbb{K}$ is an infinite field, $\Omega = \coprod_{m=1}^{\infty} \mathcal{M}^{sm\times sm}$ and $f|_{\mathcal{M}^{sm\times sm}}$ is polynomial on slices for every $m \in \mathbb{N}$, then $\Omega_{\text{d.s.e.}} = \mathcal{M}_{\text{nc}}$ and $f_{\text{d.s.e.}}$ is polynomial on slices; if, furthermore, the degrees of $f|_{\mathcal{M}^{sm\times sm}}$, $m = 1, 2, \ldots$, are bounded, then so are the degrees of $f_{\text{d.s.e.}}|_{\mathcal{M}^{n\times n}}$, $n = 1, 2, \ldots$, so that $f_{\text{d.s.e.}}$ is a nc polynomial over $\mathcal{M}$.*
(3) *If $\mathcal{M} = \mathcal{V}$ is a vector space over $\mathbb{C}$, $\mathcal{N} = \mathcal{W}$ is a Banach space equipped with an admissible system of matrix norms over $\mathcal{W}$, $\Omega \subseteq \mathcal{V}_{\text{nc}}$ is a finitely open nc set, and $f$ is G-differentiable on $\Omega$, then $f_{\text{d.s.e.}}$ is G-differentiable on $\Omega_{\text{d.s.e.}}$; moreover, if $X \in (\Omega_{\text{d.s.e.}})_n$, $Z \in \mathcal{V}^{n\times n}$, and $X \oplus Y \in \Omega_{n+m}$, then*
$$\delta f(X \oplus Y)(Z \oplus 0_{m\times m}) = \delta f_{\text{d.s.e.}}(X)(Z) \oplus f_{\text{d.s.e.}}(Y).$$
(4) *If $\mathcal{M} = \mathcal{V}$ and $\mathcal{N} = \mathcal{W}$ are Banach spaces equipped with admissible systems of matrix norms over $\mathcal{V}$ and over $\mathcal{W}$, $\Omega \subseteq \mathcal{V}_{\text{nc}}$ is an open nc set, and $f$ is analytic on $\Omega$, then $f_{\text{d.s.e.}}$ is analytic on $\Omega_{\text{d.s.e.}}$.*
(5) *If $\mathcal{M} = \mathcal{V}$ and $\mathcal{N} = \mathcal{W}$ are operator spaces, $\Omega \subseteq \mathcal{V}_{\text{nc}}$ is a uniformly-open nc set, and $f$ is uniformly analytic on $\Omega$, then $f_{\text{d.s.e.}}$ is uniformly analytic on $\Omega_{\text{d.s.e.}}$.*

PROOF. (1) Let $X \in (\Omega_{\text{d.s.e.}})_n$. Then there exist $m \in \mathbb{N}$ and $Y \in \mathcal{M}^{m\times m}$ such that $X \oplus Y \in \Omega_{n+m}$. The matrix $\mu I_n \oplus \nu I_m \in \mathcal{R}^{(n+m)\times(n+m)}$, with arbitrary distinct $\mu, \nu \in \mathcal{R}$, commutes with $X \oplus Y$. Hence, $\mu I_n \oplus \nu I_m$ commutes with $f(X \oplus Y)$. This is possible only if $f(X \oplus Y) = A \oplus B$, with some $A \in \mathcal{N}^{n\times n}$ and $B \in \mathcal{N}^{m\times m}$. We will show now that $A$ is independent of $Y$, i.e., is determined by $f$ and $X$ only. Suppose $X \oplus Y' \in \Omega_{n+m'}$ for some $m' \in \mathbb{N}$ and $Y' \in \mathcal{M}^{m'\times m'}$. Then, using the same argument as above, we conclude that $f(X \oplus Y') = A' \oplus B'$, for some $A' \in \mathcal{N}^{n\times n}$ and $B' \in \mathcal{N}^{m'\times m'}$. Since
$$(X \oplus Y)(I_n \oplus 0_{m\times m'}) = (I_n \oplus 0_{m\times m'})(X \oplus Y'),$$
we must have
$$(A \oplus B)(I_n \oplus 0_{m\times m'}) = (I_n \oplus 0_{m\times m'})(A' \oplus B'),$$
so that $A = A'$. We define $f_{\text{d.s.e.}}(X) := A$. On the other hand, since $\Omega$ is similarity invariant, we have that $Y \in (\Omega_{\text{d.s.e.}})_m$. An analogous argument then shows that $B$ is uniquely determined by $f$ and $Y$, and we define $f_{\text{d.s.e.}}(Y) := B$.

The mapping $f_{\text{d.s.e.}}\colon \Omega_{\text{d.s.e.}} \to N_{\text{nc}}$ is correctly defined. Indeed, by the similarity invariance of $\Omega$, if $X \oplus Y \in \Omega$ and $f(X \oplus Y) = A \oplus B$, then $Y \oplus X \in \Omega$ and $f(Y \oplus X) = B \oplus A$.

Clearly, $f_{\text{d.s.e.}}((\Omega_{\text{d.s.e.}})_n) \subseteq \mathcal{N}^{n\times n}$ for every $n \in \mathbb{N}$. Since, by Proposition 9.1, $\Omega_{\text{d.s.e.}}$ is a nc set, we have for arbitrary $X \in (\Omega_{\text{d.s.e.}})_n$ and $X' \in (\Omega_{\text{d.s.e.}})_{n'}$

that $X \oplus X' \in (\Omega_{\rm d.s.e.})_{n+n'}$ and $X \oplus X' \oplus Y \in \Omega_{n+n'+m}$, with some $m \in \mathbb{N}$ and $Y \in \mathcal{M}^{m\times m}$. Using the similarity invariance of $\Omega$, we observe that

$$f(X \oplus X' \oplus Y) = f_{\rm d.s.e.}(X) \oplus f_{\rm d.s.e.}(X') \oplus f_{\rm d.s.e.}(Y) = f_{\rm d.s.e.}(X \oplus X') \oplus f_{\rm d.s.e.}(Y),$$

thus $f_{\rm d.s.e.}(X \oplus X') = f_{\rm d.s.e.}(X) \oplus f_{\rm d.s.e.}(X')$, i.e., $f_{\rm d.s.e.}$ respects direct sums.

If $X \in (\Omega_{\rm d.s.e.})_n$ and $S \in \mathcal{R}^{n\times n}$ is invertible, then, by Proposition 9.1, $SXS^{-1} \in (\Omega_{\rm d.s.e.})_n$. We have $X \oplus Y \in \Omega_{n+m}$ for some $m \in \mathbb{N}$ and $Y \in \mathcal{M}^{m\times m}$. Since $\Omega$ is similarity invariant, we also have that

$$SXS^{-1} \oplus Y = (S \oplus I_m)(X \oplus Y)(S \oplus I_m)^{-1} \in \Omega_{n+m}.$$

Then

$$\begin{aligned} f_{\rm d.s.e.}(SXS^{-1}) \oplus f_{\rm d.s.e.}(Y) &= f(SXS^{-1} \oplus Y) \\ &= f\Big((S \oplus I_m)(X \oplus Y)(S \oplus I_m)^{-1}\Big) = (S \oplus I_m) f(X \oplus Y)(S \oplus I_m)^{-1} \\ &= (S \oplus I_m)(f_{\rm d.s.e.}(X) \oplus f_{\rm d.s.e.}(Y))(S \oplus I_m)^{-1} = S f_{\rm d.s.e.}(X) S^{-1} \oplus f_{\rm d.s.e.}(Y), \end{aligned}$$

thus $f_{\rm d.s.e.}(SXS^{-1}) = S f_{\rm d.s.e.}(X) S^{-1}$, i.e., $f_{\rm d.s.e.}$ respects similarities.

We conclude that $f_{\rm d.s.e.}$ is a nc function. By the construction, $f_{\rm d.s.e.}$ extends $f$ to $\Omega_{\rm d.s.e.}$ and such an extended nc function is unique.

(2) It is obvious that $\Omega_{\rm d.s.e.} = \mathcal{M}_{\rm nc}$. Let $X \in \mathcal{M}^{n\times n}$. Then $X \oplus Y \in \mathcal{M}^{sm\times sm}$ for some $m \in \mathbb{N}$ and $Y \in \mathcal{M}^{(sm-n)\times(sm-n)}$. Since $f|_{\mathcal{M}^{sm\times sm}}$ is polynomial on slices of some finite degree $M_{sm}$, one has that, for an arbitrary $Z \in \mathcal{M}^{n\times n}$,

$$f_{X\oplus Y, Z\oplus 0_{(sm-n)\times(sm-n)}}(t) = f((X + tZ) \oplus Y) = f_{\rm d.s.e.}(X + tZ) \oplus f_{\rm d.s.e.}(Y)$$

is a polynomial in $t$ of degree at most $M_{sm}$. Then so is $(f_{\rm d.s.e.})_{X,Z}(t) = f_{\rm d.s.e.}(X + tZ)$, and therefore, $f_{\rm d.s.e.}$ is polynomial on slices. If, in addition, the degrees $M_{sm}$ of $f|_{\mathcal{M}^{sm\times sm}}$, $m = 1, 2, \ldots$, are bounded, then the degrees of $f_{\rm d.s.e.}|_{\mathcal{M}^{n\times n}}$, $n = 1, 2, \ldots$, are bounded by the same constant. By Theorem 6.8, $f|_{\rm d.s.e.}$ is a nc polynomial over $\mathcal{M}$ with coefficients in $\mathcal{N}$.

(3) Let $X \in (\Omega_{\rm d.s.e.})_n$ and $Z \in \mathcal{V}^{n\times n}$. Then $X \oplus Y \in \Omega_{n+m}$ for some $m \in \mathbb{N}$ and $Y \in \mathcal{V}^{m\times m}$. Since $\Omega$ is finitely open, so is $\Omega_{\rm d.s.e.}$ by Proposition 9.1. Then both $X + tZ \in (\Omega_{\rm d.s.e.})_n$ and $(X + tZ) \oplus Y \in \Omega_{n+m}$ for a sufficiently small scalar $t$. Since $f$ is G-differentiable at $X \oplus Y$, we have

$$\begin{aligned} \delta f(X \oplus Y)(Z \oplus 0_{m\times m}) &= \lim_{t\to 0} \frac{f((X \oplus Y) + t(Z \oplus 0_{m\times m})) - f(X \oplus Y)}{t} \\ &= \lim_{t\to 0} \frac{f((X + tZ) \oplus Y) - f(X \oplus Y)}{t} \\ &= \lim_{t\to 0} \frac{(f_{\rm d.s.e.}(X + tZ) \oplus f_{\rm d.s.e.}(Y)) - (f_{\rm d.s.e.}(X) \oplus f_{\rm d.s.e.}(Y))}{t} \\ &= \lim_{t\to 0} \frac{(f_{\rm d.s.e.}(X + tZ) - f_{\rm d.s.e.}(X)) \oplus f_{\rm d.s.e.}(Y)}{t} \\ &= \lim_{t\to 0} \frac{f_{\rm d.s.e.}(X + tZ) - f_{\rm d.s.e.}(X)}{t} \oplus f_{\rm d.s.e.}(Y) \\ &= \delta f_{\rm d.s.e.}(X)(Z) \oplus f_{\rm d.s.e.}(Y), \end{aligned}$$

i.e., $\delta f_{\rm d.s.e.}(X)(Z)$ exists and is determined by $\delta f(X \oplus Y)(Z \oplus 0_{m\times m})$. Clearly, $\delta f_{\rm d.s.e.}(X)(Z)$ is independent of the choice of $Y$, since it is unique as a limit. We conclude that $f_{\rm d.s.e.}$ is G-differentiable on $\Omega_{\rm d.s.e.}$.

(4) It suffices to show, either by the result of part (4) and the definition of an analytic nc function or by Theorem 7.4, that $f_{\rm d.s.e.}$ is locally bounded on $\Omega_{\rm d.s.e.}$. Let $X \in (\Omega_{\rm d.s.e.})_n$. Then $X \oplus Y \in \Omega_{n+m}$ for some $m \in \mathbb{N}$ and $Y \in \mathcal{V}^{m\times m}$. Since $\Omega_{n+m}$ is open and $f$ is locally bounded on $\Omega_{n+m}$, there exists a constant $K > 0$ such that $\|f(W)\|_{n+m} \le K$ for every $W$ in some open ball $B(X \oplus Y, \epsilon) \subseteq \Omega_{n+m}$. By Proposition 9.1, $(\Omega_{\rm d.s.e.})_n$ is open. Let $\delta < C_1'(n,m)^{-1}\epsilon$ be such that $B(X,\delta) \subseteq (\Omega_{\rm d.s.e.})_n$, where $C_1'(n,m) = C_1'(\mathcal{V};n,m)$ is the constant on the right-hand side of (7.1) for $\mathcal{V}$. Let $Z \in B(X,\delta)$ be arbitrary. Then $Z \oplus Y \in B(X \oplus Y,\epsilon)$ and

$$\begin{aligned}\|f_{\rm d.s.e.}(Z)\|_n &\le \max\{\|f_{\rm d.s.e.}(Z)\|_n, \|f_{\rm d.s.e.}(Y)\|_m\}\\ &\le C_1(n,m)\|f_{\rm d.s.e.}(Z) \oplus f_{\rm d.s.e.}(Y)\|_{n+m}\\ &= C_1(n,m)\|f(Z \oplus Y)\|_{n+m} \le C_1(n,m)K,\end{aligned}$$

where $C_1(n,m) = C_1(\mathcal{W};n,m)$ is the constant on the left-hand side of (7.1) for $\mathcal{W}$. Thus, $f_{\rm d.s.e.}$ is locally bounded on $\Omega_{\rm d.s.e.}$.

(5) It suffices to show, either by the result of part (4) and the definition of a uniformly analytic nc function or by Corollary 7.28, that $f_{\rm d.s.e.}$ is uniformly locally bounded on $\Omega_{\rm d.s.e.}$. Let $X \in (\Omega_{\rm d.s.e.})_n$. Then $X \oplus Y \in \Omega_{n+m}$ for some $m \in \mathbb{N}$ and $Y \in \mathcal{V}^{m\times m}$. Since $\Omega$ is uniformly-open and $f$ is uniformly locally bounded on $\Omega$, there exists a constant $K > 0$ such that $\|f(W)\|_{n_W} \le K$ for every $W$ in some open nc ball $B_{\rm nc}(X \oplus Y, \epsilon) \subseteq \Omega$ (here $n_W$ is the size of the matrix $W$). By Proposition 9.1, $\Omega_{\rm d.s.e.}$ is uniformly-open. Moreover, the argument in the proof of Proposition 9.1 implies that $B_{\rm nc}(X,\epsilon) \subseteq \Omega_{\rm d.s.e.}$: for an arbitrary $Z \in B_{\rm nc}(X,\epsilon)_{nk}$, one has (9.3) with the permutation matrix $U$ defined by (9.2). By the properties of operator space norms,

$$\begin{aligned}\|f_{\rm d.s.e.}(Z)\|_{nk} &\le \Big\|f_{\rm d.s.e.}(Z) \oplus \bigoplus_{\beta=1}^{k} f_{\rm d.s.e.}(Y)\Big\|_{nk} = \Big\|f\Big(Z \oplus \bigoplus_{\beta=1}^{k} Y\Big)\Big\|_{nk}\\ &= \Big\|U^{-1} f\Big(Z \oplus \bigoplus_{\beta=1}^{k} Y\Big)U\Big\|_{nk} = \Big\|f\Big(U^{-1}\Big(Z \oplus \bigoplus_{\beta=1}^{k} Y\Big)U\Big)\Big\|_{nk} \le K.\end{aligned}$$

We conclude that $f_{\rm d.s.e.}$ is uniformly locally bounded on $\Omega_{\rm d.s.e.}$. □

Proposition 9.2 can be extended to higher order nc functions as follows.

PROPOSITION 9.3. *Let $\mathcal{M}_0, \ldots, \mathcal{M}_k, \mathcal{N}_0, \ldots, \mathcal{N}_k$ be modules over a commutative unital ring $\mathcal{R}$, let $\Omega^{(0)} \subseteq \mathcal{M}_{0,\rm nc}, \ldots, \Omega^{(k)} \subseteq \mathcal{M}_{k,\rm nc}$ be similarity invariant nc sets, and let $f \in \mathcal{T}^k(\Omega^{(0)},\ldots,\Omega^{(k)};\mathcal{N}_{0,\rm nc},\ldots,\mathcal{N}_{k,\rm nc})$. Then:*

(1) *There exists a uniquely determined* direct summands extension *of $f$, i.e., a nc function of order $k$, $f_{\rm d.s.e.} \in \mathcal{T}^k(\Omega^{(0)}_{\rm d.s.e.},\ldots,\Omega^{(k)}_{\rm d.s.e.};\mathcal{N}_{0,\rm nc},\ldots,\mathcal{N}_{k,\rm nc})$, such that $f_{\rm d.s.e.}|_{\Omega^{(0)}\times\cdots\times\Omega^{(k)}} = f$.*

(2) *If $\mathcal{M}_0 = \mathcal{V}_0, \ldots, \mathcal{M}_k = \mathcal{V}_k$ are vector spaces over $\mathbb{C}$, $\mathcal{N}_0 = \mathcal{W}_0, \ldots, \mathcal{N}_k = \mathcal{W}_k$ are Banach spaces equipped with admissible systems of rectangular matrix norms over $\mathcal{W}_0, \ldots, \mathcal{W}_k$, $\Omega^{(0)} \subseteq \mathcal{V}_{0,\rm nc}, \ldots, \Omega^{(k)} \subseteq \mathcal{V}_{k,\rm nc}$ are finitely open nc sets, and $f$ is $G_W$-differentiable on $\Omega^{(0)}\times\cdots\times\Omega^{(k)}$, then $f_{\rm d.s.e.}$ is $G_W$-differentiable on $\Omega^{(0)}_{\rm d.s.e.}\times\cdots\times\Omega^{(k)}_{\rm d.s.e.}$; moreover, if $X^0 \in (\Omega^{(0)}_{\rm d.s.e.})_{n_0}, \ldots, X^k \in (\Omega^{(k)}_{\rm d.s.e.})_{n_k}$, $Z^0 \in {\mathcal{V}_0}^{n_0\times n_0}, \ldots, Z^k \in {\mathcal{V}_k}^{n_k\times n_k}$,*

$W^1 \in \mathcal{W}_1{}^{n_0\times n_1}$, ..., $W^k \in \mathcal{W}_k{}^{n_{k-1}\times n_k}$, *and* $X^0 \oplus Y^0 \in \Omega^{(0)}_{n_0+m_0}$, ..., $X^k \oplus Y^k \in \Omega^{(k)}_{n_k+m_k}$, *then*

$$\frac{d}{dt} f\Big((X^0 \oplus Y^0) + t(Z^0 \oplus 0_{m_0\times m_0}), \dots, (X^k \oplus Y^k) + t(Z^k \oplus 0_{m_k\times m_k})\Big)$$
$$(W^1 \oplus 0_{m_0\times m_1}, \dots, W^k \oplus 0_{m_{k-1}\times m_k})\Big|_{t=0}$$
$$= \frac{d}{dt} f_{\rm d.s.e.}(X^0 + tZ^0, \dots, X^k + tZ^k)(W^1, \dots, W^k)\Big|_{t=0} \oplus 0_{m_0\times m_k}.$$

(3) *If* $\mathcal{M}_0 = \mathcal{V}_0$, ..., $\mathcal{M}_k = \mathcal{V}_k$, $\mathcal{N}_0 = \mathcal{W}_0$, ..., $\mathcal{N}_k = \mathcal{W}_k$ *are Banach spaces equipped with admissible systems of rectangular matrix norms over* $\mathcal{V}_0$, ..., $\mathcal{V}_k$, $\mathcal{W}_0$, ..., $\mathcal{W}_k$, $\Omega^{(0)} \subseteq \mathcal{V}_{0,\rm nc}$, ..., $\Omega^{(k)} \subseteq \mathcal{V}_{k,\rm nc}$ *are open nc sets, and* $f$ *is analytic on* $\Omega^{(0)} \times \cdots \times \Omega^{(k)}$, *then* $f_{\rm d.s.e.}$ *is analytic on* $\Omega^{(0)}_{\rm d.s.e.} \times \cdots \times \Omega^{(k)}_{\rm d.s.e.}$.

(4) *If* $\mathcal{M}_0 = \mathcal{V}_0$, ..., $\mathcal{M}_k = \mathcal{V}_k$, $\mathcal{N}_0 = \mathcal{W}_0$, ..., $\mathcal{N}_k = \mathcal{W}_k$ *are operator spaces,* $\Omega^{(0)} \subseteq \mathcal{V}_{0,\rm nc}$, ..., $\Omega^{(k)} \subseteq \mathcal{V}_{k,\rm nc}$ *are uniformly-open nc sets, and* $f$ *is uniformly analytic on* $\Omega^{(0)} \times \cdots \times \Omega^{(k)}$, *then* $f_{\rm d.s.e.}$ *is uniformly analytic on* $\Omega^{(0)}_{\rm d.s.e.} \times \cdots \times \Omega^{(k)}_{\rm d.s.e.}$.

We omit the proof which is essentially similar to that of Proposition 9.2.

Our next goal is to define the direct summands extension of a sequence of $\ell$-linear mappings $f_\ell\colon (\mathcal{M}^{s\times s})^\ell \to \mathcal{N}^{s\times s}$, $\ell = 0, 1, \dots$ satisfying conditions (4.4)–(4.7).

PROPOSITION 9.4. *Let* $\mathcal{M}$ *and* $\mathcal{N}$ *be modules over a commutative unital ring* $\mathcal{R}$, *let* $s_1$, ..., $s_m \in \mathbb{N}$ *be such that* $s = s_1 + \cdots + s_m$, *let* $Y = Y_{11} \oplus \cdots \oplus Y_{mm} \in \mathcal{M}^{s\times s}$, *with* $Y_{ii} \in \mathcal{M}^{s_i\times s_i}$, $i = 1, \dots, m$, *and let a sequence of* $\ell$-*linear mappings* $f_\ell\colon (\mathcal{M}^{s\times s})^\ell \to \mathcal{N}^{s\times s}$, $\ell = 0, 1, \dots$ *satisfy* (4.4)–(4.7). *Then:*

(1) *With respect to the corresponding block decomposition of* $s \times s$ *matrices,*

$$f^{\rm d.s.e.}_{0;\alpha,\beta} := (f_0)_{\alpha\beta} \in \mathcal{N}^{s_\alpha\times s_\beta}, \quad \alpha, \beta = 1, \dots, m, \tag{9.4}$$

*satisfy* $f^{\rm d.s.e.}_{0;\alpha,\beta} = 0$ *if* $\alpha \neq \beta$,

$$f_\ell(Z^1, \dots, Z^\ell)_{\alpha\beta} = \sum_{1\le\alpha_1,\dots,\alpha_{\ell-1}\le m} f^{\rm d.s.e.}_{\ell;\alpha_0,\dots,\alpha_\ell}(Z^1_{\alpha_0\alpha_1}, \dots, Z^\ell_{\alpha_{\ell-1}\alpha_\ell}), \tag{9.5}$$
$$\text{where } \alpha_0 = \alpha,\ \alpha_\ell = \beta,\ \alpha, \beta = 1, \dots, m,\ \ell = 1, 2, \dots,$$

*with uniquely determined* $\ell$-*linear mappings*

$$f^{\rm d.s.e.}_{\ell;\alpha_0,\dots,\alpha_\ell}\colon \mathcal{M}^{s_{\alpha_0}\times s_{\alpha_1}} \times \cdots \times \mathcal{M}^{s_{\alpha_{\ell-1}}\times s_{\alpha_\ell}} \to \mathcal{N}^{s_{\alpha_0}\times s_{\alpha_\ell}}$$

*satisfying*

$$Sf^{\rm d.s.e.}_{0;\beta,\beta} - f^{\rm d.s.e.}_{0;\alpha,\alpha}S = f^{\rm d.s.e.}_{1;\alpha,\beta}(SY_{\beta\beta} - Y_{\alpha\alpha}S), \quad S \in \mathcal{R}^{s_\alpha\times s_\beta}, \tag{9.6}$$

*and for* $\ell = 1, 2, \dots$,

$$Sf^{\rm d.s.e.}_{\ell;\gamma,\alpha_1,\dots,\alpha_\ell}(W^1, \dots, W^\ell) - f^{\rm d.s.e.}_{\ell;\alpha_0,\dots,\alpha_\ell}(SW^1, W^2, \dots, W^\ell) \tag{9.7}$$
$$= f^{\rm d.s.e.}_{\ell+1;\alpha_0,\gamma,\alpha_1,\dots,\alpha_\ell}(SY_{\gamma\gamma} - Y_{\alpha_0\alpha_0}S, W^1, \dots, W^\ell), \quad S \in \mathcal{R}^{s_{\alpha_0}\times\gamma},$$

$$(9.8)\quad f^{\text{d.s.e.}}_{\ell;\alpha_0,\ldots,\alpha_{j-1},\gamma,\alpha_{j+1},\ldots,\alpha_\ell}(W^1,\ldots,W^{j-1},W^jS,W^{j+1},\ldots,W^\ell) - f^{\text{d.s.e.}}_{\ell;\alpha_0,\ldots,\alpha_\ell}(W^1,\ldots,W^j,SW^{j+1},W^{j+2},\ldots,W^\ell) = f^{\text{d.s.e.}}_{\ell+1;\alpha_0,\ldots,\alpha_j,\gamma,\alpha_{j+1},\ldots,\alpha_\ell}(W^1,\ldots,W^j,SY_{\gamma\gamma}-Y_{\alpha_j\alpha_j}S,W^{j+1},\ldots,W^\ell),\quad S\in\mathcal{R}^{s_{\alpha_j}\times\gamma},$$

$$(9.9)\quad f^{\text{d.s.e.}}_{\ell;\alpha_0,\ldots,\alpha_{\ell-1},\gamma}(W^1,\ldots,W^{\ell-1},W^\ell S) - f^{\text{d.s.e.}}_{\ell;\alpha_0,\ldots,\alpha_\ell}(W^1,\ldots,W^\ell)S = f^{\text{d.s.e.}}_{\ell+1;\alpha_0,\ldots,\alpha_\ell,\gamma}(W^1,\ldots,W^\ell,SY_{\gamma\gamma}-Y_{\alpha_\ell\alpha_\ell}S),\quad S\in\mathcal{R}^{\alpha_\ell\times\gamma}.$$

*In particular, conditions (9.6)–(9.9) for* $f^{\text{d.s.e.}}_{\ell;\alpha,\ldots,\alpha}$, $\ell=0,1,\ldots$, *with a fixed* $\alpha$, *coincide with (4.4)–(4.7) for* $s$ *replaced by* $s_\alpha$ *and* $Y$ *replaced by* $Y_{\alpha\alpha}$.

(2) *If* $s_1=\cdots=s_m$ *and* $Y_{11}=\cdots=Y_{mm}$, *then the mappings* $f^{\text{d.s.e.}}_{\ell;\alpha_0,\ldots,\alpha_\ell}$, $1\le\alpha_0,\ldots,\alpha_\ell\le m$, *coincide for every fixed* $\ell$.

(3) *If* $\mathcal{R}=\mathbb{C}$, $\mathcal{M}=\mathcal{V}$ *and* $\mathcal{N}=\mathcal{W}$ *are Banach spaces equipped with admissible systems of rectangular matrix norms over* $\mathcal{V}$ *and* $\mathcal{W}$, *and the* $\ell$*-linear mappings* $f_\ell$ *are bounded, then so are* $f^{\text{d.s.e.}}_{\ell;\alpha_0,\ldots,\alpha_\ell}$.

(4) *If* $\mathcal{R}=\mathbb{C}$, $\mathcal{M}=\mathcal{V}$ *and* $\mathcal{N}=\mathcal{W}$ *are Banach spaces equipped with admissible systems of rectangular matrix norms over* $\mathcal{V}$ *and* $\mathcal{W}$, *(7.39) hold for* $\mathcal{V}$ *and* $\mathcal{W}$ *with the constants* $C^{\mathcal{V}}$ *and* $C^{\mathcal{W}}$ *which are independent of* $n,p,q$, *and* $m$, *and the* $\ell$*-linear mappings* $f_\ell$ *are completely bounded, then so are* $f^{\text{d.s.e.}}_{\ell;\alpha_0,\ldots,\alpha_\ell}$, *and*

$$(9.10)\qquad \|f^{\text{d.s.e.}}_{\ell;\alpha_0,\ldots,\alpha_\ell}\|_{\mathcal{L}^\ell_{\text{cb}}}\le C^{\mathcal{W}}(C^{\mathcal{V}})^\ell\|f_\ell\|_{\mathcal{L}^\ell_{\text{cb}}};$$

*in particular, if* $\mathcal{V}$ *and* $\mathcal{W}$ *are operator spaces, then*

$$(9.11)\qquad \|f^{\text{d.s.e.}}_{\ell;\alpha_0,\ldots,\alpha_\ell}\|_{\mathcal{L}^\ell_{\text{cb}}}\le\|f_\ell\|_{\mathcal{L}^\ell_{\text{cb}}}.$$

PROOF. (1) Let $S=E_\alpha E_\alpha^\top\in\mathcal{R}^{s\times s}$. Since $SY=YS$, we obtain from (4.4) that $(Sf_0)_{\alpha\beta}=(f_0S)_{\alpha\beta}$, i.e., $(f_0)_{\alpha\beta}=(f_0)_{\alpha\alpha}\delta_{\alpha\beta}$, hence $(f_0)_{\alpha\beta}=0$ if $\alpha\ne\beta$.

For any $\alpha,\beta=1,\ldots,m$, and for an arbitrary $S\in\mathcal{R}^{s_\alpha\times s_\beta}$, we set $T=E_\alpha SE_\beta^\top$. Define the linear mapping $f^{\text{d.s.e.}}_{1;\alpha,\beta}\colon\mathcal{M}^{s_\alpha\times s_\beta}\to\mathcal{N}^{s_\alpha\times s_\beta}$ by

$$f^{\text{d.s.e.}}_{1;\alpha,\beta}(W):=f_1(E_\alpha WE_\beta^\top)_{\alpha\beta}.$$

Then (4.4) with $T$ in the place of $S$ implies

$$Sf^{\text{d.s.e.}}_{0;\beta,\beta}-f^{\text{d.s.e.}}_{0;\alpha,\alpha}S=S(f_0)_{\beta\beta}-(f_0)_{\alpha\alpha}S=(Tf_0-f_0T)_{\alpha\beta}=f_1(TY-YT)_{\alpha\beta}$$
$$=f_1(E_\alpha(SY_{\beta\beta}-Y_{\alpha\alpha}S)E_\beta^\top)_{\alpha\beta}=f^{\text{d.s.e.}}_{1;\alpha,\beta}(SY_{\beta\beta}-Y_{\alpha\alpha}S),$$

i.e., (9.6) holds. Similarly, for every $\ell=2,3,\ldots$, $\alpha_0,\ldots,\alpha_\ell=1,\ldots,m$, one defines the $\ell$-linear mapping

$$f^{\text{d.s.e.}}_{\ell;\alpha_0,\ldots,\alpha_\ell}\colon\mathcal{M}^{\alpha_0\times\alpha_1}\times\cdots\times\mathcal{M}^{\alpha_{\ell-1}\times\alpha_\ell}\to\mathcal{N}^{\alpha_0\times\alpha_\ell}$$

by

$$(9.12)\qquad f^{\text{d.s.e.}}_{\ell;\alpha_0,\ldots,\alpha_\ell}(W^1,\ldots,W^\ell):=f_\ell(E_{\alpha_0}W^1E_{\alpha_1}^\top,\ldots,E_{\alpha_{\ell-1}}W^\ell E_{\alpha_\ell}^\top)_{\alpha_0\alpha_\ell}$$

and checks that (9.7)–(9.9) hold.

Let $1 \leq \alpha, \beta \leq m$ be arbitrary. Since $E_i E_i^\top$ commutes with $Y$ for every $i = 1, \ldots, m$, it follows by linearity from (4.5)–(4.7) that

$$\begin{aligned}
f_\ell(Z^1, \ldots, Z^\ell)_{\alpha\beta} &= f_\ell\Big( \sum_{1 \leq \gamma_0, \alpha_1 \leq m} E_{\gamma_0} Z^1_{\gamma_0 \alpha_1} E_{\alpha_1}^\top, \ldots, \\
&\quad \sum_{1 \leq \gamma_{\ell-2}, \alpha_{\ell-1} \leq m} E_{\gamma_{\ell-2}} Z^{\ell-1}_{\gamma_{\ell-2}\alpha_{\ell-1}} E^\top_{\alpha_{\ell-1}}, \sum_{1 \leq \gamma_{\ell-1}, \gamma_\ell \leq m} E_{\gamma_{\ell-1}} Z^\ell_{\gamma_{\ell-1}\gamma_\ell} E^\top_{\gamma_\ell} \Big)_{\alpha\beta} \\
&= \Bigg( E_\alpha E_\alpha^\top f_\ell\Big( \sum_{1 \leq \gamma_0, \alpha_1 \leq m} E_{\gamma_0} Z^1_{\gamma_0\alpha_1} E^\top_{\alpha_1} E_{\alpha_1} E^\top_{\alpha_1}, \ldots, \\
&\quad \sum_{1 \leq \gamma_{\ell-2}, \alpha_{\ell-1} \leq m} E_{\gamma_{\ell-2}} Z^{\ell-1}_{\gamma_{\ell-2}\alpha_{\ell-1}} E^\top_{\alpha_{\ell-1}} E_{\alpha_{\ell-1}} E^\top_{\alpha_{\ell-1}}, \\
&\quad \sum_{1 \leq \gamma_{\ell-1}, \gamma_\ell \leq m} E_{\gamma_{\ell-1}} Z^\ell_{\gamma_{\ell-1}\gamma_\ell} E^\top_{\gamma_\ell} \Big) E_\beta E_\beta^\top \Bigg)_{\alpha\beta} \\
&= f_\ell\Big( E_\alpha E_\alpha^\top \sum_{1 \leq \gamma_0, \alpha_1 \leq m} E_{\gamma_0} Z^1_{\gamma_0\alpha_1} E^\top_{\alpha_1}, \, E_{\alpha_1} E^\top_{\alpha_1} \sum_{1 \leq \gamma_1, \alpha_2 \leq m} E_{\gamma_1} Z^1_{\gamma_1\alpha_2} E^\top_{\alpha_2}, \ldots, \\
&\quad E_{\alpha_{\ell-2}} E^\top_{\alpha_{\ell-2}} \sum_{1 \leq \gamma_{\ell-2}, \alpha_{\ell-1} \leq m} E_{\gamma_{\ell-2}} Z^{\ell-1}_{\gamma_{\ell-2}\alpha_{\ell-1}} E^\top_{\alpha_{\ell-1}}, \\
&\quad E_{\alpha_{\ell-1}} E^\top_{\alpha_{\ell-1}} \sum_{1 \leq \gamma_{\ell-1}, \gamma_\ell \leq m} E_{\gamma_{\ell-1}} Z^\ell_{\gamma_{\ell-1}\gamma_\ell} E^\top_{\gamma_\ell} E_\beta E_\beta^\top \Big)_{\alpha\beta} \\
&= \sum_{1 \leq \alpha_1, \ldots, \alpha_{\ell-1} \leq m} f_\ell(E_\alpha Z^1_{\alpha\alpha_1} E^\top_{\alpha_1}, E_{\alpha_1} Z^2_{\alpha_1\alpha_2} E^\top_{\alpha_2}, \ldots, \\
&\quad E_{\alpha_{\ell-2}} Z^{\ell-1}_{\alpha_{\ell-2}\alpha_{\ell-1}} E^\top_{\alpha_{\ell-1}}, E_{\alpha_{\ell-1}} Z^\ell_{\alpha_{\ell-1}\beta} E^\top_\beta)_{\alpha\beta} \\
&= \sum_{1 \leq \alpha_1, \ldots, \alpha_{\ell-1} \leq m} f^{\text{d.s.e.}}_{\ell;\alpha,\alpha_1,\ldots,\alpha_{\ell-1},\alpha_\ell}(Z^1_{\alpha\alpha_1}, Z^2_{\alpha_1\alpha_2}, \ldots, Z^{\ell-1}_{\alpha_{\ell-2}\alpha_{\ell-1}}, Z^\ell_{\alpha_{\ell-1}\beta}),
\end{aligned}$$

i.e., (9.5) holds. On the other hand, if (9.5) is satisfied with some $\ell$-linear mappings

$$f^{\text{d.s.e.}}_{\ell;\alpha_0,\ldots,\alpha_\ell} \colon \mathcal{M}^{s_{\alpha_0} \times s_{\alpha_1}} \times \cdots \times \mathcal{M}^{s_{\alpha_{\ell-1}} \times s_{\alpha_\ell}} \to \mathcal{N}^{s_{\alpha_0} \times s_{\alpha_\ell}},$$

then

$$\begin{aligned}
f_\ell(E_{\alpha_0} Z^1_{\alpha_0\alpha_1} E^\top_{\alpha_1}, \ldots, E_{\alpha_{\ell-1}} Z^\ell_{\alpha_{\ell-1}\alpha_\ell} E^\top_{\alpha_\ell})_{\alpha_0\alpha_\ell} \\
= f^{\text{d.s.e.}}_{\ell;\alpha_0,\ldots,\alpha_\ell}\Big( (E_{\alpha_0} Z^1_{\alpha_0\alpha_1} E^\top_{\alpha_1})_{\alpha_0\alpha_1}, \ldots, (E_{\alpha_{\ell-1}} Z^\ell_{\alpha_{\ell-1}\alpha_\ell} E^\top_{\alpha_\ell})_{\alpha_{\ell-1}\alpha_\ell} \Big) \\
= f^{\text{d.s.e.}}_{\ell;\alpha_0,\ldots,\alpha_\ell}(Z^1_{\alpha_0\alpha_1}, \ldots, Z^\ell_{\alpha_{\ell-1}\alpha_\ell}),
\end{aligned}$$

which coincides with our original definition of $f^{\text{d.s.e.}}_{\ell;\alpha_0,\ldots,\alpha_\ell}$ as in (9.12).

(2) follows from (9.6)–(9.9) with $S = I_{s/m}$.

(3) By (9.12) and (7.43), we have for $W^1 \in \mathcal{V}^{s_{\alpha_0} \times s_{\alpha_1}}, \ldots, W^\ell \in \mathcal{V}^{s_{\alpha_{\ell-1}} \times s_{\alpha_\ell}}$:

$$
\begin{aligned}
\|f^{\text{d.s.e.}}_{\ell;\alpha_0,\ldots,\alpha_\ell}&(W^1,\ldots,W^\ell)\|_{s_{\alpha_0},s_{\alpha_1}} \\
&= \|\pi^{(s_1,\ldots,s_m),\mathcal{W}}_{\alpha_0,\alpha_\ell} f^\ell(E_{\alpha_0}W^1E^\top_{\alpha_1},\ldots,E_{\alpha_{\ell-1}}W^\ell E^\top_{\alpha_\ell})\|_{s_{\alpha_0},s_{\alpha_1}} \\
&\le C^{\mathcal{W}}(s_{\alpha_0},s,s,s_{\alpha_\ell})\|f_\ell\|\|E_{\alpha_0}W^1E^\top_{\alpha_1}\|_s\cdots\|E_{\alpha_{\ell-1}}W^\ell E^\top_{\alpha_\ell}|_s \\
&\le C^{\mathcal{W}}(s_{\alpha_0},s,s,s_{\alpha_\ell})C^{\mathcal{V}}(s,s_{\alpha_0},s_{\alpha_1},s)\cdots C^{\mathcal{V}}(s,s_{\alpha_0},s_{\alpha_1},s) \\
&\qquad\qquad \cdot \|f_\ell\|\,\|W^1\|_{s_{\alpha_0},s_{\alpha_1}}\cdots\|W^\ell\|_{s_{\alpha_{\ell-1}},s_{\alpha_\ell}},
\end{aligned}
$$

where we indicate the corresponding Banach space in the superscript, for the corresponding rectangular block projections and constants. Thus the $\ell$-linear mappings $f^{\text{d.s.e.}}_{\ell;\alpha_0,\ldots,\alpha_\ell}$ are bounded.

(4) By (9.12) and (7.43), we have for every $n_0, \ldots, n_\ell \in \mathbb{N}$ and for every $W^1 \in (\mathcal{V}^{s_{\alpha_0}\times s_{\alpha_1}})^{n_0\times n_1} \cong \mathcal{V}^{s_{\alpha_0}n_0\times s_{\alpha_1}n_1}, \ldots, W^\ell \in \left(\mathcal{V}^{s_{\alpha_{\ell-1}}\times s_{\alpha_\ell}}\right)^{n_{\ell-1}\times n_\ell} \cong \mathcal{V}^{s_{\alpha_{\ell-1}}n_{\ell-1}\times s_{\alpha_\ell}n_\ell}$:

$$
\begin{aligned}
&\|f^{\text{d.s.e.}(n_0,\ldots,n_\ell)}_{\ell;\alpha_0,\ldots,\alpha_\ell}(W^1,\ldots,W^\ell)\|_{s_{\alpha_0}n_0,s_{\alpha_1}n_1} \\
&= \|f_\ell^{(n_0,\ldots,n_\ell)}((\mathrm{id}_{\mathbb{C}^{n_0\times n_0}}\otimes E_{\alpha_0})W^1(\mathrm{id}_{\mathbb{C}^{n_1\times n_1}}\otimes E^\top_{\alpha_1}),\ldots, \\
&\qquad (\mathrm{id}_{\mathbb{C}^{n_{\ell-1}\times n_{\ell-1}}}\otimes E_{\alpha_{\ell-1}})W^\ell(\mathrm{id}_{\mathbb{C}^{n_\ell\times n_\ell}}\otimes E^\top_{\alpha_\ell}))_{s_{\alpha_0}n_0,s_{\alpha_1}n_1}\|_{s_{\alpha_0}n_0,s_{\alpha_1}n_1} \\
&\qquad \le C^{\mathcal{W}}\|f_\ell\|_{\mathcal{L}^\ell_{\text{cb}}}\|(\mathrm{id}_{\mathbb{C}^{n_0\times n_0}}\otimes E_{\alpha_0})W^1(\mathrm{id}_{\mathbb{C}^{n_1\times n_1}}\otimes E^\top_{\alpha_1})\|_{sn_0,sn_1}\cdots \\
&\qquad\qquad \cdot\|(\mathrm{id}_{\mathbb{C}^{n_{\ell-1}\times n_{\ell-1}}}\otimes E_{\alpha_{\ell-1}})W^\ell(\mathrm{id}_{\mathbb{C}^{n_\ell\times n_\ell}}\otimes E^\top_{\alpha_\ell})\|_{sn_{\ell-1},sn_\ell} \\
&\qquad\qquad \le C^{\mathcal{W}}(C^{\mathcal{V}})^\ell\|f_\ell\|_{\mathcal{L}^\ell_{\text{cb}}}\|W^1\|_{s_{\alpha_0}n_0,s_{\alpha_1}n_1}\cdots\|W^\ell\|_{s_{\alpha_{\ell-1}}n_{\ell-1},s_{\alpha_\ell}n_\ell},
\end{aligned}
$$

which implies (9.10). Clearly, in the case of operator spaces, $C^{\mathcal{V}} = C^{\mathcal{W}} = 1$, and (9.10) becomes (9.11). □

REMARK 9.5. Let $\Omega \subseteq \mathcal{M}_{\text{nc}}$ be a similarity invariant nc set, let $f\colon \Omega \to \mathcal{N}_{\text{nc}}$ be a nc function, and let $Y = \bigoplus_{i=1}^m Y_{ii} \in \Omega_s$, with $Y_{ii} \in \mathcal{M}^{s_i\times s_i}$. Then the sequence of $\ell$-linear mappings $f_\ell := \Delta_R^\ell f(Y,\ldots,Y)\colon (\mathcal{M}^{s\times s})^\ell \to \mathcal{N}^{s\times s}$, $\ell = 0,1,\ldots$, satisfies (4.4)–(4.7); see Remark 4.3. By Proposition 9.1, $\Omega_{\text{d.s.e.}} \subseteq \mathcal{M}_{\text{nc}}$ is a similarity invariant nc set extending $\Omega$, and by Proposition 9.2, there exists a nc function $f_{\text{d.s.e.}}\colon \Omega_{\text{d.s.e.}} \to \mathcal{N}_{\text{nc}}$ extending $f$. We then have (3.6) for $k = \ell$ and $X^0 = \cdots = X^\ell = Y$, with $f$ in the left-hand side replaced by $f_\ell$, and $f$ in the right-hand side replaced by $\Delta_R^\ell f_{\text{d.s.e.}}$. Comparing this equality with (9.5), we conclude that

$$
f^{\text{d.s.e.}}_{\ell;\alpha_0,\ldots,\alpha_\ell} = \Delta_R^\ell f_{\text{d.s.e.}}(Y_{\alpha_0\alpha_0},\ldots,Y_{\alpha_\ell\alpha_\ell}). \tag{9.13}
$$

On the other hand, given a sequence of $\ell$-linear mappings $f_\ell\colon (\mathcal{M}^{s\times s})^\ell \to \mathcal{N}^{s\times s}$, $\ell = 0,1,\ldots$, satisfying conditions (4.4)–(4.7) for $Y = \bigoplus_{i=1}^m Y_{ii} \in \mathcal{M}^{s\times s}$, by Theorem 5.13 one can define a nc function

$$
f(X) = \sum_{\ell=0}^\infty \Big(X - \bigoplus_{\alpha=1}^m Y\Big)^{\odot_s \ell} f_\ell \tag{9.14}
$$

on $\mathrm{Nilp}(\mathcal{M},Y)$ with values in $\mathcal{N}_{\text{nc}}$. Moreover, by Theorem 5.9, $f_\ell = \Delta_R^\ell f(Y,\ldots,Y)$. Extending $f$ to $\mathrm{Nilp}(\mathcal{M},Y)_{\text{d.s.e.}}$ and writing the TT expansion for $f_{\text{d.s.e.}}$ (which is a finite sum at every point of $\mathrm{Nilp}(\mathcal{M},Y)_{\text{d.s.e.}}$), we can define the direct summands extensions $f^{\text{d.s.e.}}_{\ell;\alpha_0,\ldots,\alpha_\ell}$ of the $\ell$-linear mappings $f_\ell$ by (9.13). This gives an alternative way of proving parts (1) and (2) of Proposition 9.4 using Proposition 9.2. Parts (3) and (4) of Proposition 9.4 can also be derived from Proposition 9.2 provided the the series in (9.14) defines an analytic (resp., uniformly analytic) nc function $f$ on some open (resp., uniformly open) neighborhood of $Y$.

In a similar way, we can define the direct summands extension of a sequence of $\ell$-linear mappings $f_w\colon (\mathcal{M}^{s\times s})^\ell \to \mathcal{N}^{s\times s}$, $w \in \mathcal{G}_d$, satisfying conditions (4.14)–(4.17). The next proposition can be proved similarly to Proposition 9.4, or can be derived from the result of Proposition 9.4 via relations (5.5), (5.7).

PROPOSITION 9.6. *Let $\mathcal{R}$ be a commutative unital ring, let $\mathcal{N}$ be a module over $\mathcal{R}$, let $s_1, \ldots, s_m \in \mathbb{N}$ be such that $s = s_1+\cdots+s_m$, let $Y = (Y_1,\ldots,Y_d) \in (\mathcal{R}^{s\times s})^d$ and $Y_k = (Y_k)_{11} \oplus \cdots \oplus (Y_k)_{mm} \in \mathcal{R}^{s\times s}$, with $(Y_k)_{ii} \in \mathcal{R}^{s_i\times s_i}$, $k = 1,\ldots,d$, $i = 1,\ldots,m$, and let a sequence of $\ell$-linear mappings $f_w\colon (\mathcal{R}^{s\times s})^\ell \to \mathcal{N}^{s\times s}$, $w \in \mathcal{G}_d\colon |w| = \ell$, $\ell = 0,1,\ldots$, satisfy (4.14)–(4.17). Then:*

(1) *With respect to the corresponding block decomposition of $s \times s$ matrices,*

$$f^{\text{d.s.e.}}_{\emptyset;\alpha,\beta} := (f_\emptyset)_{\alpha\beta} \in \mathcal{N}^{s_\alpha\times s_\beta}, \quad \alpha,\beta = 1,\ldots,m, \tag{9.15}$$

*satisfy $f^{\text{d.s.e.}}_{\emptyset;\alpha,\beta} = 0$ if $\alpha \neq \beta$,*

$$f_w(A^1,\ldots,A^\ell)_{\alpha\beta} = \sum_{1\le\alpha_1,\ldots,\alpha_{\ell-1}\le m} f^{\text{d.s.e.}}_{w;\alpha_0,\ldots,\alpha_\ell}(A^1_{\alpha_0\alpha_1},\ldots,A^\ell_{\alpha_{\ell-1}\alpha_\ell}), \tag{9.16}$$

$$\text{where } |w| = \ell,\ \alpha_0 = \alpha,\ \alpha_\ell = \beta,\ \alpha,\beta = 1,\ldots,m,\ \ell = 1,2,\ldots,$$

*with uniquely determined $\ell$-linear mappings*

$$f^{\text{d.s.e.}}_{w;\alpha_0,\ldots,\alpha_\ell}\colon \mathcal{R}^{s_{\alpha_0}\times s_{\alpha_1}} \times \cdots \times \mathcal{R}^{s_{\alpha_{\ell-1}}\times s_{\alpha_\ell}} \to \mathcal{N}^{s_{\alpha_0}\times s_{\alpha_\ell}}$$

*satisfying*

$$Sf^{\text{d.s.e.}}_{\emptyset;\beta,\beta} - f^{\text{d.s.e.}}_{\emptyset;\alpha,\alpha}S = \sum_{k=1}^d f^{\text{d.s.e.}}_{g_k;\alpha,\beta}(S(Y_k)_{\beta\beta} - (Y_k)_{\alpha\alpha}S), \quad S \in \mathcal{R}^{s_\alpha\times s_\beta}, \tag{9.17}$$

*and for $\ell = 1,2,\ldots$,*

$$\begin{aligned} &Sf^{\text{d.s.e.}}_{w;\gamma,\alpha_1,\ldots,\alpha_\ell}(A^1,\ldots,A^\ell) - f^{\text{d.s.e.}}_{w;\alpha_0,\ldots,\alpha_\ell}(SA^1,A^2,\ldots,A^\ell) \\ &= \sum_{k=1}^d f^{\text{d.s.e.}}_{g_kw;\alpha_0,\gamma,\alpha_1,\ldots,\alpha_\ell}(S(Y_k)_{\gamma\gamma} - (Y_k)_{\alpha_0\alpha_0}S, A^1,\ldots,A^\ell), \quad S \in \mathcal{R}^{s_{\alpha_0}\times\gamma}, \end{aligned} \tag{9.18}$$

$$\begin{aligned} &f^{\text{d.s.e.}}_{w;\alpha_0,\ldots,\alpha_{j-1},\gamma,\alpha_{j+1},\ldots,\alpha_\ell}(A^1,\ldots,A^{j-1},A^jS,A^{j+1},\ldots,A^\ell) \\ &\qquad - f^{\text{d.s.e.}}_{w;\alpha_0,\ldots,\alpha_\ell}(A^1,\ldots,A^j,SA^{j+1},A^{j+2},\ldots,A^\ell) \\ &= \sum_{k=1}^d f^{\text{d.s.e.}}_{g_{i_1}\cdots g_{i_j}g_kg_{i_{j+1}}\cdots g_{i_\ell};\alpha_0,\ldots,\alpha_j,\gamma,\alpha_{j+1},\ldots,\alpha_\ell}(A^1,\ldots,A^j, \\ &\qquad S(Y_k)_{\gamma\gamma} - (Y_k)_{\alpha_j\alpha_j}S, A^{j+1},\ldots,A^\ell), \\ &\qquad\qquad w = g_{i_1}\cdots g_{i_\ell} \in \mathcal{G}_d,\ S \in \mathcal{R}^{s_{\alpha_j}\times\gamma}, \end{aligned} \tag{9.19}$$

$$\begin{aligned} &f^{\text{d.s.e.}}_{w;\alpha_0,\ldots,\alpha_{\ell-1},\gamma}(A^1,\ldots,A^{\ell-1},A^\ell S) - f^{\text{d.s.e.}}_{w;\alpha_0,\ldots,\alpha_\ell}(A^1,\ldots,A^\ell)S \\ &= \sum_{k=1}^d f^{\text{d.s.e.}}_{wg_k;\alpha_0,\ldots,\alpha_\ell,\gamma}(A^1,\ldots,A^\ell,S(Y_k)_{\gamma\gamma} - (Y_k)_{\alpha_\ell\alpha_\ell}S), \quad S \in \mathcal{R}^{\alpha_\ell\times\gamma}. \end{aligned} \tag{9.20}$$

*In particular, conditions* (9.17)–(9.20) *for* $f^{\text{d.s.e.}}_{w;\alpha,\ldots,\alpha}$, $w \in \mathcal{G}_d$, *with a fixed* $\alpha$, *coincide with* (4.14)–(4.17) *for* $s$ *replaced by* $s_\alpha$ *and* $Y$ *replaced by* $Y_{\alpha\alpha}$.

(2) *If* $s_1 = \cdots = s_m$ *and* $Y_{11} = \cdots = Y_{mm}$, *then the mappings* $f^{\text{d.s.e.}}_{w;\alpha_0,\ldots,\alpha_\ell}$, $1 \le \alpha_0, \ldots, \alpha_\ell \le m$, *coincide for every fixed* $w$.

(3) *If* $\mathcal{R} = \mathbb{C}$, $\mathcal{M} = \mathcal{V}$ *and* $\mathcal{N} = \mathcal{W}$ *are Banach spaces equipped with admissible systems of rectangular matrix norms over* $\mathcal{V}$ *and* $\mathcal{W}$, *and the* $\ell$*-linear mappings* $f_w$ *are bounded, then so are* $f^{\text{d.s.e.}}_{w;\alpha_0,\ldots,\alpha_\ell}$.

(4) *If* $\mathcal{R} = \mathbb{C}$, $\mathcal{M} = \mathcal{V}$ *and* $\mathcal{N} = \mathcal{W}$ *are Banach spaces equipped with admissible systems of rectangular matrix norms over* $\mathcal{V}$ *and* $\mathcal{W}$, (7.39) *hold for* $\mathcal{V}$ *and* $\mathcal{W}$ *with the constants* $C^{\mathcal{V}}$ *and* $C^{\mathcal{W}}$ *which are independent of* $n, p, q$, *and* $m$, *and the* $\ell$*-linear mappings* $f_w$ *are completely bounded, then so are* $f^{\text{d.s.e.}}_{w;\alpha_0,\ldots,\alpha_\ell}$, *and*

$$\|f^{\text{d.s.e.}}_{w;\alpha_0,\ldots,\alpha_\ell}\|_{\mathcal{L}^\ell_{\text{cb}}} \le C^{\mathcal{W}}(C^{\mathcal{V}})^\ell \|f_w\|_{\mathcal{L}^\ell_{\text{cb}}}; \tag{9.21}$$

*in particular, if* $\mathcal{V}$ *and* $\mathcal{W}$ *are operator spaces, then*

$$\|f^{\text{d.s.e.}}_{w;\alpha_0,\ldots,\alpha_\ell}\|_{\mathcal{L}^\ell_{\text{cb}}} \le \|f_w\|_{\mathcal{L}^\ell_{\text{cb}}}. \tag{9.22}$$

# (Some) earlier work on nc functions

We concentrate here on the works most directly related to our main theme, namely the theory of nc functions viewed as functions on square matrices of all sizes, and how they compare to our results; we apologize in advance for any omissions.

Taylor [**99**, Section 6] introduced the algebra that he denoted by $\mathfrak{D}(U)$, where $U$ is (in our terminology) an open nc subset of $(\mathbb{C}^n)_{\rm nc}$, that consists of functions $f\colon U \to \mathbb{C}_{\rm nc}$ that preserve matrix size and respect intertwining, and are *assumed* to be analytic when restricted to $U \cap \left(\mathbb{C}^{k\times k}\right)^n$ for all $k$. He proved, using the analyticity assumption, the existence of the nc difference-differential $\Delta_R f$ (to be specific, the analyticity assumption is used to show that $\Delta_R f(X, Y)$ is a linear mapping between appropriate matrix spaces), and established the generalized finite difference formula (our (1.8) in the Introduction and Section 2.4) as well as the properties of $\Delta_R f(X, Y)$ as a function of $X$ and $Y$ (our Section 2.5). He also introduced explicitly the corresponding space $\mathfrak{D}(U, U)$ of (in our terminology) first order nc functions on $U \times U$ — again, with an analyticity assumption when restricted to each (pair) of matrix sizes, and showed the canonical isomorphism $\mathfrak{D}(U)\widehat{\otimes}\mathfrak{D}(U) \cong \mathfrak{D}(U, U)$ (compare our Remark 3.6); here all the function spaces are given the topology of uniform convergence on compacta in every matrix size, and $\widehat{\otimes}$ denotes the completed projective tensor product.

Taylor also considered in [**99**, Section 6] the algebra of nc power series with a given multiradius of convergence. He noticed that it is not a nuclear Frechet space, whereas $\mathfrak{D}(U)$ is always nuclear, implying (in our terminology and without an explicit example, compare our Example 8.14) that not every analytic nc function on a nc polydisc is uniformly analytic.

In [**100**], Taylor considered quite generally a locally convex topological algebra $A$ together with a unital algebra homomorphism from the free associative algebra $\mathbb{C}\langle z_1, \ldots, z_n\rangle$ to $A$, equipped with continuous linear maps $\Delta_i\colon A \to A\widehat{\otimes}A$[1], $i = 1, \ldots, n$ — playing the role of partial nc difference-differential operators — that satisfy the Leibnitz rule (compare our Section 2.3.4) and a version of the generalized finite difference formula. Taylor called $\mathbb{C}\langle z_1, \ldots, z_n\rangle \to A$ a localization (he introduced this notion in [**99**] in a more general context from certain homological considerations). By considering representations of $A$, he showed that each element of $A$ yields a function on a class of $n$-tuples of operators on Banach spaces that respects intertwining, much like nc functions except that infinite-dimensional Banach spaces may be allowed. With respect to such an infinite-dimensional setting, we notice also the work of Hadwin [**46**] and Hadwin–Kaonga–Mathes [**47**] for infinite-dimensional Hilbert spaces and functions respecting orthogonal direct sums and

[1]To be precise, Taylor assumes that the multiplication on $A$ is separately continuous and looks at the completed inductive tensor product; this makes no difference if $A$ is a Fréchet space.

unitary similarity, and the work of Muhly–Solel [**69, 71**] for representations of a $W^*$-correspondence.

Taylor showed in [**100**, Section 4] that for any localization $\mathbb{C}\langle z_1, \dots, z_n\rangle \to A$, $(1 \otimes \Delta) \circ \Delta = (\Delta \otimes 1) \circ \Delta$, where $\Delta\colon A \to \bigoplus_{i=1}^n A\widehat{\otimes} A$ with the coordinates $\Delta_i$ (in particular, each $\Delta_i$ is a coassociative comultiplication), compare our Section 3.4. He used this to define $\Delta^k\colon A \to \bigoplus_{w\in\mathcal{G}_n,\,|w|=k} A^{\widehat{\otimes}k}$ (here we use our notation for the free monoid, see Section 1.1) iteratively, compare our formulae (3.34) and (3.39). He then derived (in our terminology) a finite TT expansion as in our Theorem 4.1, and showed that it converges under some additional conditions on the localization $\mathbb{C}\langle z_1, \dots, z_n\rangle \to A$ (essentially, refined continuity requirements for $\Delta_i$) that he referred to as a free analytic function algebra on $n$ generators and that he verified for the algebra of nc power series with a given multiradius of convergence.

In this connection, Taylor also considered nc power series with coefficients in a Banach algebra (in the case of a matrix algebra these are our nc power series (1.19) and(8.1), though Taylor never considers their evaluations) and posed the question of when they form a free analytic function algebra; while he was unable to answer this question, he did derive a version of our conditions (1.15)–(1.18) and (4.4)–(4.7). Among other noteworthy results in [**100**], we mention a version of the implicit function theorem.

Voiculescu introduced in [**106**] (which is a continuation of [**105**]) what he called fully matricial sets and fully matricial functions. A fully matricial $G$-set $\Omega$, for $G$ a Banach space, is in our terminology a nc set in the nc space over $G$ that is similarity invariant and contains all direct summands of matrices from $\Omega$ (so that $\Omega$ coincides with its direct summands extension as in Chapter 9). A fully matricial $H$-valued function on $\Omega$, for $H$ another Banach space, is in our terminology a nc function on $\Omega$ with values in the nc space over $H$. Voiculescu called such a function continuous or analytic if for all $n$, $f|_{\Omega_n}$ is a continuous or analytic function on $\Omega_n \subseteq G^{n\times n}$ (assumed to be open) with values in $H^{n\times n}$. He introduced $p$-variable fully matricial analytic functions — in our terminology, these are nc functions of order $p-1$ that are assumed to be analytic when restricted to fixed matrix sizes, and established the existence of (in our terminology, see Sections 2.6 and 3.5) directional nc difference-differential operator, that he denoted by $\partial$, from scalar fully matricial analytic functions to 2-variable scalar fully matricial analytic functions[2]; the analyticity assumption is used essentially in the proof. Voiculescu proved that $\partial$ satisfies the Leibnitz rule (see our Section 2.3.4) and a coassociativity property (see our Section 3.4), so that the algebra of scalar fully matricial analytic functions becomes a topological version of what he called a generalized difference quotient ring (an infinitesimal bialgebra in the terminology of [**90, 3**]). One of the main results of [**106**] is a duality transform relating the generalized difference quotient ring $B\langle Y\rangle$ generated freely by an element $Y$ over a Banach algebra $B$ with a unit and the generalized difference quotient ring of scalar fully matricial analytic functions on what Voiculescu called the full $B$-resolvent set of $Y$ (which is $\coprod_{n=1}^{\infty} \{b \in B^{n\times n}\colon \bigoplus_{\alpha=1}^{n} Y - b$ is invertible in $E^{n\times n}\}$, where $E$ is a Banach algebra containing $B$ and $Y$ and some additional technical assumptions are satisfied).

[2]To be more precise, Voiculescu considered the case where $G$ is an operator system hence has a distinguished vector $1_G$, and looked at the nc difference-differential operator in the direction $1_G$.

In [**107**, Section 11], Voiculescu introduced what he called stably matricial sets and stably matricial functions by relaxing the similarity invariance property for a fully matricial set (so that, using our terminology, nc balls in the nc space over an operator space and naturally defined nc half planes in the nc space over an operator system are included), and showed that an analytic stably matricial function naturally extends to an analytic fully matricial function (compare our canonical extension of a nc function to the similarity invariant envelope of the original domain of definition in the Appendix, and the direct summands extension in Chapter 9). We mention also an inverse function theorem in [**107**, Section 11.5]. In [**107**, Section 13], Voiculescu established that an analytic fully matricial function on a fully matricial set containing the origin admits, in our terminology, a nc power series expansion at the origin (see our (7.15) in Theorem 7.8, except for the difference of notation for nc power series and the fact that Voiculescu does not identify the multlinear forms appearing as coefficients as the values at the origin of the higher order nc difference-differential operator applied to the function). It is interesting to notice that unlike Taylor's approach in [**100**] mentioned above and our approach in Chapters 4 and 7, Voiculescu does not obtain the nc power series expansion via the nc difference-differential calculus; he rather starts with the usual power series expansion of $f|_{\Omega_n}$ at $0_{n\times n}$ for every matrix size $n$, and uses the fact that $f$ respects direct sums and similarities to deduce that for all $N$ (using our notation, compare Remark 7.3) $\delta^N(f|_{\Omega_n})(0_{n\times n})(Z) = Z^{\odot N}\alpha_N$ for some bounded $N$-linear mapping $\alpha_N$. In [**107**, Sections 14–16], Voiculescu combines the nc power series expansion with asymptotic freeness results for random matrices to study boundary values for (using our terminology) bounded nc functions on the open nc unit ball in $(\mathbb{C}^{k\times k})_{\rm nc}$ with the operator space structure $\|X\|_n = \|X\|$ for $X \in \left(\mathbb{C}^{k\times k}\right)^{n\times n} \cong \mathbb{C}^{nk\times nk}$ and in $(\mathbb{C}^k)_{\rm nc}$ with the operator space structure $\|X\|_n = \max\{\|X_1\|, \ldots, \|X_k\|\}$ for $X = (X_1, \ldots, X_k) \in (\mathbb{C}^{n\times n})^k$ (here $\|\cdot\|$ denotes the usual Euclidean norm of a complex matrix), and to obtain asymptotic integral formulae for the coefficients in terms of integration over the corresponding distinguished boundary.

The main thrust of [**107**] is to develop nc function theory on the Grassmanian completion of (using our terminology) the nc space over a Banach algebra $B$, which is to play the role of the Riemann sphere completion of the complex plane in the case $B = \mathbb{C}$.

We finish this review with the mention of the two papers [**93**] (see also [**94**]) and [**76**] dealing with functions of a single complex or real nc variable. More precisely, Schanuel considered in [**93**] a family $f$ of functions $f_A\colon A \to A$, where $A$ runs over all finite-dimensional $\mathbb{K}$-algebras with unit in the case $\mathbb{K} = \mathbb{C}$ and over all finite-dimensional $\mathbb{K}$-algebras with unit that are embedded in upper triangular matrices in the case $\mathbb{K} = \mathbb{R}$, and the functions $f_A$ respect algebra homomorphisms. Schanuel called $f$ a global function; $f$ is essentially the same as (in our terminology) a nc function from $\mathbb{C}_{\rm nc}$ to $\mathbb{C}_{\rm nc}$ in the case $\mathbb{K} = \mathbb{C}$, and as a nc function from $\Lambda$ to $\mathbb{R}_{\rm nc}$ in the case $\mathbb{K} = \mathbb{R}$, where $\Lambda \subseteq \mathbb{R}_{\rm nc}$ is the nc set consisting of matrices with only real eigenvalues (notice that $\Lambda$ is not an open nc set). Schanuel remarked that every entire function in the case $\mathbb{K} = \mathbb{C}$ and every $C^\infty(\mathbb{R})$ function in the case $\mathbb{K} = \mathbb{R}$ extend naturally to a global function. He proved that if

- either for every $A$ and every $\lambda \in \mathbb{K}$, $f_A$ is bounded on an open neighbourhood of $\lambda$ in $A$,

- or there exists an integer $n \geq 0$ so that for every $A$ and every $a, b \in A$ the function $t \mapsto t^n f_A(a+tb)$ from $\mathbb{K}$ to $A$ is bounded in a neignhourhood of 0,

then for every $A$, $f_A\colon A \to A$ is an analytic function in the case $\mathbb{K} = \mathbb{C}$ and a $C^\infty$ function in the case $\mathbb{K} = \mathbb{R}$, and the global function $f$ is obtained from $f_{\mathbb{K}}$ by the natural extension (in the case $\mathbb{K} = \mathbb{C}$, compare our Section 7.1 — notice that for nc functions on $\mathbb{C}_{\rm nc}$ local boundedness at scalar points implies local boundedness everywhere because of the decomposition (5.8)). The proof uses evaluation on upper triangular matrices and a formal power series expansion for a global function in terms of a sequence of scalar functions (see our (5.9)–(5.10) — notice that the desired conclusion is that $f_0$ is an entire function in the case $\mathbb{K} = \mathbb{C}$ and a $C^\infty$ function in the case $\mathbb{K} = \mathbb{R}$, and $f_\ell' = f_{\ell+1}$ for $\ell = 0, 1, \ldots$).

Niemiec [**76**] considered the sets that he denoted by $\mathcal{D}_k(\Omega)$ — of $k \times k$ diagonalizable matrices with all the eigenvalues in $\Omega$ — and by $\mathcal{M}_k(\Omega)$ — of all $k \times k$ matrices with all the eigenvalues in $\Omega$; here $\Omega \subseteq \mathbb{R}$ or $\Omega \subseteq \mathbb{C}$ and the matrices are over $\mathbb{R}$ or over $\mathbb{C}$, respectively. For any function $f\colon \Omega \to \mathbb{C}$, there is a natural extension to (in our terminolgy) nc function from the nc set $\coprod_{k=1}^\infty \mathcal{D}_k(\Omega)$ to $\mathbb{C}_{\rm nc}$ (here we view $\mathbb{C}$ as a $\mathbb{R}$-vector space in the case $\Omega \subseteq \mathbb{R}$) that Niemiec denoted by $f_{\rm op}$. He showed that for an open set $\Omega \subseteq \mathbb{C}$ and for any $k \geq 2$, $f_{\rm op}$ is continuous on $\mathcal{D}_k(\Omega)$ if and only if $f_{\rm op}$ extends to to a continuous function from $\mathcal{M}_k(\Omega)$ to $\mathcal{M}_k(\mathbb{C})$ if and only if $f$ is analytic. He also showed that for any set $\Omega \subseteq \mathbb{C}$, $f_{\rm op}$ is bounded on the intersection of some nc ball around $\lambda$ (in our terminology) with $\coprod_{k=1}^\infty \mathcal{D}_k(\Omega)$ for all $\lambda \in \Omega$ if and only if for all $\lambda \in \Omega$ and all $\epsilon > 0$ there exists a $\delta > 0$ such that $\|f_{\rm op}(X) - I_k f(\lambda)\| < \epsilon$ for all $k = 1, 2, \ldots$ and all $X \in \mathcal{D}_k(\Omega)$: $\|X - I_k\lambda\| < \delta$, if and only if $f$ extends to an analytic function on an open set in $\mathbb{C}$ containing $\Omega$. For both of these results, compare our Section 7.1. Niemiec also proved that for an open set (or an interval) $\Omega \subseteq \mathbb{R}$ and for any $k \geq 3$, $f_{\rm op}$ is continuous on $\mathcal{D}_k(\Omega)$ if and only if $f$ is of class $C^{k-2}$ and $f^{(k-2)}$ is locally Lipschitz, while $f_{\rm op}$ extends to to a continuous function from $\mathcal{M}_k(\Omega)$ to $\mathcal{M}_k(\mathbb{C})$ if and only if $f$ is of class $C^{k-1}$. The proofs use Taylor's formula with remainder and the calculus of divided differences.

APPENDIX A

# Similarity invariant envelopes and extension of nc functions

(by Shibananda Biswas[1], Dmitry S. Kaliuzhnyi-Verbovetskyi, and Victor Vinnikov)

PROPOSITION A.1. *Let $\mathcal{M}$ be a module over a commutative unital ring $\mathcal{R}$, and let $\Omega \subseteq \mathcal{M}_{\rm nc}$ be a nc set. Then*

$$\widetilde{\Omega} := \{SXS^{-1} \colon X \in \Omega_n, S \in \mathcal{R}^{n\times n} \text{ invertible}, n \in \mathbb{N}\}$$

*is a nc set.*

Clearly, $\widetilde{\Omega}$ is the smallest nc set that contains $\Omega$ and that is invariant under similarities. We shall call $\widetilde{\Omega}$ the *similarity invariant envelope of the nc set* $\Omega$.

PROOF OF PROPOSITION A.1. Let $X \in \Omega_n$, $Y \in \Omega_m$, and let $S \in \mathcal{R}^{n\times n}$, $T \in \mathcal{R}^{m\times m}$ be invertible, $n, m \in \mathbb{N}$. Then $S \oplus T \in \mathcal{R}^{(n+m)\times(n+m)}$ is invertible, $X \oplus Y \in \Omega_{n+m}$, and

$$SXS^{-1} \oplus TYT^{-1} = (S \oplus T)(X \oplus Y)(S \oplus T)^{-1} \in \widetilde{\Omega}_{n+m}.$$

□

PROPOSITION A.2. *If $\Omega \subseteq \mathcal{M}_{\rm nc}$ is a right (resp., left) admissible nc set, then so is its similarity invariant envelope $\widetilde{\Omega}$. Moreover, for any $\widetilde{X} \in \widetilde{\Omega}_n$, $\widetilde{Y} \in \widetilde{\Omega}_m$ and $Z \in \mathcal{M}^{n\times m}$, one has*

$$\begin{bmatrix} \widetilde{X} & Z \\ 0 & \widetilde{Y} \end{bmatrix} \in \widetilde{\Omega}_{n+m},$$

*and, respectively, for any $\widetilde{X} \in \widetilde{\Omega}_n$, $\widetilde{Y} \in \widetilde{\Omega}_m$ and $Z \in \mathcal{M}^{m\times n}$, one has*

$$\begin{bmatrix} \widetilde{X} & 0 \\ Z & \widetilde{Y} \end{bmatrix} \in \widetilde{\Omega}_{n+m},$$

PROOF. We shall give the proof for a right admissible nc set $\Omega$. A similar argument works for a left admissible nc set $\Omega$. Let $\widetilde{X} \in \widetilde{\Omega}_n$, $\widetilde{Y} \in \widetilde{\Omega}_m$ and $Z \in \mathcal{M}^{n\times m}$. Then there exist $X \in \Omega_n$, $Y \in \Omega_m$, and invertible $S \in \mathcal{R}^{n\times n}$, $T \in \mathcal{R}^{m\times m}$, and $r \in \mathcal{R}$ such that $\widetilde{X} = SXS^{-1}$, $\widetilde{Y} = TYT^{-1}$, and

$$\begin{bmatrix} X & rS^{-1}ZT \\ 0 & Y \end{bmatrix} \in \Omega_{n+m}.$$

[1]Department of Mathematics, Ben-Gurion University of the Negev, Beer-Sheva, Israel, 84105; current address: Department of Mathematics and Statistics, Indian Institute of Science Education and Research Kolkata, Mohanpur – 741246, Nadia, West Bengal, India; e-mail address: shibananda@gmail.com.

Then

$$\begin{bmatrix} \widetilde{X} & Z \\ 0 & \widetilde{Y} \end{bmatrix} = \begin{bmatrix} r^{-1}S & 0 \\ 0 & T \end{bmatrix} \begin{bmatrix} X & rS^{-1}ZT \\ 0 & Y \end{bmatrix} \begin{bmatrix} r^{-1}S & 0 \\ 0 & T \end{bmatrix}^{-1} \in \widetilde{\Omega}_{n+m}.$$

□

PROPOSITION A.3. *Let $\mathcal{M}$, $\mathcal{N}$ be modules over a commutative unital ring $\mathcal{R}$, let $\Omega \subseteq \mathcal{M}_{\rm nc}$ be a nc set, and let $f\colon \Omega \to \mathcal{N}_{\rm nc}$ be a nc function. Then there exists a unique nc function $\widetilde{f}\colon \widetilde{\Omega} \to \mathcal{N}_{\rm nc}$ such that $\widetilde{f}|_\Omega = f$.*

PROOF. For any $n \in \mathbb{N}$ and $\widetilde{X} \in \widetilde{\Omega}_n$, there exist $X \in \Omega_n$ and an invertible $S \in \mathcal{R}^{n\times n}$ such that $\widetilde{X} = SXS^{-1}$, and we set $\widetilde{f}(\widetilde{X}) = Sf(X)S^{-1}$. This definition is correct, since $SXS^{-1} = TYT^{-1}$ for $X, Y \in \Omega_n$ and for invertible $S, T \in \mathcal{R}^{n\times n}$ implies that $Y = (T^{-1}S)X(T^{-1}S)^{-1}$, hence $f(Y) = (T^{-1}S)f(X)(T^{-1}S)^{-1}$ and $Tf(Y)T^{-1} = Sf(X)S^{-1}$. It is straightforward to check that $\widetilde{f}\colon \widetilde{\Omega} \to \mathcal{N}_{\rm nc}$ is a nc function and that $\widetilde{f}|_\Omega = f$. If $\widetilde{g}\colon \widetilde{\Omega} \to \mathcal{N}_{\rm nc}$ is another nc function such that $\widetilde{g}|_\Omega = f$, then for any $X \in \Omega_n$ and invertible $S \in \mathcal{R}^{n\times n}$ we have

$$\widetilde{g}(SXS^{-1}) = S\widetilde{g}(X)S^{-1} = Sf(X)S^{-1} = \widetilde{f}(SXS^{-1}),$$

i.e., we have necessarily $\widetilde{g} = \widetilde{f}$. □

We will call the nc function $\widetilde{f}\colon \widetilde{\Omega} \to \mathcal{N}_{\rm nc}$ from Proposition A.3 the *canonical extension of the nc function* $f\colon \Omega \to \mathcal{N}_{\rm nc}$.

EXAMPLE A.4. Let $\mathcal{V} = \mathcal{W} = \mathbb{C}$, with the canonical operator space structure. The nc unit ball $\Omega = B_{\rm nc}(0_{1\times 1}, 1)$ is a nc set. Then its similarity invariant envelope is

$$\widetilde{\Omega} = \{Z \in \mathbb{C}_{\rm nc} \colon r_{\rm spec}(Z) < 1\}.$$

Indeed, let $Z \in \mathbb{C}^{n\times n}$ be such that $r_{\rm spec}(Z) < 1$. We need to show that $Z = SXS^{-1}$ for some $X \in \mathbb{C}^{n\times n}$ with $\|X\| < 1$ and some invertible $S \in \mathbb{C}^{n\times n}$. It suffices to show this for $Z$ a Jordan cell, i.e., for

$$Z = \begin{bmatrix} \lambda & 1 & & \\ & \ddots & \ddots & \\ & & \ddots & 1 \\ & & & \lambda \end{bmatrix}$$

with $|\lambda| < 1$. Let $\epsilon > 0$ be such that $\|X\| < 1$, where

$$X = \begin{bmatrix} \lambda & \epsilon & & \\ & \ddots & \ddots & \\ & & \ddots & \epsilon \\ & & & \lambda \end{bmatrix}.$$

Then $Z = SXS^{-1}$ with $S = \operatorname{diag}(\epsilon^{-1}, \ldots, \epsilon^{-n})$.

Clearly the nc function $f\colon \Omega \to \mathcal{W}_{\rm nc}$ defined by $f(Z) = (I - Z)^{-1}$ can be canonically extended to a nc function on $\widetilde{\Omega}$ using the same formula.

Proposition A.3 can be extended to higher order nc functions as follows.

PROPOSITION A.5. *Let $\mathcal{M}_0, \ldots, \mathcal{M}_k, \mathcal{N}_0, \ldots, \mathcal{N}_k$ be modules over a commutative unital ring $\mathcal{R}$, let $\Omega^{(0)} \subseteq \mathcal{M}_{0,\mathrm{nc}}, \ldots, \Omega^{(k)} \subseteq \mathcal{M}_{k,\mathrm{nc}}$ be nc sets, and let $f \in \mathcal{T}^k(\Omega^{(0)}, \ldots, \Omega^{(k)}; \mathcal{N}_{0,\mathrm{nc}}, \ldots, \mathcal{N}_{k,\mathrm{nc}})$. Then there exists a unique nc function of order $k$, $\widetilde{f} \in \mathcal{T}^k(\widetilde{\Omega}^{(0)}, \ldots, \widetilde{\Omega}^{(k)}; \mathcal{N}_{0,\mathrm{nc}}, \ldots, \mathcal{N}_{k,\mathrm{nc}})$, such that $\widetilde{f}|_{\Omega^{(0)} \times \cdots \times \Omega^{(k)}} = f$.*

PROOF. For any $n_0, \ldots, n_k \in \mathbb{N}$ and $\widetilde{X}^j \in \widetilde{\Omega}^{(j)}_{n_j}$, there exist $X^j \in \Omega^{(j)}_{n_j}$ and invertible $S_j \in \mathcal{R}^{n_j \times n_j}$ such that $\widetilde{X}^j = S_j X^j S_j^{-1}$, $j = 0, \ldots, k$. We set, for any $Z^j \in \mathcal{N}_j^{n_{j-1} \times n_j}$, $j = 1, \ldots, k$,

$$\widetilde{f}(\widetilde{X}^0, \ldots, \widetilde{X}^k)(Z^1, \ldots, Z^k) = S_0 f(X^0, \ldots, X^k)(S_0^{-1} Z^1 S_1, \ldots, S_{k-1}^{-1} Z^k S_k) S_k^{-1}.$$

This definition is correct, since the equalities $S_j X^j S_j^{-1} = T_j Y^j T_j^{-1}$ for $X^j, Y^j \in \Omega^{(j)}_{n_j}$ and for invertible $S_j, T_j \in \mathcal{R}^{n_j \times n_j}$ imply $Y^j = (T_j^{-1} S_j) X^j (T_j^{-1} S_j)^{-1}$, hence by (3.3)

$$\begin{aligned} f(Y^0, \ldots, Y^k)&(T_0^{-1} S_0 Z^1 S_1^{-1} T_1, \ldots, T_{k-1}^{-1} S_{k-1} Z^k S_k^{-1} T_k) \\ &= T_0^{-1} S_0 f(X^0, \ldots, X^k)(Z^1, \ldots, Z^k) S_k^{-1} T_k, \end{aligned}$$

and for $W_j = S_{j-1} Z^j S_j^{-1}$ we have

$$\begin{aligned} T_0 f(Y^0, \ldots, Y^k)&(T_0^{-1} W^1 T_1, \ldots, T_{k-1}^{-1} W^k T_k) T_k^{-1} \\ &= S_0 f(X^0, \ldots, X^k)(S_0^{-1} W^j S_1, \ldots, S_{k-1}^{-1} W^1 S_k) S_k^{-1}. \end{aligned}$$

It is obvious that $\widetilde{f}$ respects similarities and that $\widetilde{f}|_{\Omega^{(0)} \times \cdots \times \Omega^{(k)}} = f$. We also have

$$\begin{aligned} &\widetilde{f}(\widetilde{X}^{0\prime} \oplus \widetilde{X}^{0\prime\prime}, \ldots, \widetilde{X}^{k\prime} \oplus \widetilde{X}^{k\prime\prime}) \left( \begin{bmatrix} Z^{1\prime,\prime} & Z^{1\prime,\prime\prime} \\ Z^{1\prime\prime,\prime} & Z^{1\prime\prime,\prime\prime} \end{bmatrix}, \ldots, \begin{bmatrix} Z^{k\prime,\prime} & Z^{k\prime,\prime\prime} \\ Z^{k\prime\prime,\prime} & Z^{k\prime\prime,\prime\prime} \end{bmatrix} \right) \\ &= (S_0' \oplus S_0'') f(X^{0\prime} \oplus X^{0\prime\prime}, \ldots, X^{k\prime} \oplus X^{k\prime\prime}) \left( \begin{bmatrix} (S_0')^{-1} Z^{1\prime,\prime} S_1' & (S_0')^{-1} Z^{1\prime,\prime\prime} S_1'' \\ (S_0'')^{-1} Z^{1\prime\prime,\prime} S_1' & (S_0'')^{-1} Z^{1\prime\prime,\prime\prime} S_1'' \end{bmatrix}, \right. \\ &\qquad \left. \ldots, \begin{bmatrix} (S_{k-1}')^{-1} Z^{k\prime,\prime} S_k' & (S_{k-1}')^{-1} Z^{k\prime,\prime\prime} S_k'' \\ (S_{k-1}'')^{-1} Z^{k\prime\prime,\prime} S_k' & (S_k'')^{-1} Z^{k\prime\prime,\prime\prime} S_k'' \end{bmatrix} \right) (S_k' \oplus S_k'')^{-1} \\ &\qquad\qquad = \begin{bmatrix} S_0' f^{\prime,\prime} (S_k')^{-1} & S_0' f^{\prime,\prime\prime} (S_k'')^{-1} \\ S_0'' f^{\prime\prime,\prime} (S_k')^{-1} & S_0'' f^{\prime\prime,\prime\prime} (S_k'')^{-1} \end{bmatrix}, \end{aligned}$$

where according to (3.1) and (3.2), for $\alpha, \beta \in \{\prime, \prime\prime\}$,

$$\begin{aligned} f^{\alpha,\beta} = \sum_{\alpha_0, \ldots, \alpha_k \in \{\prime,\prime\prime\} \colon \alpha_0 = \alpha, \alpha_k = \beta} & f(X^{0\alpha_0}, \ldots, X^{k\alpha_k}) \\ & \left( (S_0^{\alpha_0})^{-1} Z^{1\alpha_0,\alpha_1} S_1^{\alpha_1}, \ldots, (S_{k-1}^{\alpha_{k-1}})^{-1} Z^{k\alpha_{k-1},\alpha_k} S_k^{\alpha_k} \right). \end{aligned}$$

Here $n_j' \in \mathbb{N}$, $n_j'' \in \mathbb{Z}_+$, $X^{j\alpha} \in \Omega^{(j)}_{n_j^\alpha}$, $\widetilde{X}^{j\alpha} \in \widetilde{\Omega}^{(j)}_{n_j^\alpha}$ for $j = 0, \ldots, k$ are such that $\widetilde{X}^{j\alpha} = S_j^\alpha X^{j\alpha} (S_j^\alpha)^{-1}$ for an invertible $S_j \in \mathcal{R}^{n_j^\alpha \times n_j^\alpha}$, $\alpha \in \{\prime, \prime\prime\}$; $Z^{j\alpha,\beta} \in \mathcal{N}_j^{n_{j-1}^\alpha \times n_j^\beta}$ for $j = 1, \ldots, k$, $\alpha, \beta \in \{\prime, \prime\prime\}$, the block entry $f^{\alpha,\beta}$ is void if either $n_0^\alpha$ or $n_k^\beta$ is 0, and a summand in the right-hand side sum is 0 if at least one of $n_j^{\alpha_j}$, $j = 1, \ldots, k-1$, is 0. Therefore,

$$\widetilde{f}(\widetilde{X}^{0\prime} \oplus \widetilde{X}^{0\prime\prime}, \ldots, \widetilde{X}^{k\prime} \oplus \widetilde{X}^{k\prime\prime}) \left( \begin{bmatrix} Z^{1\prime,\prime} & Z^{1\prime,\prime\prime} \\ Z^{1\prime\prime,\prime} & Z^{1\prime\prime,\prime\prime} \end{bmatrix}, \ldots, \begin{bmatrix} Z^{k\prime,\prime} & Z^{k\prime,\prime\prime} \\ Z^{k\prime\prime,\prime} & Z^{k\prime\prime,\prime\prime} \end{bmatrix} \right) = \begin{bmatrix} \widetilde{f}^{\prime,\prime} & \widetilde{f}^{\prime,\prime\prime} \\ \widetilde{f}^{\prime\prime,\prime} & \widetilde{f}^{\prime\prime,\prime\prime} \end{bmatrix},$$

with

$$\begin{aligned} \widetilde{f}^{\alpha,\beta} = S_0^\alpha f^{\alpha,\beta} (S_k^\beta)^{-1} &= \sum_{\alpha_0,\ldots,\alpha_k \in \{\prime,\prime\prime\}:\ \alpha_0=\alpha,\alpha_k=\beta} S_0^\alpha f\left(X^{0\alpha_0}, \ldots, X^{k\alpha_k}\right) \\ &\qquad \left((S_0^{\alpha_0})^{-1} Z^{1\alpha_0,\alpha_1} S_1^{\alpha_1}, \ldots, (S_{k-1}^{\alpha_{k-1}})^{-1} Z^{k\alpha_{k-1},\alpha_k} S_k^{\alpha_k}\right) (S_k^\beta)^{-1} \\ &= \sum_{\alpha_0,\ldots,\alpha_k \in \{\prime,\prime\prime\}:\ \alpha_0=\alpha,\alpha_k=\beta} \widetilde{f}(\widetilde{X}^{0\alpha_0}, \ldots, \widetilde{X}^{k\alpha_k})(Z^{1\alpha_0,\alpha_1}, \ldots, Z^{k\alpha_{k-1},\alpha_k}). \end{aligned}$$

Thus, $\widetilde{f}$ respects direct sums. Next, if $\widetilde{g} \in \mathcal{T}^k(\widetilde{\Omega}^{(0)}, \ldots, \Omega^{(k)}; \mathcal{N}_{0,\text{nc}}, \ldots, \mathcal{N}_{k,\text{nc}})$ is another nc function of order $k$ such that $\widetilde{g}\,|_{\Omega^{(0)} \times \cdots \times \Omega^{(k)}} = f$, then for any $X^j \in \Omega^{(j)}_{n_j}$ and invertible $S_j \in \mathcal{R}^{n_j \times n_j}$, $j = 0, \ldots, k$, we have

$$\begin{aligned} &\widetilde{g}(S_0 X^0 S_0^{-1}, \ldots, S_k X^k S_k^{-1})(S_0 Z^1 S_1^{-1}, \ldots, S_{k-1} Z^k S_k^{-1}) \\ &\quad = S_0 \widetilde{g}(X^0, \ldots, X^k)(Z^1, \ldots, Z^k) S_k^{-1} = S_0 f(X^0, \ldots, X^k)(Z^1, \ldots, Z^k) S_k^{-1} \\ &\qquad = \widetilde{f}(S_0 X^0 S_0^{-1}, \ldots, S_k X^k S_k^{-1})(S_0 Z^1 S_1^{-1}, \ldots, S_{k-1} Z^k S_k^{-1}), \end{aligned}$$

i.e., we have necessarily $\widetilde{g} = \widetilde{f}$. □

We will call the nc function $\widetilde{f} \in \mathcal{T}^k(\widetilde{\Omega}^{(0)}, \ldots, \widetilde{\Omega}^{(k)}; \mathcal{N}_{0,\text{nc}}, \ldots, \mathcal{N}_{k,\text{nc}})$ the *canonical extension of the nc function* $f \in \mathcal{T}^k(\Omega^{(0)}, \ldots, \Omega^{(k)}; \mathcal{N}_{0,\text{nc}}, \ldots, \mathcal{N}_{k,\text{nc}})$.

# Bibliography


[1] G. Abduvalieva and D. S. Kaliuzhnyi-Verbovetskyi. Fixed point theorems for noncommutative functions. *J. Math. Anal. Appl.* 401 (2013), no. 1, 436–446.

[2] V. M. Adamjan and D. Z. Arov. On unitary coupling of semiunitary operators. Russian, *Mat. Issled.* **1** (1966), vyp. 2, 3–64. English translation: *Amer. Math. Soc. Transl. (2)* (95) (1970), 75–129.

[3] M. Aguiar. Infinitesimal Hopf algebras. *New trends in Hopf algebra theory (La Falda, 1999)*, 1–29, *Contemp. Math.*, 267, Amer. Math. Soc., Providence, RI, 2000.

[4] J. Agler. On the representation of certain holomorphic functions defined on a polydisc. *Oper. Theory Adv. Appl.* 48, pp. 47–66, Birkhäuser Verlag, Basel, 1990.

[5] J. Agler and J. E. McCarthy. Global holomorphic functions in several non-commuting variables. Preprint, arXiv:1305.1636.

[6] J. Agler and J. E. McCarthy. Pick Interpolation for free holomorphic functions. Preprint, arXiv:1308.3730.

[7] J. Agler and N. J. Young. Symmetric functions of two noncommuting variables. Preprint, arXiv:1307.1588.

[8] D. Alpay and D. S. Kaliuzhnyi-Verbovetskyi. On the intersection of null spaces for matrix substitutions in a non-commutative rational formal power series. *C. R. Math. Acad. Sci. Paris* 339 (2004), no. 8, 533–538.

[9] D. Alpay and D. S. Kaliuzhnyi-Verbovetskyi. Matrix-J-unitary non-commutative rational formal power series. *Oper. Theory Adv. Appl.* 161, Birkhäuser, Basel, 2006, 49–113.

[10] C.-G.Ambrozie and D. Timotin. A von Neumann type inequality for certain domains in $C^n$. *Proc. Amer. Math. Soc.* 131:859–869, 2003.

[11] S. A. Amitsur. Rational identities and applications to algebra and geometry. *J. Algebra*, 3:304–359, 1966.

[12] J. A. Ball, A. Biswas, Q. Fang, and S. ter Horst. Multivariable generalizations of the Schur class: Positive kernel characterization and transfer function realization. *Recent Advances in Operator Theory and Applicaions*, pp. 17–79, OT 187, Birkhäuser, Basel, 2008.

[13] J. A. Ball and V. Bolotnikov. Realization and interpolation for Schur-Agler-class functions on domains with matrix polynomial defining function in $C^n$. *J. Funct. Anal.* 213:45–87, 2004.

[14] J.A. Ball and N. Cohen. de Branges–Rovnyak operator models and systems theory: a survey. *Topics in Matrix and Operator Theory* (ed. by H. Bart, I. Gohberg, and M.A. Kaashoek), pp. 93–136, *Operator Theory: Adv. Appl.* 50, Birkhäuser-Verlag, Boston, 1991.

[15] J. A. Ball, G. Groenewald, and T. Malakorn. Structured noncommutative multidimensional linear systems. *SIAM J. Control Optim.*, 44(4):1474–1528 (electronic), 2005.

[16] J. A. Ball, G. Groenewald, and T. Malakorn. Bounded real lemma for structured noncommutative multidimensional linear systems and robust control. *Multidimens. Syst. Signal Process.*, 17(2-3):119–150, 2006.

[17] J. A. Ball, G. Groenewald, and T. Malakorn. Conservative structured noncommutative multidimensional linear systems. In *The state space method generalizations and applications*, volume 161 of *Oper. Theory Adv. Appl.*, pages 179–223. Birkhäuser, Basel, 2006.

[18] J. A. Ball and D. S. Kaliuzhnyi-Verbovetskyi. Conservative dilations of dissipative multidimensional systems: The commutative and non-commutative settings. *Multidim. Syst. Signal Processing*, 19:79–122, 2008.

[19] J. A. Ball and T. T. Trent. Unitary colligations, reproducing kernel Hilbert spaces, and Nevanlinna-Pick interpolation in several variables. *J. Funct. Anal.* 157:1–61, 1998.

[20] J. A. Ball and V. Vinnikov, Lax-Phillips scattering and conservative linear systems: a Cuntz-algebra multidimensional setting, *Memoirs Amer. Math. Soc.* 178, no. 837, 2005.

[21] S. T. Belinschi, M. Popa, and V. Vinnikov. Infinite divisibility and a non-commutative Boolean-to-free Bercovici-Pata bijection. *J. Funct. Anal.* 262 (2012), no. 1, 94–123.

[22] S. T. Belinschi, M. Popa, and V. Vinnikov. On the operator-valued analogues of the semicircle, arcsine and Bernoulli laws. em J. Oper. Theory 70 (2013), 239–258.

[23] H. Bercovici and D.-V. Voiculescu. Free convolutions of measures with unbounded support. *Indiana Univ. Math. J.* 42:733–773, 1993.

[24] G. M. Bergman. Skew fields of noncommutative rational functions, after Amitsur. In *Séminaire Schützenberger–Lentin–Nivat, Année 1969/70, No. 16.* Paris, 1970.

[25] G. M. Bergman and W. Dicks. On Universal Derivations. *J. Algebra* 36 (1975), 193–211.

[26] J. Berstel and C. Reutenauer. *Rational series and their languages*, volume 12 of *EATCS Monographs on Theoretical Computer Science.* Springer-Verlag, Berlin, 1988.

[27] O. Bratelli and P. E. T. Jorgensen, Iterated function systems and permutation representations of the Cuntz algebra, *Memoirs Amer. Math. Soc.* no. 139, 1999.

[28] P. M. Cohn. The embedding of firs in skew fields. *Proc. London Math. Soc.*, 23:193–213, 1971.

[29] P. M. Cohn. *Free rings and their relations.* Academic Press, London, 1971. London Mathematical Society Monographs, No. 2.

[30] P. M. Cohn. Universal skew fields of fractions. *Symposia Math.*, 8:135–148, 1972.

[31] P. M. Cohn. *Free ideal rings and localization in general rings.* Cambridge University Press, Cambridge, 2006. New Mathematical Monographs 3.

[32] K. R. Davidson and D. R. Pitts, Invariant subspaces and hyper-reflexivity for the free semigroup algebras, *Proc. London Math. Soc.* 78:401–430, 1999.

[33] W. Dicks and J. Lewin. A Jacobian conjecture for free associative algebras. *Comm. Algebra* 10 (1982), no. 12, 1285–1306.

[34] M. A. Dritschel, S. Marcantognini, and S. McCullough. Interpolation in semigroupoid algebras. *J. Reine Angew. Math.* 606:1–40, 2007.

[35] M. A. Dritschel and S. McCullough. Test functions, kernels, realizations and interpolation. *Operator Theory, Structured Matrices, and Dilation*, pp. 153–179, Theta Series in Advanced Mathematics, Bucharest, 2007.

[36] E. B. Dynkin. Calculation of the coefficients in the Campbell-Hausdorff formula. *Doklady Akad. Nauk SSSR (N.S.)* 57:323–326, 1947 (Russian).

[37] E. G. Effros and Zh.-J. Ruan. *Operator spaces.* London Mathematical Society Monographs. New Series, **23**. The Clarendon Press, Oxford University Press, New York, 2000.

[38] M. Fliess. Sur le plongement de l'algèbre des séries rationnelles non commutatives dans un corps gauche. *C. R. Acad. Sci. Paris, Ser. A*, 271:926–927, 1970.

[39] M. Fliess. Matrices de Hankel. *J. Math. Pures Appl.*, 53(9):197–222, 1974.

[40] M. Fliess. Sur divers produits de séries formelles. *Bull. Soc. Math. France*, 102:181–191, 1974.

[41] E. Formanek. *The polynomial identities and invariants of $n \times n$ matrices.* CBMS Regional Conference Series in Mathematics, **78**. The American Mathematical Society, Providence, RI, 1991.

[42] I. Gelfand, S. Gelfand, V. Retakh, and R. L. Wilson. Quasideterminants. *Adv. Math.* 193(1):56–141, 2005.

[43] I. M. Gelfand, D. Krob, A. Lascoux, B. Leclerc, V. S. Retakh, and J.-Y. Thibon. Noncommutative symmetric functions. *Adv. Math.* 112(2):218–348, 1995.

[44] I. M. Gelfand and V. S. Retakh. Determinants of matrices over noncommutative rings. *Funktsional. Anal. i Prilozhen.)* 25(2):13–25, 96, 1991 (Russian); translation in *Funct. Anal. Appl.* 25(2):91–102, 1991.

[45] I. M. Gelfand and V. S. Retakh. Theory of noncommutative determinants, and characteristic functions of graphs. *Funktsional. Anal. i Prilozhen.* 26(4):1–20, 96, 1992 (Russian); translation in *Funct. Anal. Appl.* 26(4):231–246, 1992.

[46] D. W. Hadwin. Continuous functions of operators; a functional calculus. *Indiana Univ. Math. J.* 27 (1978), no. 1, 113–125.

[47] D. W. Hadwin, L. Kaonga, and B. Mathes. Noncommutative continuous functions. *J. Korean Math. Soc.* 40 (2003), no. 5, 789–830.

[48] J. W. Helton. Discrete time systems, operator models, and scattering theory. *J. Funct. Anal.* (16) (1974), 15–38.

[49] J. W. Helton, "Positive" noncommutative polynomials are sums of squares. *Ann. of Math.* (2) 156 (2002), no. 2, 675–694.

[50] J. W. Helton. Manipulating matrix inequalities automatically. In *Mathematical systems theory in biology, communications, computation, and finance (Notre Dame, IN, 2002)*, volume 134 of *IMA Vol. Math. Appl.*, pages 237–256. Springer, New York, 2003.

[51] J. W. Helton, I. Klep, and S. McCullough. Analytic mappings between noncommutative pencil balls. *J. Math. Anal. Appl.* 376 (2011), no. 2, 407–428.

[52] J. W. Helton, I. Klep, and S. McCullough. Proper Analytic Free Maps. *J. Funct. Anal.* 260 (2011), no. 5, 1476–1490.

[53] J. W. Helton, I. Klep, and S. McCullough. Free analysis, convexity and LMI domains. *Mathematical methods in systems, optimization and control* (ed. by H. Dym, M. de Oliveira, and M. Putinar), pp. 195–219, *Operator Theory: Adv. Appl.* 222, Birkhauser, 2012.

[54] J. W. Helton, I. Klep, and S. McCullough. Free Convex Algebraic Geometry. *Semidefinite Optimization and Convex Algebraic Geometry* (ed. by G. Blekherman, P. Parrilo, and R. Thomas), pp. 341–405, SIAM, 2013.

[55] J. W. Helton, I. Klep, S. McCullough, and N. Slinglend. Non-commutative ball maps. *J. Funct. Anal.* 257(2009), 47–87.

[56] J. W. Helton and S. McCullough. Convex noncommutative polynomials have degree two or less. *SIAM J. Matrix Anal. Appl.* 25 (2004), no. 4, 1124–1139.

[57] J. W. Helton and S. McCullough. Free convex sets defined by rational expressions have LMI representations. Preprint, arXiv:1209.3412.

[58] J. W. Helton, S. A. McCullough, M. Putinar, and V. Vinnikov. Convex Matrix Inequalities versus Linear Matrix Inequalities, *IEEE Trans. Aut. Control*, 54(5):952–964, 2009.

[59] J. W. Helton, S. A. McCullough, and V. Vinnikov. Noncommutative convexity arises from linear matrix inequalities. *J. Funct. Anal.*, 240(1):105–191, 2006.

[60] R. G. Heyneman and M. E. Sweedler. Affine Hopf algebras. I. *J. Algebra*, **13**:192–241, 1969.

[61] E. Hille and R. S. Phillips. *Functional analysis and semi-groups.* Third printing of the revised edition of 1957. American Mathematical Society Colloquium Publications, Vol. XXXI. American Mathematical Society, Providence, R. I., 1974.

[62] R. A. Horn and C. R. Johnson. *Topics in matrix analysis.* Corrected reprint of the 1991 original. Cambridge University Press, Cambridge, 1994.

[63] J. E. Humphreys. *Introduction to Lie algebras and representation theory.* Springer-Verlag, New-York–Heidelberg–Berlin, 1972.

[64] D. S. Kaliuzhnyi-Verbovetskyi and V. Vinnikov. Singularities of Noncommutative Rational Functions and Minimal Factorizations. *Lin. Alg. Appl.* 430:869–889, 2009.

[65] D. S. Kaliuzhnyi-Verbovetskyi and V. Vinnikov. Noncommutative rational functions, their difference-differential calculus and realizations. *Multidimens. Syst. Signal Process.* 23 (2012), no. 1–2, 49–77.

[66] S. C. Kleene. Representation of events in nerve nets and finite automata. In *Automata studies*, Annals of mathematics studies, no. 34, pages 3–41. Princeton University Press, Princeton, N. J., 1956.

[67] J. Lewin. A Matrix Representation for Associative Algebras I, II. *Trans. Amer. Math. Soc.* 188 (1977), 293–308, 309–317.

[68] P. S. Muhly and B. Solel. Hardy algebras, $W^*$-correspondences and interpolation theory. *Math. Ann.* 330:353–415, 2004.

[69] P. S. Muhly and B. Solel. Progress in noncommutative function theory. *Sci. China Math.* 54 (2011), no. 11, 2275–2294.

[70] P. S. Muhly and B. Solel. Schur class operator functions and automorphisms of Hardy algbras. *Doc. Math.* 13:365–411, 2008.

[71] P. S. Muhly and B. Solel. Tensorial function theory: From Berezin transforms to Taylor's Taylor series and back. Integral Equations Operator Theory 76 (2013), no. 4, 463–508.

[72] J. Mujica. *Complex analysis in Banach spaces. Holomorphic functions and domains of holomorphy in finite and infinite dimensions.* North-Holland Mathematics Studies, **120**. Notas de Matema'tica [Mathematical Notes], **107**. North-Holland Publishing Co., Amsterdam, 1986.

[73] A. Nemirovskii. Advances in convex optimization: conic programming. *Plenary Lecture, International Congress of Mathematicians (ICM)*, Madrid, Spain, 2006.

[74] Yu. Nesterov and A. Nemirovskii. *Interior-point polynomial algorithms in convex programming*, volume 13 of *SIAM Studies in Applied Mathematics.* Society for Industrial and Applied Mathematics (SIAM), Philadelphia, PA, 1994.

[75] A. Nica and R. Speicher. *Lectures on combinatorics of free probability*, London Math. Soc. Lecture Note Series 335, Cambridge Univ. Press, 2006.
[76] P. Niemiec. Functional calculus for diagonalizable matrices. *Linear Multilinear Algebra* 62 (2014), no. 3, 297–321.
[77] V. I. Paulsen. Representations of function algebras, abstract operator spaces, and Banach space geometry. *J. Funct. Anal.* **109**(1):113–129, 1992.
[78] V. Paulsen. *Completely bounded maps and operator algebras.* Cambridge Studies in Advanced Mathematics, **78**. Cambridge University Press, Cambridge, 2002.
[79] G. Pisier. *Introduction to operator space theory.* London Mathematical Society Lecture Note Series, **294**. Cambridge University Press, Cambridge, 2003.
[80] M. Popa and V. Vinnikov. Non-Commutative Functions and Non-Commutative Free Levy-Hincin Formula. *Adv. Math.* 236 (2013), 131–157.
[81] G. Popescu, Models for infinite sequences of noncommuting operators, *Acta Sci. Math.* 53:355-368, 1989.
[82] G. Popescu, Characteristic functions for infinite sequences of noncommuting operators, *J. Operator Theory* 22:51-71, 1989.
[83] G. Popescu, Isometric dilations for infinite sequences of noncommuting operators, *Trans. Amer. Math. Soc.* 316:523-536, 1989.
[84] G. Popescu, Multi-analytic operators on Fock spaces, *Math. Ann.* 303:31-46, 1995.
[85] G. Popescu. Free holomorphic functions on the unit ball of $B(H)^n$. *J. Funct. Anal.* 241:268–333, 2006.
[86] G. Popescu. Free holomorphic automorphisms of the unit ball of $B(H)^n$. *J. reine angew. Math.* 638:119–168, 2010.
[87] G. Popescu. Operator theory on noncommutative domains. *Mem. Amer. Math. Soc.* 205 (2010), no. 964.
[88] G. Popescu. Free biholomorphic functions and operator model theory. *J. Funct. Anal.* 262 (2012), no. 7, 3240–3308.
[89] G. Popescu. Free biholomorphic functions and operator model theory, II. *J. Funct. Anal.* 265 (2013), no. 5, 786–836.
[90] S. A. Joni and G.-C. Rota. Coalgebras and bialgebras in combinatorics. *Stud. Appl. Math.* 61 (1979), no. 2, 93–139.
[91] L. H. Rowen. *Polynomial identities in ring theory*, volume **84** of *Pure and Applied Mathematics.* Academic Press Inc. [Harcourt Brace Jovanovich Publishers], New York, 1980.
[92] Zh.-J. Ruan. On real operator spaces. International Workshop on Operator Algebra and Operator Theory (Linfen, 2001). *Acta Math. Sin. (Engl. Ser.)*, **19**(3):485–496, 2003.
[93] S. H. Schanuel. Continuous extrapolation to triangular matrices characterizes smooth functions. *J. Pure Appl. Algebra* 24 (1982), no. 1, 59–71.
[94] S. H. Schanuel and W. R. Zame. Naturality of the functional calculus. *Bull. Lond. Math. Soc.* 14 (1982), 218–220.
[95] M. P. Schützenberger. On the definition of a family of automata. *Information and Control*, 4:245–270, 1961.
[96] M. P. Schützenberger. Certain elementary families of automata. In *Proc. Sympos. Math. Theory of Automata (New York, 1962)*, pages 139–153. Polytechnic Press of Polytechnic Inst. of Brooklyn, Brooklyn, New York, 1963.
[97] B. V. Shabat. *Introduction to Complex Analysis. Part II.* Translations of Mathematical Monographs 110, Functions of several variables, Translated from the third (1985) Russian edition by J. S. Joel, AMS, Providence, RI, 1992.
[98] R. E. Skelton, T. Iwasaki, and K. M. Grigoriadis. *A Unified Algebraic Approach to Linear Control Design.* Taylor & Francis, 1997.
[99] J. L. Taylor. A general framework for a multi-operator functional calculus. *Advances in Math.*, **9**:183–252, 1972.
[100] J. L. Taylor. Functions of several noncommuting variables. *Bull. Amer. Math. Soc.*, **79**:1–34, 1973.
[101] I. M. Vinogradov, Elements of Number Theory, Mineola, NY: Dover Publications, 2003, reprint of the 1954 edition.
[102] D.-V. Voiculescu. Symmetries of some reduced free product $C^*$-algebras. *Operator algebras and their connections with topology and ergodic theory (Busçteni, 1983)*, Lecture Notes in Math. 1132, Springer, Berlin, 1985, pp. 556–588.

[103] D.-V. Voiculescu. Addition of certain non-commutative random variables. *J. Funct. Anal.* 66:323–346, 1986.

[104] D.-V. Voiculescu. Operations on certain non-commutative operator-valued random variables. *Astérisque* 232:243–275, 1995.

[105] D.-V. Voiculescu. The coalgebra of the free difference quotient and free probability. *Internat. Math. Res. Not.* 2000 (2000), no. 2, 79–106.

[106] D.-V. Voiculescu. Free Analysis Questions I: Duality Transform for the Coalgebra of $\partial_{X:B}$ *International Math. Res. Notices* 16:793–822, 2004.

[107] D.-V. Voiculescu. Free analysis questions. II: The Grassmannian completion and the series expansion at the origin. *J. Reine Angew. Math.* 645 (2010), 155–236.

[108] D.-V. Voiculescu, K. J. Dykema, and A. Nica. Free random variables. A noncommutative probability approach to free products with applications to random matrices, operator algebras and harmonic analysis on free groups. CRM Monograph Series, 1. American Mathematical Society, Providence, RI, 1992. vi+70 pp.

[109] J. H. M. Wedderburn. On continued fractions in non-commutative quantities. *Ann. of Math.* 15:101–105, 1913.

[110] M. A. Zorn. Derivatives and Fréchet differentials. *Bull. Amer. Math. Soc.*, **52**:133-137, 1946.

# Index

# Selected Published Titles in This Series

199 **Dmitry S. Kaliuzhnyi-Verbovetskyi and Victor Vinnikov,** Foundations of Free Noncommutative Function Theory, 2014

198 **Jörg Jahnel,** Brauer Groups, Tamagawa Measures, and Rational Points on Algebraic Varieties, 2014

197 **Richard Evan Schwartz,** The Octagonal PETs, 2014

196 **Silouanos Brazitikos, Apostolos Giannopoulos, Petros Valettas, and Beatrice-Helen Vritsiou,** Geometry of Isotropic Convex Bodies, 2014

195 **Ching-Li Chai, Brian Conrad, and Frans Oort,** Complex Multiplication and Lifting Problems, 2013

194 **Samuel Herrmann, Peter Imkeller, Ilya Pavlyukevich, and Dierk Peithmann,** Stochastic Resonance, 2014

193 **Robert Rumely,** Capacity Theory with Local Rationality, 2013

192 **Messoud Efendiev,** Attractors for Degenerate Parabolic Type Equations, 2013

191 **Grégory Berhuy and Frédérique Oggier,** An Introduction to Central Simple Algebras and Their Applications to Wireless Communication, 2013

190 **Aleksandr Pukhlikov,** Birationally Rigid Varieties, 2013

189 **Alberto Elduque and Mikhail Kochetov,** Gradings on Simple Lie Algebras, 2013

188 **David Lannes,** The Water Waves Problem, 2013

187 **Nassif Ghoussoub and Amir Moradifam,** Functional Inequalities: New Perspectives and New Applications, 2013

186 **Gregory Berkolaiko and Peter Kuchment,** Introduction to Quantum Graphs, 2013

185 **Patrick Iglesias-Zemmour,** Diffeology, 2013

184 **Frederick W. Gehring and Kari Hag,** The Ubiquitous Quasidisk, 2012

183 **Gershon Kresin and Vladimir Maz'ya,** Maximum Principles and Sharp Constants for Solutions of Elliptic and Parabolic Systems, 2012

182 **Neil A. Watson,** Introduction to Heat Potential Theory, 2012

181 **Graham J. Leuschke and Roger Wiegand,** Cohen-Macaulay Representations, 2012

180 **Martin W. Liebeck and Gary M. Seitz,** Unipotent and Nilpotent Classes in Simple Algebraic Groups and Lie Algebras, 2012

179 **Stephen D. Smith,** Subgroup complexes, 2011

178 **Helmut Brass and Knut Petras,** Quadrature Theory, 2011

177 **Alexei Myasnikov, Vladimir Shpilrain, and Alexander Ushakov,** Non-commutative Cryptography and Complexity of Group-theoretic Problems, 2011

176 **Peter E. Kloeden and Martin Rasmussen,** Nonautonomous Dynamical Systems, 2011

175 **Warwick de Launey and Dane Flannery,** Algebraic Design Theory, 2011

174 **Lawrence S. Levy and J. Chris Robson,** Hereditary Noetherian Prime Rings and Idealizers, 2011

173 **Sariel Har-Peled,** Geometric Approximation Algorithms, 2011

172 **Michael Aschbacher, Richard Lyons, Stephen D. Smith, and Ronald Solomon,** The Classification of Finite Simple Groups, 2011

171 **Leonid Pastur and Mariya Shcherbina,** Eigenvalue Distribution of Large Random Matrices, 2011

170 **Kevin Costello,** Renormalization and Effective Field Theory, 2011

169 **Robert R. Bruner and J. P. C. Greenlees,** Connective Real $K$-Theory of Finite Groups, 2010

168 **Michiel Hazewinkel, Nadiya Gubareni, and V. V. Kirichenko,** Algebras, Rings and Modules, 2010

For a complete list of titles in this series, visit the
AMS Bookstore at **www.ams.org/bookstore/survseries/**.